Systems of Nonlinear Partial Differential Equations

NATO ASI Series

Advanced Science Institutes Series

A series presenting the results of activities sponsored by the NATO Science Committee, which aims at the dissemation of advanced scientific and technological knowledge, with a view to strengthening links between scientific communities.

The series is published by an international board of publishers in conjunction with the NATO Scientific Affairs Division

A	Life Sciences	Plenum Publishing Corporation
B	Physics	London and New York
C	Mathematical and Physical Sciences	D. Reidel Publishing Company Dordrecht, Boston and Lancaster
D	Behavioural and Social Sciences	Martinus Nijhoff Publishers
E	Engineering and Materials Sciences	The Hague, Boston and Lancaster
F	Computer and Systems Sciences	Springer Verlag
G	Ecological Sciences	Heidelberg

Series C: Mathematical and Physical Sciences No. 111

Systems of Nonlinear Partial Differential Equations

edited by

J. M. Ball

Department of Mathematics,
Heriot-Watt University, Edinburgh, U.K.

D. Reidel Publishing Company

Dordrecht / Boston / Lancaster

Published in cooperation with NATO Scientific Affairs Division

Proceedings of the NATO Advanced Study Institute on
Systems of Nonlinear Partial Differential Equations
Oxford, U.K.
July 25-August 7, 1982

Library of Congress Cataloging in Publication Data

Main entry under title:

Systems of nonlinear partial differential equations.

(NATO ASI series. Series C, Mathematical and physical sciences; no. 111)
"Published in cooperation with NATO Scientific Affairs Division."
"Proceedings of the NATO Advanced Study Institute on Systems of Nonlinear Partial Differential Equations, Oxford, U.K., July 25—August 7, 1982" — Verso t.p.
Includes index.
1. Differential equations, Partial—Addresses, essays, lectures. 2. Differential equations, Nonlinear—Addresses, essays, lectures. I. Ball, J. M. (John MacLeod), 1948– II. North Atlantic Treaty Organization. Scientific Affairs Division. III. NATO Advanced Study Institute on Systems of Nonlinear Partial Differential Equations (1982: Oxford, Oxfordshire). IV. Series.
QA377.S967 1983 515.3'53 83–11070
ISBN 90–277–1629–3

Published by D. Reidel Publishing Company
P.O. Box 17, 3300 AA Dordrecht, Holland

Sold and distributed in the U.S.A. and Canada
by Kluwer Academic Publishers,
190 Old Derby Street, Hingham, MA 02043, U.S.A.

In all other countries, sold and distributed
by Kluwer Academic Publishers Group,
P.O. Box 322, 3300 AH Dordrecht, Holland

D. Reidel Publishing Company is a member of the Kluwer Academic Publishers Group

---

Printed in The Netherlands

TABLE OF CONTENTS

PART II
SPECIAL SESSIONS

Lectures not appearing in this volume

Expository lectures: T. B. Benjamin (University of Oxford), *A Novel Hamiltonian System deriving from Hydrodynamics*

Special sessions

a. R. V. Kohn (Courant Institute), *The Incompressible Limit in Nonlinear Elasticity*

d. M. G. Crandall (University of Wisconsin), *Solutions of the Porous Medium Equation in* $\mathbb{R}^n$ *under Optimal Conditions on Initial Values*

e. Analysis and Computational Fluid Dynamics, organized by A.J. Chorin (University of California, Berkeley)

A. J. Chorin (University of California, Berkeley), *Vortex Methods for Boundary Layer Analysis*

O. Hald (University of California, Berkeley), *Inverse Eigenvalue Problems for the Mantle*

PREFACE

This volume contains the proceedings of a NATO/London Mathematical Society Advanced Study Institute held in Oxford from 25 July - 7 August 1982. The institute concerned the theory and applications of systems of nonlinear partial differential equations, with emphasis on techniques appropriate to systems of more than one equation. Most of the lecturers and participants were analysts specializing in partial differential equations, but also present were a number of numerical analysts, workers in mechanics, and other applied mathematicians.

The organizing committee for the institute was J.M. Ball (Heriot-Watt), T.B. Benjamin (Oxford), J. Carr (Heriot-Watt), C.M. Dafermos (Brown), S. Hildebrandt (Bonn) and J.S. Pym (Sheffield).

The programme of the institute consisted of a number of courses of expository lectures, together with special sessions on different topics.

It is a pleasure to thank all the lecturers for the care they took in the preparation of their talks, and S.S. Antman, A.J. Chorin, J.K. Hale and J.E. Marsden for the organization of their special sessions.

The institute was made possible by financial support from NATO, the London Mathematical Society, the U.S. Army Research Office, the U.S. Army European Research Office, and the U.S. National Science Foundation.

The lectures were held in the Mathematical Institute of the University of Oxford, and residential accommodation was provided at Hertford College.

Valuable assistance with the organization of the institute and with the typing of the proceedings was given by Maureen Gardiner, Anne Hodgson and Janice McClelland.

J.M. BALL

PART I

EXPOSITORY LECTURES

ALGEBRAIC AND TOPOLOGICAL INVARIANTS FOR REACTION-DIFFUSION EQUATIONS

Charles C. Conley and Joel A. Smoller[1]

University of Wisconsin and University of Michigan

§ 1 INTRODUCTION

In the last few years, a good deal of progress has been made in the understanding of the qualitative properties of solutions of reaction-diffusion equations. This has been due to the introduction of new topological techniques into the field; in particular, we mention the concept of an isolated invariant set, and its index, as developed in [2].

This index is a homotopy invariant associated with an isolated invariant set. It is a generalization of the classical Morse index of an isolated rest point for a gradient flow in $\mathbb{R}^n$, but it is applicable to a far wider class of invariant sets; in fact, it is even relevant for partial differential equations, as we shall show here.

Under some very reasonable and natural hypotheses, one can decompose an isolated invariant set into its so-called Morse sets. This immediately implies the existence of certain exact sequences of cohomology groups defined on these Morse sets. Thus one obtains algebraic invariants associated with the given isolated invariant set. As is the usual case with algebraic invariants, they are discrete, and since they "depend continuously on the topology", a change in the algebraic structure reflects a change in the underlying topology - e.g., a bifurcation can occur. Moreoever, the exact sequence can also be used to compute indices. Finally, we point out that the aforementioned index actually has a "dynamic" aspect to it, in that it carries stability information. That is, one can use it to obtain precise statements on the dimensions of unstable manifolds.

J. M. Ball (ed.), Systems of Nonlinear Partial Differential Equations, 3–24.

In this paper, we want to point out the relevance of these ideas in the study of reaction-diffusion equations. The reader will see that there is involved only a bare minimum of algebraic topology, and that the "rewards" one obtains far exceed the effort involved in learning the topology.

§ 2 THE ABSTRACT FRAMEWORK

A. Algebraic Topology

We begin with some (very elementary) ideas from algebraic topology. The reader can consult any standard textbook in the field for motivation, background, and details.

We denote by (A,B) a pair of topological spaces where $A \supset B$. To any such pair, there is associated a sequence of Abelian groups $\{H^n(A,B) : n = 0,1,2,\dots\}$, and $H^n(A,B)$ is called the nth (integral) cohomology group of the pair (A,B). If S^k is a k-sphere then it turns out that

$$H^n(S^k, \text{pt.}) = \begin{cases} \mathbb{Z}, & k = n \\ 0, & k \neq n \end{cases}$$

If A, B, and C are topological spaces, and $A \supset B \supset C$, then there is a so called exact sequence of cohomology groups:

$$\dots \to H^{n-1}(B,C) \xrightarrow{\phi_{n-1}} H^n(A,B) \xrightarrow{\psi_n} H^n(A,C) \xrightarrow{\theta_n} H^n(B,C) \xrightarrow{\phi_n} \dots .$$

This means that the above maps are all linear, and the image of any map is the kernel of the following map. For example,

$$\ker \psi_n = \operatorname{im} \phi_{n-1} .$$

This is basically all of the algebraic topology that we shall need.

B. Isolated Invariant Sets and Morse Decompositions

Let Γ be a metric space, and let F be a flow on Γ; $F : \Gamma \times \mathbb{R} \to \Gamma$. We denote $F(\gamma,t)$ by $\gamma \cdot t$. For $N \subset \Gamma$, set

$$I(N) = \{\gamma \in N : \gamma \cdot \mathbb{R} \subset N\} ;$$

I(N) is called the <u>invariant set</u> in N. For $\gamma \in \Gamma$, we define the <u>alpha</u>- and <u>omega-limit sets</u> of γ by

$$\alpha(\gamma) = \bigcap_{t<0} \mathrm{cl}\{\gamma\cdot(-\infty,t)\}\ ,\quad \omega(\gamma) = \bigcap_{t>0} \mathrm{cl}\{\gamma\cdot(t,\infty)\}\ .$$

<u>Definition</u> 1. (<u>Morse Decomposition</u>): Let I be a compact invariant set in Γ. A Morse decomposition of I is a finite collection $\{M_i\}_1^n$ of disjoint, compact invariant sets in I which can be ordered $(M_1, M_2, \ldots, M_n)$ such that if $\gamma \in I \setminus \bigcup_1^n M_i$, then there are indices $i<j$ with $\alpha(\gamma) \in M_j$ and $\omega(\gamma) \in M_i$. The above ordering will be termed admissible.

For example, if I consists of two rest points M_1 and M_2 which are connected by an orbit (see figure 1), then (M_1, M_2) is a Morse decomposition of I.

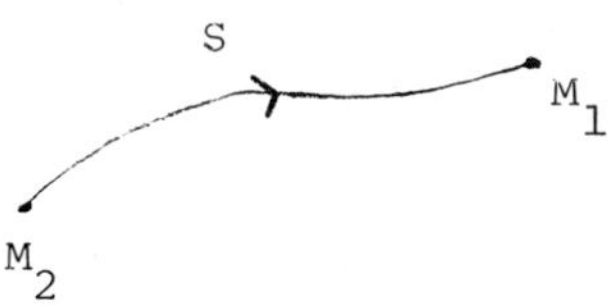

<u>FIGURE 1</u>

The reason that Morse decompositions are of interest to us is that many "sensible" properties of the equation are reflected in the Morse decomposition of its (isolated) invariant sets. Here sensible means stable under small perturbations.

In order to make our abstract theory applicable to partial differential equations, we introduce the concept of a local flow.

<u>Definition</u> 2. (<u>Local Flow</u>). A locally compact subset $X \subset \Gamma$ is called a local flow if for each $\gamma \in X$ there exists an $\varepsilon>0$, and a neighborhood U of γ such that $(X \cap U)\cdot[0,\varepsilon) \subset X$.

For example, we may think of X as a Sobolev space W^s; then our definition means that if the data is in W^s, the same is true for the corresponding solutions for small time.

From now on, we will always assume that X is a local flow in Γ.

<u>Definition</u> 3. (<u>Isolated Invariant Set</u>): Let N be a compact subset of X. If $I(N) \subset \mathrm{int}(N)$, relative to X, then I(N) is called an isolated invariant set and N is called an isolating neighborhood.

The proof of the next lemma follows easily from the definitions, see [5].

Lemma 4. If S is an isolated invariant set, and $\{M_i\}_1^n$ is a Morse decomposition of S, then each M_i is an isolated invariant set.

For a given isolated invariant set S, we can associate to it an "index". To this end we require the following notion. Recall that by a pair of subsets (A,B) we mean $A \supset B$.

Definition 5. (Index Pair): Let S be an isolated invariant set and N an isolating neighborhood for S. A compact pair of subsets (N_1, N_0) of N is called an index pair for S provided the following three conditions hold:

a) $cl(N_1 \setminus N_0)$ is an isolating neighborhood of S.

b) N_0 is positively invariant relative to N_1; ie if $\gamma \in N_0$ and $\gamma \cdot [0,t] \subset N_1$, then $\gamma \cdot [0,t] \subset N_0$.

c) If $\gamma \in N_1$ and $\gamma \cdot \mathbb{R}_+ \not\subset N$, then there is a $t>0$ with $\gamma \cdot [0,t) \subset N_1$, and $\gamma \cdot t \in N_0$.

The definition takes into account that we have only a local flow. Note too that c) implies that the orbit through γ exits N only through N_0.

To illustrate this definition, consider the picture in the figure below. Here we can take the square B

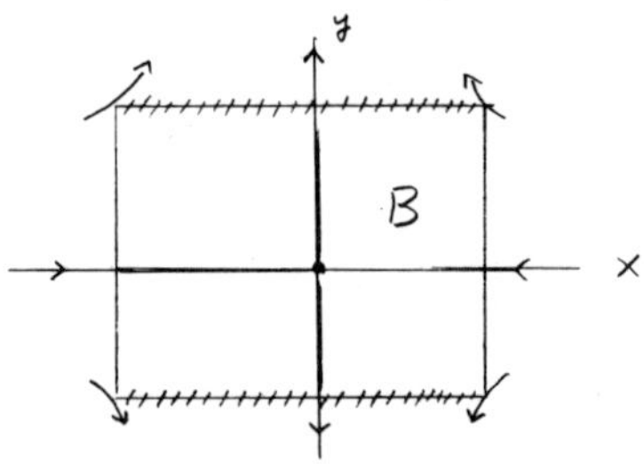

FIGURE 2

to be N_1 and the "top" and "bottom" sides to be N_0. Of course S is the hyperbolic rest point at the origin.

The desired topological invariants will be described in terms of index pairs. Before giving this description, we need to recall a few standard topological notions. First, we say that two topological spaces X and Y are homotopic if we can continuously deform one into the other. Next, if A is a closed subset of X, the quotient space X/A is the topological space obtained from X by identifying all points $a \in A$ to a single point. We can now give the main theorem.

Theorem 6. i) Index pairs exist; ie, given an isolated invariant set S, one can always prove the existence of an index pair for S.

ii) For any index pair (N_1, N_0) of S, the homotopy type of the quotient space N_1/N_0 is independent of N, N_1, N_0, and depends only on S. It is thus a topological invariant associated with the isolated invariant set S.

In view of ii), we may unambiguously define the (homotopy) index of S, by

$$h(S) = [N_1/N_2] .$$

where the brackets denote the homotopy equivalence class. Observe that for the above example $N_1/N_0 = \Sigma^1$, a (pointed)[2] 1-sphere.

Now it is not difficult to check that this index extends the Morse index in the sense that if x_0 is a non-degenerate rest point of a gradient system of ordinary differential equations, and k is the dimension of the unstable manifold at x_0, then $h(x_0) = \Sigma^k$. Furthermore, there is an addition formula for the index. Namely, if S_1 and S_2 are disjoint isolated invariant sets, then $S_1 \cup S_2$ is an isolated invariant set, and

$$h(S) = h(S_1) \vee h(S_2) , \qquad (*)$$

where by v, we denote the operation of "glueing" the two (quotient) spaces together at their distinguished point.

Note that if $S = \phi$, then it is easy to see $h(S) = \bar{0}$, the one point (pointed) space. Thus if $h(S) \neq \bar{0}$, then $S \neq \phi$. This is how the index is used as an existence tool.

Next, we turn to a useful result which can be viewed as a generalization of the first part of the last theorem.

Theorem 7. ([2,5]). Let S be an isolated invariant set and let $(M_1, M_2, \ldots, M_n)$ be an admissible ordering of a Morse decomposition of S. Then there exists an increasing sequence of compact sets $N_0 \subset N_1 \subset \ldots \subset N_n$ such that i) (N_n, N_o) is an index pair for S, and ii) (N_k, N_{k-1}) is an index pair for M_K.

We say that the sets N_0, N_1, ..., N_n form a Morse filtration of S.

Now if $N_0 \subset \dots \subset N_n$ is a Morse filtration of S, we consider the triple (N_{p+1}, N_p, N_{p-1}). As we have seen above, there is an exact sequence,

$$\dots \to H^{n-1}(N_p, N_{p-1}) \to H^n(N_{p+1}, N_p) \to H^n(N_{p+1}, N_{p-1}) \to \dots \quad .$$

Now under reasonable hypotheses, there is an isomorphism $H^n(A,B) \approx H^n(A/B,\phi) \equiv H^n(A/B)$, provided B is closed in A. Applying this to the above exact sequence gives

$$\dots \to H^{n-1}(N_p/N_{p-1}) \to H^n(N_{p+1}/N_p) \to H^n(N_{p+1}/N_{p-1}) \to \dots \quad .$$

In particular, referring back to our previous example of an orbit connecting to rest points (see figure 1), we have the important exact sequence in terms of indices:

$$\dots \to H^{n-1}(h(M_1)) \to H^n(h(M_2)) \to H^n(h(S)) \to \dots \quad . \qquad (1)$$

It is useful to have a criterion for deciding when Morse decompositions exist, as well as one for being able to find isolated invariant sets. Both of these are difficult problems, in general, but here is one useful result. Before stating it recall that a differential equation is called <u>gradient-like</u> in a region if there is a function which strictly increases on orbits in the region.

<u>Theorem</u> 8. Let X be a gradient-like local flow.

1) If the flow has a finite rest point set C contained in an isolating neighborhood N, then the rest points form a Morse decomposition for I(N).

2) If γ_0 is an isolated rest point, then γ_0 is an isolated invariant set.

<u>Proof</u>. 1) Let ϕ be the gradient-like function. If $\gamma \in I(N) \setminus C$ and $t>0$, then $\phi(\gamma \cdot t) > \phi(\gamma)$, and so ϕ is constant on $\omega(\gamma)$. Consider the flow on the compact set I(N). The flow is gradient-like on I(N); thus $\omega(\gamma)$ and $\alpha(\gamma)$ both must be rest points. If we now order $C = \{\gamma_{i_1}, \dots, \gamma_{i_n}\}$ in such a way that if $k<j$, $F(x_{i_k}) > F(x_{i_j})$, then 1) easily follows. Let N be a compact neighborhood of γ_0 such that γ_0 is the only rest point in N. Let U be any neighborhood of γ_0, $U \subsetneq N$. Since $N \setminus U$ is compact, there is a $\delta>0$ such that $F(\gamma \cdot 1) - F(\gamma) > \delta > 0$ for all $\gamma \in N \setminus U$. If $\gamma \in N$ and $\gamma \cdot \mathbb{R} \subset N$, then $\omega(\gamma)$ and $\alpha(\gamma)$ are rest points in I(N);

thus $\omega(\gamma) = \gamma_o = \alpha(x_o)$, and $y \cdot \mathbb{R} \subset U$. Since U was arbitrary, $\gamma \cdot \mathbb{R} = \{\gamma_o\}$. Thus $\{\gamma_o\}$ is the only invariant set in N, and this proves (2).

We turn now to the last general notion, that of continuation. Like every "index" worthy of the name, our index is a homotopy invariant. Thus, consider a parametrised family of equations $\dot{u} = f(\lambda,u)$. Suppose that N is an isolating neighborhood for all parameter values λ in a closed connected set, say an arc. Then the various so determined isolated invariant sets are said to be _related by continuation_. By definition, if S' is so related to S, and S to S", then S' is so related to S". In this case one can prove the following theorem.

Theorem 9. If S and S' are isolated invariant sets which are related by continuation, then $h(S) = h(S')$.

Of course this result is often used to compute indices; namely one continues the given equation to one in which it is easier to compute indices.

§ 3 LOCAL FLOWS AND REACTION-DIFFUSION EQUATIONS

In order to be able to apply these general topological methods to partial differential equations, it is necessary to show that solutions generate local flows. We shall prove a general theorem on the existence of local flows, and we then shall show that the hypotheses can be verified for systems of reaction-diffusion equations.

Again let Γ be a metric space with metric ρ; by a _curve_ on Γ is meant a continuous function $\gamma: I \to \Gamma$, where $I = \mathrm{dom}(\gamma)$ is an open interval in $\mathbb{R}$. A curve is called _regular_ if its graph is closed in $\mathbb{R}\times\Gamma$. $C(I',\Gamma')$ denotes the set of regular curves γ on Γ such that $I' \subset \mathrm{dom}(\gamma)$ and $\gamma(I') \subset \Gamma'$.

Let F be a flow on Γ. The local flow X is called _two sided_ if for all $\gamma \in X$, there is an $\varepsilon > 0$ and a neighborhood U of γ such that $(U \cap X)\cdot(-\varepsilon,\varepsilon) \subset X$. If $\gamma \in X$ implies $\gamma \cdot \mathbb{R}_+ \subset X$, X is called a _semi-flow_. If $\gamma \in X$ implies $\gamma \cdot \mathbb{R} \subset X$, X is called an invariant set; or more consistently, a _sub flow_. We give some examples. Let $C(\Gamma)$ denote the regular curves on Γ.

1. For any $t_o \in \mathbb{R}$, $C(t_o,\Gamma)$ is a two-sided local flow.
2. $C(\mathbb{R}_-,\Gamma)$ is a semi-flow.
3. Let C^- consist of those curves which have an interval of the form $(-\infty,t)$ in their domain. Then C^- is an invariant set.

It is easy to check that the local flows in the above examples are all Hausdorff spaces (in the relative topology).

A local flow X is called <u>locally compact</u> if X is locally compact in the relative topology.

<u>Definition</u> (<u>Equicontinuity</u>): A set of curves $\Phi \subset C(\Gamma)$ is equicontinuous in the set $U \subset \Gamma$ if the following conditions are satisfied:

i) there is a positive function $\varepsilon(\delta)$ of the positive real variable δ such that $\varepsilon(\delta) \to 0$ as $\delta \to 0$, and

ii) If $\gamma \in \Phi$ and I is an interval with $\gamma(I) \subset U$, then for $s, t \in I$, $\rho(\gamma(s), \gamma(t)) \leq \varepsilon(|s-t|)$.

Here is the main theorem.

<u>Theorem</u> 10. Let Γ be a locally compact metric space, and let $X \subset C(\Gamma)$ be a local flow which is equicontinuous in any compact set. Then for any $\bar{t} \in \mathbb{R}$, and compact set $\bar{K} \subset \Gamma$, $cl(C(\bar{t},\bar{k}) \cap X)$ is compact. In particular, if $C(\bar{t},\bar{k})$ is closed for all $\bar{t}$ and $\bar{K}$, X is locally compact.

<u>Proof</u>. Since Γ is locally compact, any compact set $K_o \subset \Gamma$ admits a compact neighborhood K_1. Since X is equicontinuous on K_1, it follows that there is some $\varepsilon > 0$ such that for $\gamma \in X$ and $\gamma([t_o,t_1]) \subset K_o$, $\gamma([t_o-\varepsilon,t_1+\varepsilon]) \subset K_1$.

Suppose that $\{\gamma_n : n \in \mathbb{Z}_+\}$ is a sequence of curves in $C(\bar{t},\bar{K})$. Since X is equicontinuous on (a neighborhood of) $\bar{K}$, and since there is an interval $[t_o,t_1]$ containing $\bar{t}$ such that all the curves in $X \cap C(\bar{t},\bar{K})$ have $[t_o,t_1]$ in their domain, there is a curve $\tilde{\gamma}$ which is a limit point of the sequence $\{\gamma_n\}$ and contains $[t_o,t_1]$ in its domain.

Consider now all such curves which are limit points of $\{\gamma_n\}$, and (partially) order them by their graphs. Let $\hat{\gamma}$ be a maximal element whose domain contains $[t_o,t_1]$. If $\hat{\gamma}$ did not have closed graph, its domain could be enlarged. Namely, if the domain is $(\hat{t}_o,\hat{t}_1)$, and $\hat{\gamma}(t) \to \bar{\gamma}$ as $t \downarrow \hat{t}_o$ (or as $t \uparrow \hat{t}_1$), then all curves in X near $\hat{\gamma}$ have an interval of uniform size about $\hat{t}_o$ $(\hat{t}_1)$ in their domain. It follows from this that $\hat{\gamma}$ is not a maximal element in the limit set. Consequently, $\hat{\gamma}$ must have closed graph, and thus $\hat{\gamma} \in C(\Gamma)$.

Since $C(\Gamma)$ is second countable, the sequential compactness of the closure of X in $C(\bar{t},\bar{K})$ implies its compactness, and the theorem follows.

Consider now the reaction-diffusion equation on the set $\Omega \times \mathbb{R}_+$, where Ω is a bounded domain in $\mathbb{R}^m$ having "nice" boundary:

$$u_t = D\Delta u + f(u), \; bu = 0 \text{ on } \partial\Omega, \; u(x,0) = u_o(x), \; x \varepsilon \Omega. \tag{2}$$

Here bu represents the boundary conditions, say homogeneous Neumann or Dirichlet conditions, $u = u(x,t)$ is in $\mathbb{R}^n$, D is a non-negative diagonal matrix, and f, as well as u_o are smooth functions. We always assume that <u>the system</u> (2) <u>admits arbitrarily large bounded invariant regions</u> (rectangles), see [1].

We shall find it convenient to rewrite (2) in an abstract form as

$$u' = Au + f(u), \; u(0) = u_o, \tag{3}$$

where u is in Γ, the C^1 curves on some space of functions satisfying the boundary conditions, in the domain of A. Of course it is well-known that (2) does not define a (2-sided) flow. In order to get around this difficulty, we shall append past histories to solutions of (3). This is made possible by the fact that solutions of (3) satisfy a uniform modulus of continuity estimate up to $t = 0$; namely we have an estimate of the form ([6]):

$$|u(t)-u(s)|_0 \leq C(1+|u_o|_1)|t-s|^{\nu}, \quad t, s \; \varepsilon \; \mathbb{R}_+ \tag{4}$$

where $0 < \nu \leq 1/2$, C and ν are independent of t, s and u. Here $|\cdot|_o$ is the C^o-(sup)-norm, and $|\cdot|_1$ denotes the C^1-norm.

We can define a (trivial) flow on Γ by

$$(\gamma,t) \; \gamma\cdot t, \text{ where } (\gamma\cdot t)(s) = \gamma(t+s),$$

for $\gamma \; \varepsilon \; \Gamma$, t, s $\varepsilon \; \mathbb{R}$. Of course, this is merely the translation flow on Γ. We now define the candidate for a local flow in Γ; namely, consider the set

$$X = X(R) = \{u \; \varepsilon \; \Gamma\colon u(\cdot) \; \varepsilon \; C^1, \; u \text{ satisfies (4) (for } t, s \text{ in } \mathbb{R}), \; u(0) \; \varepsilon \; R, \text{ and for } t \geq 0, \; u \text{ satisfies (3)}\},$$

where R is a given invariant region. Theorem 10 shows that X is locally compact in Γ. To see that X is a local flow, let $\bar{u} \; \varepsilon_o X$. Then $\bar{u}(0)$ lies in the invariant region R.* Hence there is a C^o-neighborhood U of $\bar{u}$ such that if $u \; \varepsilon \; U$, $u(0) \; \varepsilon \; R$. (The solution exists since we have assumed the existence of bounded invariant regions). Again for t, s in $\mathbb{R}$, the

* So C and $|u_o|_1$ depend only on R (in (4)).

estimate (4) holds for u, and thus $u \in X$. We have thus shown that for any $\bar{u} \in X$, there is a C^1-neighborhood of $\bar{u}$ in Γ such that $(U \cap X) \cdot \mathbb{R}_+ \subset X$. This proves that X is, in fact, a semi-flow.

This result enables us to apply the theory discussed in §2 to systems of the form (2). Of course the difficulty will now be how to recognize isolated invariant sets and Morse decompositions. Our only general tool for these things is theorem 8. This requires that we prove that our equations admit a gradient-like structure. This will always be true for scalar equations, but we shall also show how to apply the theory to some interesting systems of reaction-diffusion equations.

§ 4 STABILITY OF STATIONARY SOLUTIONS

In this section we shall be concerned with a scalar equation in one-space variable:

$$u_t = u_{xx} + f(u), \quad |x| < L, \quad t > 0, \tag{5}$$

which satisfies the boundary conditions

$$(\alpha u + (\alpha - 1)u_x)(-L,t) = 0 = (\alpha u + (1-\alpha)u_x)(L,t), \quad t > 0, \tag{6}$$

where $0 \leq \alpha \leq 1$. Notice that if $\alpha = 0$ or $\alpha = 1$, (6) gives homogeneous Neumann or Dirichlet conditions.

If we define the functional Φ by

$$\Phi(u) = \int_{|x|<L} \left(\frac{1}{2} u u_{xx} + F(u)\right) dx, \quad F' = f,$$

then it is easy to check that $\Phi' > 0$, and $\Phi = 0$ only at the rest points (i.e., stationary solutions) of Φ. Thus our equations are all gradient-like.

A. Dirichlet Boundary Conditions. We consider a general function f(u) which (qualitatively) can be depicted as in figure 3. That is, $0 < f(u)$ if $u < 0$, $f(u) < 0$ if

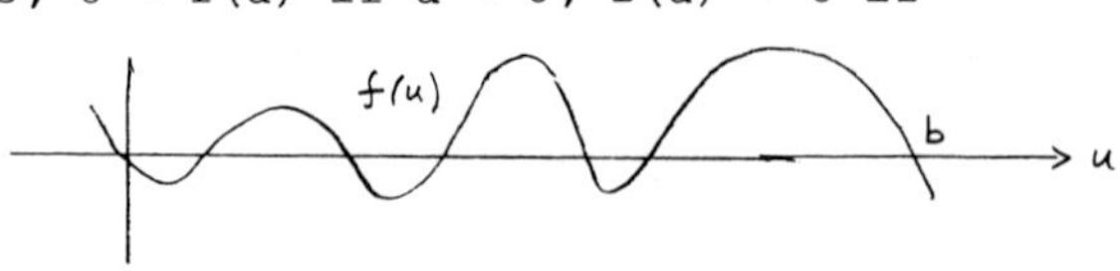

FIGURE 3

$u > b$, $f'(0) < 0$, $f'(b) < 0$ and $\int_0^b f(u) > 0$. The steady-state equation $u'' + f(u) = 0$, can be written as the first order system

$$u' = v, \; v' = -f(u), \quad u(\pm L) = 0. \tag{7}$$

Note that $u \equiv 0$ is always (i.e., for every $L > 0$), a solution of (7). For this system, we may assume that the phase plane can be depicted as in figure 4. Solutions of (7) are thus represented by orbit

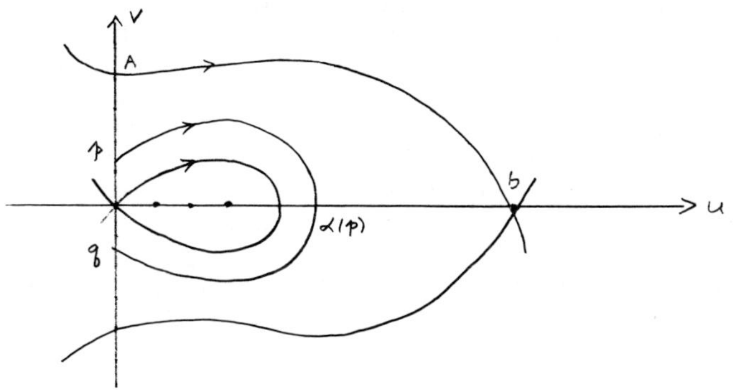

FIGURE 4

segments which begin on the line $u = 0$, where $0 < v < A$, end on the v-axis, and take "time", (parameter length), 2L. If $T(p)$ denotes the time an orbit takes to go from p to $\alpha(p)$ (see Figure 4), then we may write

$$T(p) = \int_0^{\alpha(p)} \frac{du}{\sqrt{2F(\alpha(p)) - 2F(u)}} .$$

We assume that $T(p)$ is a Morse function; that is, T has a finite number of critical points, all of which are non-degenerate[3]. Thus T has the qualitative form as depicted in figure 5. From the results in [8], we

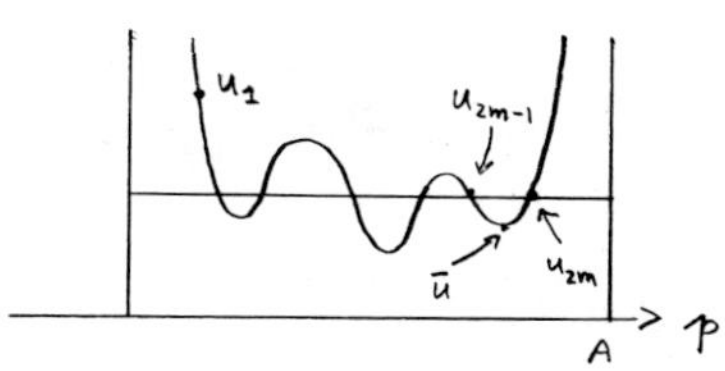

FIGURE 5

know that if $T'(p) \neq 0$, then the corresponding solution u is non-degenerate in the sense that 0 is not in the spectrum of the linearized operator. Each non-degenerate solution is an isolated rest point of (5) and (6) ($\alpha = 1$) and so by theorem 8, it is an isolated invariant set. It is also easy to see that (7) has an even number of non-degenerate non-constant solutions, $u_1, u_2, \ldots, u_{2m}$.

We shall now compute the dimensions of the unstable manifolds of these solutions. To this end, note that for large L, say $L > L_1$, there are only three solutions; namely 0, u_1, and u_{2m}. Using a method described in [3], we can show that $h(u_{2m}) = \Sigma^o$, for $L \gg 1$, and thus $h(u_{2m}) = \Sigma^o$ for all L, by the continuation property. Since u_{2m} is non-degenerate, and the spectrum of the associated linearized operator $d^2/dx^2 + f'(u_{2m}(x))$ has a finite number of positive eigenvalues, it follows that u_{2m} is an attractor, i.e., that it is a stable solution.

Now consider the solution u_{2m-1}. Note that both u_{2m} and u_{2m-1} bifurcate out of the degenerate solution $\bar{u}$ (figure 5). Since $\bar{u}$ continues to the empty set, we have $h(\bar{u}) = \bar{0}$. At this point we need a lemma.

<u>Lemma</u> 11. Suppose that $u = u_k$ and $v = u_{k+1}$ are a unique pair of solutions of (7), not both degenerate which bifurcate out of a degenerate solution $\bar{u}$. Then if $h(u) = \Sigma^n$ and $h(v) = \Sigma^k$, $n = k \pm 1$, and there are solutions of (5), (6) which connect u and v.

<u>Proof</u>. u and v are the only rest points near $\bar{u}$ contained in an isolating neighborhood N, where $h(I(N)) = \bar{0}$. If there were no other complete orbit in N, then the addition formula (*) would give the contradiction

$$\bar{0} = h(I(N)) = h(u) \vee h(v) \neq \bar{0}.$$

Thus there is a third complete orbit γ contained in N. The gradient-like nature of the flow forces γ to connect u and v. We now use (1) to compute indices. Suppose that $h(u) = \Sigma^n$ and $h(v) = \Sigma^k$, and that the orbit runs from u to v. Then if $u = M_2$, $v = M_1$ we have the exact sequence

$$\cdots \to H^{p-1} \to h(I(N)) \to H^{p-1}(h(v)) \to H^p(h(u)) \to H^p(h(I(N)) \to \cdots ,$$

so that

$$\cdots \to 0 \to H^{p-1}(\Sigma^k) \to H^p(\Sigma^n) \to 0 \to \cdots .$$

It follows that $H^{p-1}(\Sigma^k)$ is isomorphic to $H^p(\Sigma^n)$. Thus

$$H^{p-1}(\Sigma^k) \approx H^p(\Sigma^n) = \begin{cases} 0, & \text{if } p \neq n \\ \mathbb{Z}, & \text{if } p = n. \end{cases}$$

But $H^{p-1}(\Sigma^k) = \mathbb{Z}$ if $p = k + 1$, and is 0 otherwise. It follows that $n = k + 1$, and the proof is complete.

Using this result we see that $h(u_{2m-1}) = \Sigma^1$ and that $h(u_{2m-2}) = \Sigma^0$ or Σ^2. This is all that the topological techniques can give us. To remove the ambiguity, we need the following lemma, which takes advantage of the fact that all solutions of (5) are non-negative.

Lemma 12. Let u be a non-degenerate solution of (7). Then u has at most a one-dimensional unstable manifold.

Proof. Consider the eigenvalue equation

$$v'' + f'(u(x))v = \lambda v, \quad v(\pm L) = 0 .$$

Set $w = u'$; then $w'' + f'(u(x))w = 0$. If $\lambda > 0$, then $f'(u(x)) - \lambda < f'(u(x))$ so by the Sturm comparison theorem, w must have a zero between any two zeros of v. The non-negativity of u means that w has precisely one zero on $|x| < L$. It follows that λ is the principal eigenvalue, and we are done.

Using this lemma, it follows that $h(u_{2k}) = \Sigma^0$, and $h(u_{2k-1}) = \Sigma^1$, $k = 1, 2, \ldots, m$. Since 0 is easily seen to be a stable solution, we may depict (schematically), the entire set of solutions of (5),(6) ($\alpha = 1$) in the following diagram. Here u_{2k-2} and u_{2k} are connected to u_{2k-1} by orbits.

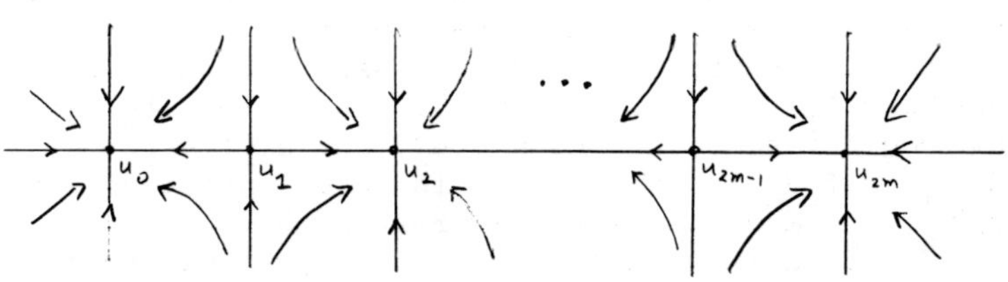

FIGURE 6

and solutions in the infinite dimensional subspace outside of this "plane", tend to this "plane" as $t \to \infty$. There are no other solutions, i.e., we have given the complete qualitative description of all solutions of (5), (6)($\alpha = 1$) .

B. Neumann Boundary Conditions. We now consider the problem (5), (6) ($\alpha = 0$). If we again consider the associated order system for the stationary solutions, we know from [9] that these solutions can all be depicted as in figure 7; (we are assuming that $f(u)$ is a cubic polynomial of the form $f(u) = -u(u+b)(u-1)$, $0 < b < 1/2$).

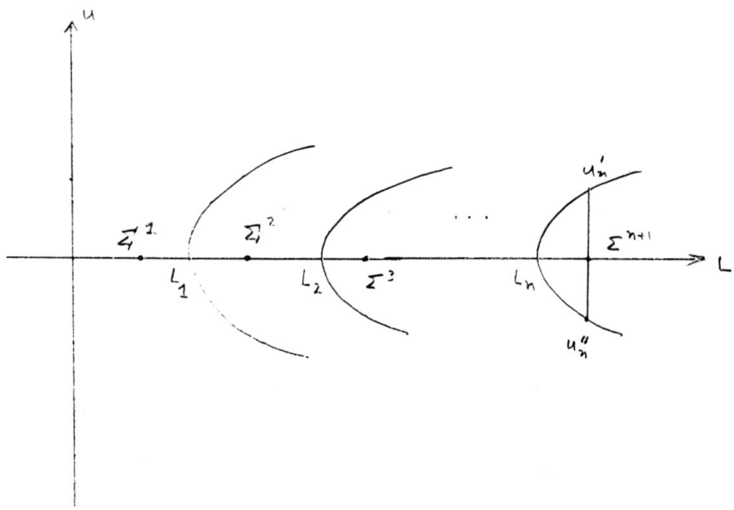

FIGURE 7

The constants $u = b$, $u = 1$ and $u = 0$ are solutions for each $L > 0$, and one can check by linearization, that $u = b$ and $u = 1$ are always stable. Furthermore, as L passes thru certain discrete numbers $L_1, L_2, \ldots$, the 0 solution bifurcates into a pair of non-degenerate solutions u_n', u_n'', and $h(0) = \Sigma^{n+1}$, if $L_n < L < L_{n+1}$. If S_n is the maximal isolated invariant set containing u_n', u_n'', and 0 as its only rest points, then $h(S_n) = \Sigma^n$. Since $u_n'(x) = u_n''(-x)$, it is easy to see that $h(u_n'') = h(u_n')$. Let M_n, M_n', and M_n'' be Morse sets in S_n containing 0, u_n' and u_n'', respectively. We know that $h(u_n') = h(u_n'') = \Sigma^k$ for some k. Here is the main theorem.

Theorem 13. $k = n$, and there exist orbits of (5), (6) $\alpha = 0$, which connect 0 to both u_n' and u_n''.

Proof. First note that exactly as in lemma 11, there must be orbits which connect the critical points. Since it is easy to check that $\Phi(u_n') = \Phi(u_n'')$, these connect 0 to u_n', and by symmetry, 0 to u_n''.

Assume that there are orbits running from 0 to u_n', u_n''. Then we have the Morse decomposition of S_n given by $(u_n' \amalg u_n'', 0)$, where $\amalg$ means disjoint union. Using (1), we find

$$\ldots \to H^k(h(S_n)) \to H^k(h(u_n')) + H^k(h(u_n'')) \to H^{k+1}(h(0)) \to \ldots$$

or

$$\ldots \to H^k(\Sigma^n) \to H^k(h(u_n')) + H^k(h(u_n'')) \to H^{k+1}(\Sigma^{n+1}) \to \ldots .$$

If $k \neq n$, we obtain

$$0 \to H^k(h(u_n')) + H^k(h(u_n'')) \to 0 ,$$

so that $H^k(h(u_n')) = H^k(h(u_n'')) = 0$, for all $k \neq n$. Since $h(u_n') = \Sigma^p$ for some p, we see $p = n$.

Now we shall show that it is not possible for the connecting orbits to run from u_n' and u_n'' to 0. For if this were true, S_n would have the Morse decomposition $(0, u_n' \amalg u_n'')$. From (1) again, we have the long exact sequence

$$\ldots \to H^{k-1}(\Sigma^n) \to H^{k-1}(\Sigma^{n+1}) \to H^k(h(u_n')) + H^k(h(u_n'')) \to H^k(\Sigma^n) \to \ldots .$$

If $k = n + 2$, we obtain

$$0 \to \mathbb{Z} \to H^{n+2}(h(u_n')) + H^{n+2}(h(u_n'')) \to 0 ,$$

which gives the contradiction $\mathbb{Z} \simeq H^{n+2}(h(u_n')) + H^{n+2}(h(u_n''))$. This completes the proof of the theorem.

C. The Fitz-Hugh Nagumo Equations. We end this section by showing how our results can be applied to the Fitz-Hugh-Nagumo equations, complete details will be given in a forthcoming publication.

Thus consider the system

$$v_t = v_{xx} + f(v) - u, \quad u_t = \delta v - \gamma u \qquad |x| < L \tag{8}$$

where $f(v) = -v(v-a)(v-1)$, $0 < a < 1/2$. Here δ and γ are positive constants. We shall study the Dirichlet problem for (8); i.e., we seek solutions of (8) which satisfy the boundary conditions

$$v(\pm L, t) = 0, \qquad t > 0. \tag{9}$$

The stationary solutions of (8), (9) are solutions of

$$v'' + f(v) - u = 0, \quad \delta v - \gamma u = 0 , \tag{10}$$

where

$$v(\pm L) = 0 . \tag{11}$$

Observe that (10) implies that v solves the scalar equation

$$v'' + \phi(v) = 0, \quad \phi(v) = f(v) - \delta v/\gamma , \tag{12}$$

subject to the same boundary conditions (11). Now in order that this problem have non-zero solutions, we assume that the function ϕ has three roots, $0 < b < c$, and $\int_0^c \phi(v) > 0$.

Since ϕ is a cubic polynomial with three non-negative roots, we may apply the results of [9] to conclude that there is an $L_o > 0$ such that if $L > L_o$, (10), (11) has precisely three solutions, (0,0), $U_o = (1,\delta/\gamma)v_o$, $U_1 = (1,\delta/\gamma)v_1$, where v_1 and v_o are solutions of (12). In view of our results in §A, we may assume that the spectrum of the operator $d^2/dx^2 + \phi'(v_i)$ contains i positive elements, i = 0, 1. Using this fact, we can show that if $\gamma^2 \geq \delta$, U_o is a stable solution of (8), (9), while U_1 has a 1-dimensional unstable manifold. Furthermore, the solution (0,0) is always stable. Now it is a remarkable fact that for this range of parameter values, the system (8), (9) is gradient-like. It follows from this that if $\gamma^2 \geq \delta$, the complete orbit structure of (8), (9) can be determined; i.e., we can describe, qualitatively, all the solutions of the Fitz-Hugh-Nagumo equations.

The relevant parameter values can be depicted in figure 8. Here Δ is chosen such that if $\delta/\gamma > \Delta$

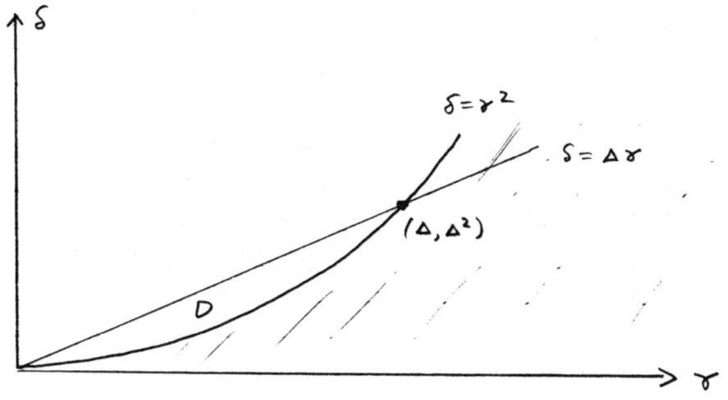

FIGURE 8

then (12), (11) has only $v = 0$ as a solution, and conversely. The Lyapunov function exists below the parabola. As δ crosses the parabola from below (with $0 < \gamma < \Delta$), the solution U_o undergoes a Hopf bifurcation, and a periodic solution appears. As δ increases further into the region D still more interesting solutions appear.

§ 5 CONNECTION INDICES

We want to briefly describe how one can use these algebraic-topological methods to find travelling wave solutions which connect critical points. We shall outline the theory, and illustrate it by an example.

Let S be an isolated invariant set and let $\{M_\pi : \pi \varepsilon P\}$ be a Morse decomposition of S. We call $I \subset P$ an <u>interval</u> if it corresponds to an interval in some admissible ordering. Then I determines a Morse set (i.e., an element in a Morse decomposition) M_I, defined by

$$M_I = \{\gamma : \omega(\alpha) \cup \alpha(\gamma) \subset \bigcup_{\pi \varepsilon I} M_\pi\} .$$

(I,J) is an <u>adjacent</u> pair of intervals if the elements of I come before those of J for some admissible ordering in which $I \cup J$ is also an interval. In particular, with the ordering $M_1, \ldots, M_n$, (j, j+1) is an adjacent pair of intervals.

If (I,J) is an adjacent pair of intervals, then there exists a coboundary map $\partial_{JI}: H_I \to H_J$, where H_I and H_J are the cohomology indices of M_I and M_J respectively. The map comes from the exact sequence

$$\cdots \to H_J^{n-1} \to H_{I\cup J}^n \to H_I^n \xrightarrow{\partial_{JI}} H_J^n \to \cdots .$$

All of these maps are uniquely defined by the flow, and are invariant under continuation of the admissible orderings. For example, if we have a Morse filtration $N_0 \subset N_1 \subset \ldots \subset N_n$ for the admissible ordering $(M_1, M_2, \ldots, M_n)$ of a Morse decomposition of S, then we have the natural inclusions

$$(N_{i-1}, \phi) \subset (N_i, \phi) \subset (N_i, N_{i-1}) ,$$

which induce exact sequences on cohomology; namely,

$$\ldots \to H^k(N_{i-1}) \to H^{k+1}(N_i, N_{i-1}) \to H^{k+1}(N_i) \xrightarrow{\partial_{i,i-1}} H^{k+1}(N_{i-1}) \to \ldots .$$

If $\partial_{i,i-1} \neq 0$ for some k, then there are solutions which go from M_i to M_{i-1}.

The object now is to define more maps which might become defined when the flow is perturbed. These are not generally uniquely defined since "bifurcation" may occur. For example, consider the picture below. Here S is a set containing two hyperbolic

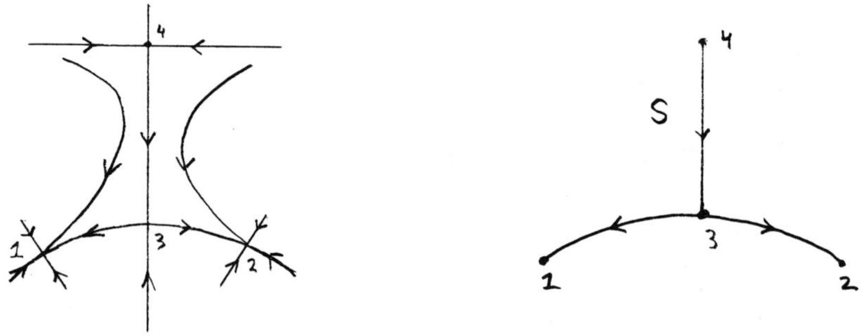

FIGURE 9

points, two attractors, and three connecting orbits. The Morse sets are the critical points, and the admissible orderings are (1,2,3,4) and (2,1,3,4). The flow defines ∂_{21}, ∂_{32}, ∂_{43}, ∂_{31}

and ∂_{12}. Of these, all are zero except ∂_{31} and ∂_{32}, which are isomorphisms. Note that $\partial_{43} = 0$ even though there is a connection - it isn't stable. Now under perturbation, one can get either of the following isolated invariant sets in figure 10.

FIGURE 10

In either case, one has the new admissible orderings (1,2,4,3) and (2,1,4,3) so that maps ∂_{42} and ∂_{41} become defined. In case A, $\partial_{42} = 0$ and ∂_{41} is an isomorphism, while in case B, ∂_{42} is an isomorphism, and $\partial_{41} = 0$.

Rules can now be given for defining the new maps so that all possibilities obtained by perturbation will be defined, and (in an "algebraic" way, at least), <u>only</u> all such possibilities.

Let $C = \oplus \{H_\pi : \pi \in P\}$; $C_I = \oplus \{H_\pi : \pi \in I\}$.

$\Delta: C \to C$ is called a cohomology map if $\Delta^2 = 0$. $H(\Delta) = \ker \Delta / \mathrm{Im}\, \Delta$, Pr I, and In I are the natural projection and injection of C_I relative to C. Here are the four rules.

1) $\Delta_I = \Pr I\, \Delta\, \mathrm{In}\, I$ is a cohomology map. There is an isomorphism $\Phi_I : H(\Delta_I) \to H_I$.

2) $\Delta_{JI} = \Pr J\, \Delta\, \mathrm{In}\, I$ induces a cohomology map $\tilde{\Delta}_{JI} : H(\Delta_I) \to H(\Delta_J)$.

3) If ∂_{JI} is defined by the flow, $\Phi_J \tilde{\Delta}_{JI} \Phi_I^{-1} = \partial_{JI}$.

4) If $(\ldots M_{\pi'}, \ldots M_\pi \ldots)$ is an admissible orderings, then $\Delta_{\pi'\pi} = 0$.

In the above example,

$$\Delta: \begin{pmatrix} H_4 \\ H_3 \\ H_2 \\ H_1 \end{pmatrix} \to \begin{pmatrix} H_4 \\ H_3 \\ H_2 \\ H_1 \end{pmatrix} \quad \text{is (partially) given by}$$

$$\Delta = \begin{bmatrix} 0 & 0 & - & - \\ 0 & 0 & 1 & 1 \\ 0 & 0 & 0 & 0 \\ 0 & 0 & 0 & 0 \end{bmatrix} .$$

The blanks can be filled in by 1, 0, in case B, or 0, 1, in case A. These are the only possibilities since $\operatorname{rank}(\Delta) = 2$; namely, $H(\Delta)$ is isomorphic to $H(S) = 0$. Thus dim ker = dim Im.

Now consider the equations[4]

$$\begin{aligned} u' &= v \\ v' &= -cv - \nabla F(u) \\ c' &= \delta(\bar{c}^2 - c^2) \qquad \bar{c} = \text{const.},\ \delta \text{ small},\ \delta > 0. \end{aligned} \tag{13}$$

Assume that F has finitely-many critical points, that $F(u) \to -\infty$ as $\|u\| \to \infty$ and for some constant $k < \inf F|\{\text{critical points}\}$, $F = k$ is a convex surface. Then the set of bounded solutions S of (13) is an isolated invariant set having u coordinates contained in the set $F \geq k$; the same statement is true for the equation $\dot{u} = \nabla F$. If u^* is an attracting critical point of F, then the index of the (relative) repellor in S ($\dot{u} = \nabla F$) has index $\bar{0}$ (see figure 11)

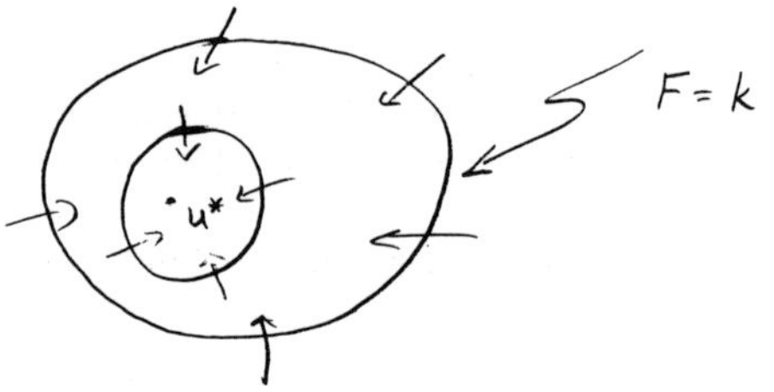

FIGURE 11

Now if $\bar{c} >> 1$, then the (two) invariant sets in $c = \pm\bar{c}$ (respectively) have the same connection structure as $\dot{u} = \nabla F$, except for index and direction. To see this, note that since $\bar{c}$ is large, (13) is well-approximated by the system $u' = -c^{-1}\nabla F(u)$, $v' = -cv$. Thus, if $\tilde{u}$ is a critical point of F with index $h(\tilde{u}, \nabla F) = \Sigma^r$, then $(\tilde{u}, 0, -\bar{c})$ has index Σ_-^{r+n+1}; the n comes from the v-equation, and the 1 from the $\bar{c}$ equation. Similarly, $(\tilde{u}, 0, \bar{c})$ has index Σ^{n-r}, the complementary index for the reversed flow. Therefore, if $\tilde{u}$ and $\tilde{\tilde{u}}$ are attractors for $\dot{u} = \nabla F$, then there is an admissible ordering of the critical points which has a Morse decomposition in the form

$$\ldots,\ (\tilde{u}, 0, -\bar{c}),\ (\tilde{\tilde{u}}, 0, \bar{c}),\ \ldots .$$

The dotted parts have index $\bar{0}$ since their correspondants for $\dot{u} = \nabla F$ do. The total index being $\bar{0}$ implies that $(\hat{u}, 0, -\bar{c})$, $(\tilde{\tilde{u}}, 0, \bar{c})$ has index $\underline{0}$. Thus $(\check{u}, 0, -\bar{c})$ is algebraically connected to $(\tilde{\tilde{u}}, 0, \bar{c})$.

With a limiting argument, one now has the following theorem concerning travelling waves; namely, there is a chain of waves from $\check{u}$ to $\tilde{\tilde{u}}$ via other attractors, along which the wave velocity increases:

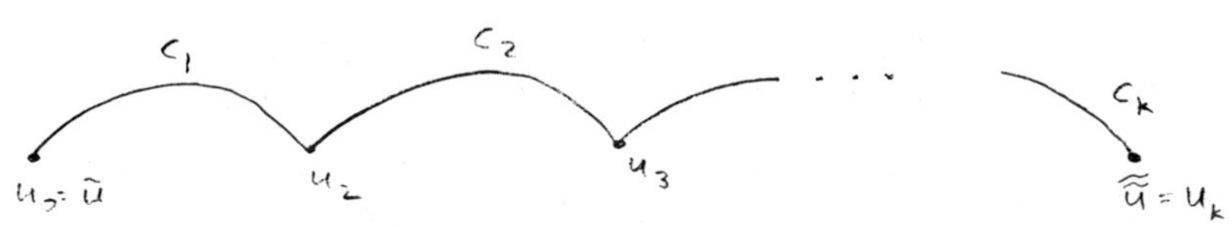

$$c_1 < c_2 < \dots < c_k.$$

A finer argument shows that all but one of the attractors must admit a homoclinic orbit. Complete details of this outline will be described in a forthcoming publication.

FOOTNOTES

[1]Research for both authors supported in part by the NSF.
[2]N_o is a distinguished point in the quotient space; hence we use the notation Σ^1 rather than S^1, in order to keep this in mind.
[3]This hypothesis is true generically; see [10]. It is also true if f is a cubic polynomial, or even if f is C^1-close to a cubic polynomial, see [9].
[4]Notice that the first two equations are the travelling wave equations for the equation $u_t = u_{xx} + \nabla F(u)$.

REFERENCES

1. Chueh, K., Conley, C., and Smoller, J., Positively invariant regions for systems of nonlinear diffusion equations, Ind. U. Math. J., 26, (1977), 373-392.

2. Conley, C., Isolated Invariant Sets and the Morse Index, Conf. Bd. Math. Sci., No.38, Amer. Math. Soc., Providence, (1978).

3. Conley, C., and Smoller, J., Remarks on the stability of steady-state solutions of reaction-diffusion equations, in Bifurcation Phenomena in Mathematical Physics and Related Phenomena, ed. by C. Bardos and D. Bessis, Reidel Publ. Co., Dordrecht, Holland, (1980), 47-56.

4. Conley, C., and Smoller, J., Topological techniques in reaction-diffusion equations, in Biological Growth and Spread, Springer Lecture Notes in Biomathematics, No. 38, (1980), 473-483.

5. Conley, C., and Zehnder, E., Morse type index theory for flows and periodic solutions for Hamiltonian equations, to appear.

6. Mora, X., Semilinear parabolic problems define semiflows on C^k-spaces, Trans. Amer. Math. Soc., to appear.

7. Smoller, J., Shock Waves and Reaction-Diffusion Equations, Springer-Verlag, New York, (1982), to appear.

8. Smoller, J., Tromba, A, and Wasserman, A., Non-degenerate solutions of boundary-value problems, Nonlinear Anal., 4, (1980), 207-215.

9. Smoller, J., and Wasserman, A., Global bifurcation of steady-state solutions, J. Diff. Eqns., 39, (1981), 269-290.

10. Smoller, J., and Wasserman, A., Generic bifurcation of steady-state solutions, to appear.

HYPERBOLIC SYSTEMS OF CONSERVATION LAWS

C. M. Dafermos

Lefschetz Center for Dynamical Systems
Division of Applied Mathematics
Brown University
Providence, Rhode Island 02912 USA

INTRODUCTION

The study of systems of quasilinear hyperbolic equations that result from the balance laws of continuum physics was initiated more than a century ago yet, despite considerable progress in recent years, most of the fundamental problems in the analytical theory remain unsolved. From the outset the student of the subject encounters obstacles such as the non-existence of globally defined smooth solutions, as a reflection of the physical phenomenon of breaking of waves and development of shocks, and the nonuniqueness of (weak) solutions in the class of discontinuous functions. Furthermore, no strong a priori estimates have yet been discovered for systems of more than one equation so that the subject is not amenable to the functional analytic techniques that have swept through the theory of partial differential equations of other types. It is only now that a functional analytic treatment of these systems is emerging, based on the delicate theory of compensated compactness which relies upon weak a priori estimates. The reader may find an account of these interesting recent developments in the article by L. Tartar in this volume.

The umbilical cord that joins the theory of systems of hyperbolic conservation laws with continuum physics is still vital for the proper development of the subject and should not be severed. In fact it is by using this connection that it has become possible to identify admissible solutions and thus resolve the issue of nonuniqueness. Furthermore, it is evident that systems arising in continuum physics are endowed with special features that have yet to be fully detected and incorporated in

J. M. Ball (ed.), Systems of Nonlinear Partial Differential Equations, 25–70.

the abstract theory.

Much of the recent progress in the area is due to the development of the random choice method which constructs solutions to the Cauchy problem for systems in one space dimension, under initial data of small total variation, by monitoring the propagation and interaction of waves. The advantage of this approach is that it is explicitly constructive so not only does it establish the existence of solutions but it also provides a very detailed picture of their rich geometric structure as well as a technique for their numerical computation. On the other hand, the explicitness of the random choice method turns into a drawback when one attempts to apply it to multidimensional problems. In fact at the present time nothing is known about the existence and uniqueness of global solutions to the Cauchy problem for systems of more than one equation in several space dimensions. Moreover, multidimensional wave interaction patterns are extremely intricate so it is anticipated that the effort to unravel the geometric structure of solutions will face serious difficulties.

The objective of these lecture notes is to give an overview of the current state of affairs in the theory of hyperbolic systems of balance laws with emphasis to those aspects of the subject that are intimately related to continuum physics. With this in mind, after formulating, in Section 1, the general problem, we introduce as models, in Section 2, the systems that result from the balance laws of mass, momentum and energy of elasticity and adiabatic thermoelasticity in one and several space dimensions. The feature of nonexistence of classical solutions is exhibited in Section 3 and the geometric structure of solutions in the class of functions of bounded variation is described in Section 4. Nonuniqueness of weak solutions is demonstrated in Section 5. Sections 6 to 9 present the motivation behind the shock, entropy, viscosity and entropy rate admissibility criterion and discuss the interconnections between these criteria. In Section 10 we treat the Riemann problem whose solution provides an enlargement of the local structure of general solutions in one space dimension. Finally, in Section 11, we sketch the construction of the solution to the Cauchy problem in one space dimension by means of the random choice method.

At the present time there is no comprehensive treatment of the subject in book form even though extensive parts are covered in the monographs by Courant and Friedrichs [11], Rozhdestvensky and Yanenko [55] and Jeffrey [31]. The forthcoming text by Smoller [57] will fulfill a need. There are earlier surveys, with different emphases, by Lax [33,34] and Dafermos [12]. The bibliographic information at the end of each section provides only a sample of relevant references and is by no means comprehensive.

In concluding, I wish to thank John Ball for his hospitality during the conference and acknowledge the support of the U.S. Army, under contract ARO-DAAG-29-79-C-0161 and the National Science Foundation, under grant MCS-79-05774-05.

1. SYSTEMS OF HYPERBOLIC BALANCE LAWS

In the framework of continuum physics, individual theories are identified by the list of quantities that satisfy equations of balance, stating that production in any region of space-time is balanced by a flux through the boundary of this region. For example, continuum mechanics is founded on the balance laws of mass and momentum while continuum thermomechanics rests upon the balance laws of mass, momentum and energy. It is generally postulated that the production of a balanced quantity in a sufficiently smooth set Ω is expressed as an integral of a production density over Ω and the flux of the quantity through the boundary $\partial\Omega$ of Ω is given by the integral of a flux density over $\partial\Omega$. It is further assumed that the flux density at any point of $\partial\Omega$ depends solely upon the point itself and the local orientation of $\partial\Omega$ at this point.

To express the above ideas in symbols, we let R^{m+1}, with typical point $x=(x_0,x_1,\ldots,x_m)$, play the role of space-time; x_0 denotes time and $(x_1,\ldots,x_m)$ are the space coordinates. A balance law over an open subset $\mathscr{B}\subset R^{m+1}$ is expressed by an equation

$$\oint_{\partial\Omega} h(x,N(x))dS = \int_{\Omega} g(x)dx \qquad (1.1)$$

which is to hold for every sufficiently smooth subset Ω of $\mathscr{B}$. In (1.1) $N(x)$ denotes the unit normal on $\partial\Omega$ at $x\in\partial\Omega$ and $h(x,N), g(x)$ take values in R^n.

Cauchy realized that an equation such as (1.1) cannot hold for arbitrary Ω unless $h(x,N)$ is linear in N. We sketch Cauchy's original proof under the (stronger than necessary) assumption that $g(x)$ is bounded and $h(x,N)$ is continuous in x for fixed N.

We fix $\bar{x}\in\mathscr{B}$ and $N\in R^{m+1}$ with $|N| = 1$. We apply (1.1) for the m-dimensional ball $\Omega = \{x\in\mathscr{B} : (x-\bar{x})\cdot N = 0, |x-\bar{x}| \le r\}$, we multiply through by r^{-m} and let $r\downarrow 0$ to deduce $h(\bar{x},N) = -h(\bar{x},-N)$. Next we pick an orthonormal basis

$e^0, e^1, \ldots, e^m$ in R^{m+1}, such that $N = \sum_{\alpha=0}^{m} N_\alpha e^\alpha$ with $N_\alpha \neq 0$, and apply (1.1) for the simplex $\Omega = \{ x \in \mathscr{B} : (x_\alpha - \bar{x}_\alpha)N_\alpha \geq 0$, $\alpha = 0,\ldots,m$, $\sum_{\alpha=0}^{m} (x_\alpha - \bar{x}_\alpha)N_\alpha \leq r\}$. Multiplying through by $m! r^{-m} N_0 N_1 \ldots N_m$, namely the inverse of the measure of the face of the simplex which is orthogonal to N, and passing to the limit as $r \downarrow 0$ we conclude that $h(\bar{x},N)$ is linear in N, that is

$$h(\bar{x},N) = \sum_{\alpha=0}^{m} N_\alpha h(\bar{x},e^\alpha). \tag{1.2}$$

As a consequence of the above result, balance laws are always written in the form

$$\oint_{\partial\Omega} \sum_{\alpha=0}^{m} f^\alpha(x) N_\alpha(x)\,dS = \int_\Omega g(x)\,dx \tag{1.3}$$

where $f^\alpha(x)$, $\alpha=0,1,\ldots,m$, are fields taking values in R^n. By the Gauss-Green theorem, whenever the $f^\alpha(x)$ are continuously differentiable and $g(x)$ is continuous, (1.3) is equivalent to the field equation

$$\sum_{\alpha=0}^{m} \partial_\alpha f^\alpha(x) = g(x), \qquad x \in \mathscr{B} \tag{1.4}$$

with ∂_α standing for $\partial/\partial x_\alpha$.

In continuum physics balance laws are supplemented with *constitutive relations*. In general terms a constitutive relation characterizes the medium by relating the values of f^α, $\alpha=0,1,\ldots,m$, and g in (1.4) at any point $x \in \mathscr{B}$ to the state of the medium on an appropriate subset of $\mathscr{B}$. The state of the medium is commonly characterized by a vector field $U(x)$ defined on $\mathscr{B}$ and taking values in R^n. Here we shall be concerned with the simplest case in which f^α, $\alpha=0,1,\ldots,m$ and g at $x \in \mathscr{B}$ are determined by the state of the medium at x alone, that is,

$$f^\alpha(x) = F^\alpha(U(x),x)\ , \ g(x) = G(U(x),x)\ , \tag{1.5}$$

where $F^\alpha(U,x)$, $\alpha=0,1,\ldots,m$, and $G(U,x)$ are known smooth functions defined on $R^n \times \mathscr{B}$ and taking values in R^n. When F^α, $\alpha=0,1,\ldots,m$, does not depend explicitly on x, the medium is called *homogeneous*. Henceforth, for the sake of simplicity we will be dealing with homogeneous media.

Combining the balance law (1.4) with the constitutive relation (1.5) we obtain the system of first order partial differential equations

$$\sum_{\alpha=0}^{m} \partial_\alpha F^\alpha(U(x)) = G(U(x),x) \tag{1.6}$$

from which the state vector field $U(x)$ is to be determined. We will call (1.6) a system of *balance laws*. Whenever $G \equiv 0$, (1.6) is called a system of *conservation laws*.

The system (1.6) is called *hyperbolic* in the x_0 direction if for every $U \in R^n$ and $\nu \in R^m$, with $|\nu| = 1$, the matrix $\nabla F^0(U)$ is nonsingular and the eigenvalue problem

$$\left[\sum_{\alpha=1}^{m} \nu_\alpha \nabla F^\alpha(U) - \lambda \nabla F^0(U)\right] r = 0 \tag{1.7}$$

has real eigenvalues $\lambda_1(U;\nu),\ldots,\lambda_n(U;\nu)$, called *characteristic speeds* in the direction ν, and n linearly independent *right characteristic eigenvectors* $r_1(U;\nu),\ldots,r_n(U;\nu)$. When the characteristic speeds are distinct, the system is called *strictly hyperbolic*.

With the set of right eigenvectors we associate a set of *left characteristic eigenvectors* $\ell_1(U;\nu),\ldots,\ell_n(U;\nu)$ normalized so that

$$\ell_i(U;\nu)^T \nabla F^0(U) r_j(U;\nu) = \delta_{ij}, \quad i,j = 1,2,\ldots,n. \tag{1.8}$$

Any system (1.6) with $\nabla F^0(U)$ nonsingular and $\nabla F^\alpha(U)$, $\alpha=0,1,\ldots,m$, symmetric is called *symmetric hyperbolic*.

Bibliographic Remarks: An exhaustive study of the balance laws of continuum physics is contained in the treatise on the classical field theories [60] by Truesdell and Toupin. For general information on hyperbolicity see, e.g., [31].

2. MODELS FROM CONTINUUM PHYSICS

We record here examples of systems of hyperbolic balance laws in continuum thermomechanics which have been providing the essential models for the development of the general theory of the subject.

We begin with the referential (commonly, though inappropriately, called Lagrangean) formulation of the set of balance laws of momentum and energy for an m-dimensional (m=1,2 or 3), homogeneous, thermoelastic nonconductor of heat with reference density $\rho_0 = 1$:

$$\begin{cases} \partial_t F_{i\alpha} - \partial_\alpha v_i = 0 & i,\alpha = 1,\dots,m \\ \partial_t v_i - \sum_{\alpha=1}^{m} \partial_\alpha T_{i\alpha}(F,\eta) = f_i & i = 1,\dots,m \\ \partial_t \{e(F,\eta) + \frac{1}{2}\sum_{i=1}^{m} v_i^2\} - \sum_{i,\alpha=1}^{m} \partial_\alpha\{v_i T_{i\alpha}(F,\eta)\} = \sum_{i=1}^{m} f_i v_i + r. \end{cases} \tag{2.1}$$

System (2.1) is written in the classical notation of mechanics so time is denoted by t rather than x_0, F is deformation gradient, v is velocity, T denotes the Piola-Kirchhoff stress tensor, η is entropy and e is internal energy. The body force f and heat supply r are prescribed, as functions of $(t,x_1,\dots,x_m)$. The state is characterized by F,v and η. Internal energy is determined by F and η through a given constitutive relation $e(F,\eta)$ which also induces the constitutive relations for stress and absolute temperature θ via the thermodynamic relations

$$T_{i\alpha}(F,\eta) = \frac{\partial e(F,\eta)}{\partial F_{i\alpha}}, \quad \theta(F,\eta) = \frac{\partial e(F,\eta)}{\partial \eta} \tag{2.2}$$

whose role will be elucidated in Section 7, when discussing the entropy admissibility criterion. We assume $\theta(F,\eta) > 0$ for all (F,η).

To (2.1) we should append

$$\partial_\beta F_{i\alpha} - \partial_\alpha F_{i\beta} = 0, \qquad i,\alpha,\beta = 1,\dots,m, \tag{2.3}$$

which expresses that F is a gradient. In view of $(2.1)_1$, (2.2) is just a restriction on initial data.

System (2.1) is hyperbolic provided $e(F,\eta)$ is strictly rank-one convex in F or, equivalently, the acoustic tensor

$$Q_{ij}(F,\eta;\nu) = \sum_{\alpha,\beta=1}^{m} \frac{\partial^2 e}{\partial F_{i\alpha}\partial F_{j\beta}} \nu_\alpha \nu_\beta \tag{2.4}$$

is positive definite, for any nonzero vector $\nu \in R^m$. The (positive and negative) square roots of the eigenvalues of Q and zero, with multiplicity m^2-m+1, are the characteristic speeds in the direction ν.

Under proper conditions, motions are assumed isentropic, that is, η = constant. Strictly speaking, an isentropic motion may be sustained only by controlling the heat supply r so that the energy equation $(2.1)_3$ remains balanced. Isentropic motions are to be determined by the first two of (2.1), namely,

$$\begin{cases} \partial_t F_{i\alpha} - \partial_\alpha v_i = 0 & i,\alpha=1, \dots, m \\ \partial_t v_i - \sum_{\alpha=1}^{m} \partial_\alpha T_{i\alpha}(F) = f_i & i = 1, \dots, m \end{cases} \tag{2.5}$$

with $T(F)$ determined through

$$T_{i\alpha}(F) = \frac{\partial e(F)}{\partial F_{i\alpha}} . \tag{2.6}$$

System (2.5) is hyperbolic when the acoustic tensor

$$Q_{ij}(F;\nu) = \sum_{\alpha,\beta=1}^{m} \frac{\partial^2 e}{\partial F_{i\alpha}\partial F_{j\beta}} \nu_\alpha \nu_\beta = \sum_{\alpha,\beta=1}^{m} \frac{\partial T_{i\alpha}}{\partial F_{j\beta}} \nu_\alpha \nu_\beta \tag{2.7}$$

is positive definite for every nonzero $\nu \in R^m$. The (positive and negative) square roots of Q and zero, with multiplicity m^2-m, are the characteristic speeds of (2.5) in the direction ν.

Next we record the one-dimensional versions of (2.1) and (2.5) in the familiar notation of the theory of gas dynamics. They read

$$\begin{cases} \partial_t u - \partial_x v = 0 \\ \partial_t v + \partial_x p(u,\eta) = 0 \\ \partial_t\{e(u,\eta) + \frac{1}{2}v^2\} + \partial_x\{vp(u,\eta)\} = 0 , \end{cases} \tag{2.8}$$

where

$$p(u,\eta) = -e_u(u,\eta) \quad , \quad \theta(u,\eta) = e_\eta(u,\eta) , \tag{2.9}$$

and

$$\begin{cases} \partial_t u - \partial_x v = 0 \\ \partial_t v + \partial_x p(u) = 0 . \end{cases} \tag{2.10}$$

In (2.8), (2.9), (2.10), u is specific volume, v is velocity, p is pressure, η is entropy, e is internal energy and θ

denotes absolute temperature.

System (2.8) is strictly hyperbolic, providing $\theta(u,\eta) > 0$ and $p_u(u,\eta) < 0$. The characteristic speeds are $\pm\sqrt{-p_u(u,\eta)}$ and 0. Similarly, (2.10) is strictly hyperbolic when $p_u(u) < 0$ and the characteristic speeds are

$$\lambda_1 = -\sqrt{-p_u(u)} \quad , \quad \lambda_2 = \sqrt{-p_u(u)} \; . \tag{2.11}$$

Bibliographic Remarks: A good source of information on the types of nonlinear evolution equations that arise in continuum physics is the encyclopedic work [59] by Truesdell and Noll.

3. BREAKING OF WAVES AND GENUINE NONLINEARITY

By a classical solution of (1.6) in $\mathscr{B} \subset R^{m+1}$ we mean a Lipschitz continuous vector field $U(x)$ which satisfies (1.6) almost everywhere in $\mathscr{B}$. It turns out that the class of classical solutions is too narrow to encompass all solutions of physical interest. The reason is that, as signals propagate at characteristic speed which depends on the state, wave profiles may get steeper and finally break in a finite time. Consequently, solutions of the Cauchy problem for (1.6) that start out from smooth initial data stay temporarily smooth but eventually develop singularities.

Blow-up of smooth solutions can be seen even in the simplest case of a single nonlinear hyperbolic equation in one space variable:

$$\partial_t u + \lambda(u)\partial_x u = 0. \tag{3.1}$$

Assuming that a smooth solution $u(x,t)$ of (3.1) exists in a strip $-\infty < x < \infty$, $0 \leq t < T$, we consider a characteristic Γ, that is, an integral curve $x = \xi(t)$ of the ordinary differential equation $\dot{\xi} = \lambda(u(\xi,t))$. By (3.1), $u(x,t)$ is constant along Γ, say $u(\xi(t),t) = \bar{u}$. Let us set $a(t) = \partial_x u(\xi(t),t)$. Differentiating (3.1) with respect to x we deduce that $a(t)$ satisfied the differential equation

$$\frac{da}{dt} + \nabla\lambda(\bar{u})a^2 = 0. \tag{3.2}$$

Thus, if $\nabla\lambda(\bar{u}) < 0$ (or > 0) and $a(0) > 0$ (or < 0) $a(t)$ will blow up in finite time. Therefore, the Cauchy problem for (3.1) will have a globally defined classical solution only when the initial data satisfy $\partial_x \lambda(u(x,0)) \geq 0$, a severe restriction unless of course (3.1) is linear.

We now turn to any homogeneous system of conservation laws (1.6) with $G \equiv 0$. We switch to the traditional notation writing t for x_0. We fix $\nu \in R^m$ with $|\nu| = 1$ and look for solutions which remain constant on m-dimensional hyperplanes orthogonal to ν; that is, $U = U(y,t)$ with $y = \sum_{\alpha=1}^{m} \nu_\alpha x_\alpha$. Any Lipschitz function of this type will be a solution of (1.6) if

$$\nabla F^0(U)\partial_t U + \sum_{\alpha=1}^{m} \nu_\alpha \nabla F^\alpha(U)\partial_y U = 0. \tag{3.3}$$

We assume Γ is a C^1 smooth curve $y = \xi(t)$ on the y-t plane defined on an interval $0 \leq t < T$ and $U(y,t)$ is a Lipschitz continuous solution of (3.3) on the strip $-\infty < x < \infty$, $0 \leq t < T$ and such that (a) on the right of Γ $U(y,t) = \bar{U} =$ constant and (b) on the left of Γ $U(y,t)$ is twice continuously differentiable and first and second derivatives $\partial_y U, \partial_t U, \partial^2_{yy} U, \partial^2_{yt} U$ attain limits $\partial_y U(\xi(t)-,t)$, $\partial_t U(\xi(t)-,t)$, $\partial^2_{yy} U(\xi(t)-,t)$, $\partial^2_{yt} U(\xi(t)-,t)$ on Γ^-. Moreover, $\partial_y U(\xi(t)-,t) \neq 0$ so that $\partial_y U$ experiences a jump discontinuity across Γ. We then say Γ is a plane *weak wave* propagating into the constant state $\bar{U}$ in the direction ν with speed $\dot{\xi}(t)$.

Since the restriction of U on Γ is constant,

$$\partial_t U(\xi(t)-,t) + \dot{\xi}(t)\, \partial_y U(\xi(t)-,t) = 0, \quad t \in [0,T). \tag{3.4}$$

At the same time (3.3) yields

$$\nabla F^0(\bar{U})\partial_t U(\xi(t)-,t) + \sum_{\alpha=1}^{m} \nu_\alpha \nabla F^\alpha(\bar{U})\partial_y U(\xi(t)-,t) = 0, \quad t \in [0,T). \tag{3.5}$$

Combining (3.4) with (3.5),

$$\{\sum_{\alpha=1}^{m} \nu_\alpha \nabla F^\alpha(\bar{U}) - \dot{\xi}(t)\nabla F^0(\bar{U})\}\partial_y U(\xi(t)-,t) = 0, \quad t \in [0,T), \tag{3.6}$$

which implies

$$\dot{\xi}(t) = \lambda(\bar{U};\nu), \qquad \partial_y U(\xi(t)-,t) = a(t) r(\bar{U};\nu) \tag{3.7}$$

where λ is a characteristic speed and r is a corresponding normalized right characteristic eigenvector. In particular, the wave travels at a constant speed $\lambda(\bar{U};\nu)$ and Γ is a straight line. The function $a(t)$ measures the *amplitude* of the wave. We proceed to determine the evolution of $a(t)$ along Γ. From (3.7),

$$\frac{da}{dt}r(\overline{U};\nu) = \partial^2_{yt}U(\xi(t)-,t) + \lambda(\overline{U};\nu)\partial^2_{yy}U(\xi(t)-,t). \tag{3.8}$$

In the region on the left of Γ, we take the inner product of (3.3) with $\ell(U;\nu)$, the left characteristic eigenvector associated with $\lambda(U;\nu)$, thus obtaining

$$\ell(U;\nu)^T\nabla F^0(U)\{\partial_t U + \lambda(U;\nu)\partial_y U\} = 0. \tag{3.9}$$

We differentiate (3.9) with respect to y, we pass to the limit as $y \uparrow \xi(t)$ and use (3.7),(3.4),(3.8) and (1.8) to get

$$\frac{da}{dt} + \{\nabla\lambda(\overline{U};\nu)\cdot r(\overline{U};\nu)\}a^2 = 0. \tag{3.10}$$

Equation (3.10) should be compared with (3.2). It follows from (3.10) that if $\nabla\lambda(\overline{U};\nu)\cdot r(\overline{U};\nu) < 0$ (or > 0) and $a(0) > 0$ (or < 0) $a(t)$ will blow up in finite time.

As we shall see in following sections, the condition

$$\nabla\lambda(U;\nu)\cdot r(U;\nu) \neq 0 \tag{3.11}$$

arises in different contexts within the theory, always distinguishing linear from nonlinear behavior. Consequently, a characteristic field that satisfies (3.11) (at a point $\overline{U}$) is called *genuinely nonlinear* (at $\overline{U}$). A characteristic field that violates (3.11) identically, i.e.,

$$\nabla\lambda(U;\nu)\cdot r(U;\nu) = 0 \quad , \quad U \in R^n , \tag{3.12}$$

is called *linearly degenerate.*

The characteristic fields with zero characteristic speed of the model systems (2.1),(2.5) and (2.8) are obviously linearly degenerate. Any characteristic field of (2.1) or (2.5) with nonzero characteristic speed, which, as we saw in Section 2, is a square root of one of the eigenvalues of the acoustic tensor Q defined by (2.4) or (2.7), is genuinely nonlinear when

$$\sum_{\substack{i,j,k\\ \alpha,\beta,\gamma}} \frac{\partial^3 e}{\partial F_{i\alpha}\partial F_{j\beta}\partial F_{k\gamma}} \nu_\alpha\nu_\beta\nu_\gamma\zeta_i\zeta_j\zeta_k \neq 0 \tag{3.13}$$

where ζ is a corresponding eigenvector of Q. In particular, the characteristic speeds of (2.10) are genuinely nonlinear if

$$p_{uu}(u) \neq 0. \tag{3.14}$$

Bibliographic Remarks: For information on the blow up of smooth solutions of (2.10) see Lax [35] and Klainerman and Majda [32]. There is voluminous literature on the time evolution of the amplitude of weak waves; see, for instance, Boilat [2] and Chen [4]. The concept of genuine nonlinearity was introduced by Lax [33].

4. GEOMETRIC STRUCTURE OF WEAK SOLUTIONS OF HYPERBOLIC SYSTEMS OF BALANCE LAWS

In view of the discussion in Section 3, it is necessary to consider weak solutions of systems of hyperbolic balance laws. Since (1.6) is in divergence form, L^∞ functions may be solutions in the sense of distributions. The article of L. Tartar in this volume reports on progress in the research program which seeks to establish the existence of L^∞ solutions by the method of compensated compactness. At the same time, from the viewpoint of continuum physics, it would be desirable to have solutions in a smaller function class in which the balance law, in its original integral form (1.3), is meaningful. For that purpose, the ideal framework is provided by the class BV of functions of bounded variation in the sense of Tonelli-Cesari.

A function $U(x)$ in $L^\infty(\mathscr{B})$ is of class BV when all its partial derivatives $\partial_\alpha U$ are Borel measures. The domain of $U(x) \in BV$ is the union of three pairwise disjoint sets $\mathscr{C}, \mathscr{J}, \mathscr{R}$ with the following properties: (i) $\mathscr{C}$ is the set of points of approximate continuity of U, in the sense of Lebesgue; (ii) $\mathscr{J}$ is the set of points of approximate jump discontinuity of U, in the sense that through each point $x \in \mathscr{J}$ there is a "tangent" m-dimensional hyperplane from either side of which U attains distinct one-sided Lebesgue approximate limits $U^-(x)$ and $U^+(x)$; $\mathscr{R}$ is the residual set. Furthermore, $\mathscr{R}$ is sparse in that its m-dimensional Hausdorff measure is zero while

$\mathscr{J} = \bigcup_{k=1}^{\infty} \mathscr{J}_k$ where $\mathscr{J}_k$ is Lipschitz homeomorphic to a compact subset of R^m.

If $U(x) \in BV$ is a solution of (1.6) in the sense of distributions then $F^\alpha(U(x))$, $\alpha=0,1,\ldots,m$, are also of class BV and (1.6) is satisfied in the sense of measures. The set $\mathscr{J}$ of points of approximate jump discontinuity of U may be visualized as a countable family of surfaces in space-time or, equally well, as a countable family of propagating surfaces in space, which are called *shock waves*.

The Gauss-Green theorem holds for any function of class BV so if $U(x) \in BV$ is a weak solution of (1.6) then (1.3) is satisfied on any smooth set Ω with $f^\alpha(x)$ and $g(x)$ given by (1.5). In fact, (1.3) holds for any set Ω whose characteristic function χ_Ω is of class BV (set of *finite perimeter*) provided $\partial\Omega$ is interpreted as the set of points of approximate jump discontinuity of χ_Ω (*essential boundary*) and $N(x)$ is selected orthogonal to the "tangent" hyperplane on $\partial\Omega$ at x. In particular, if x is on the set $\mathscr{J}$ of points of approximate jump discontinuity of U,

$$\sum_{\alpha=0}^{m} N_\alpha\{F^\alpha(U^+(x))-F^\alpha(U^-(x))\} = 0 , \tag{4.1}$$

where N is any vector orthogonal to the "tangent" plane on $\mathscr{J}$ at x. It is instructive to normalize N so that $N = (-s,\nu)$ with $\nu \in R^m$, $|\nu| = 1$, in which case ν may be interpreted as the direction and s as the speed of propagation of the shock wave at x. Then (4.1) can be rewritten in the form

$$-s\{F^0(U^+)-F^0(U^-)\} + \sum_{\alpha=1}^{m} \nu_\alpha\{F^\alpha(U^+)-F^\alpha(U^-)\} = 0. \tag{4.2}$$

Equation (4.2) is called the *Rankine-Hugoniot* jump condition.

In the literature, a bracket is commonly used to denote the jump of the enclosed quantity across a shock. In this notation, the Rankine-Hugoniot relations for our model system (2.5) read

$$\begin{cases} s[F_{i\alpha}] + \nu_\alpha[v_i] = 0 , & i = 1,\dots,m;\ \alpha = 1,\dots,m \\ s[v_i] + \sum_{\alpha=1}^{m} \nu_\alpha[T_{i\alpha}] = 0 , & i = 1,\dots,m \end{cases} \tag{4.3}$$

to which one should append

$$\nu_\beta[F_{i\alpha}]-\nu_\alpha[F_{i\beta}] = 0, \qquad i = 1,\dots,m;\ \alpha,\beta = 1,\dots m, \tag{4.4}$$

as a consequence of (2.3). Note that (4.4) follows automatically from $(4.3)_1$, when $s \neq 0$. The Rankine-Hugoniot relations for (2.10) are

$$\begin{cases} s[u] + [v] = 0 \\ s[v] - [p(u)] = 0 \end{cases} \tag{4.5}$$

whence the speed of propagation of shocks is

$$s = \pm\left\{-\frac{p(u^+)-p(u^-)}{u^+-u^-}\right\}^{1/2}. \tag{4.6}$$

It has been shown that solutions of strictly hyperbolic systems of conservation laws in one space dimension, constructed by the random choice method, are of class BV and are endowed with additional regularity, namely: (i) Solutions are continuous on $\mathscr{C}$, in the classical sense. (ii) $\mathscr{J}$ is the (at most countable) union of Lipschitz continuous arcs. The solution attains one-sided limits on either side of $\mathscr{J}$ in the classical sense. (iii) $\mathscr{R}$ is (at most) countable and contains only the endpoints of the arcs in $\mathscr{J}$ (points at which shocks are formed or shock collide.) At this point, the reader may be tempted to conjecture that any solution is piecewise smooth, that is it only contains a finite number of shocks and it is Lipschitz continuous on $\mathscr{C}$. Indeed all particular solutions of physical or engineering interest that are recorded in the literature are piecewise smooth. Unfortunately, examples of C^∞-smooth initial data have been exhibited which yield BV solutions that are not piecewise smooth! It is still being conjectured that generically solutions are piecewise smooth but this has only been established for the case of a single, genuinely nonlinear equation.

In several space dimensions, the regularity of solutions has not been determined even in the case of the single equation, for which the existence of BV solutions is known.

Bibliographic Remarks: Estimates involving "total variation" were first established by Oleinik (e.g. [52]) for the case of a single equation in one space dimension. The relevance of the space BV was first recognized by Conway and Smoller [9]. The geometric structure of BV functions was brought into the theory of conservation laws by Volpert [61]. For general information on BV functions, see Federer [23]. The regularity results in one space dimension are due to DiPerna [19], for genuinely nonlinear systems, and to Liu [38], for general systems.

5. HOPF's EQUATION. EXAMPLES OF NONUNIQUENESS

The simplest example of a hyperbolic conservation law is provided by Hopf's equation:

$$\partial_t u + \frac{1}{2}\partial_x u^2 = 0. \tag{5.1}$$

Simple as it looks, (5.1) has a rich theory of its own. Here, however, we only employ (5.1) as a model to demonstrate that the initial value problem for systems of hyperbolic conservation laws may have several solutions. To this end we fix $0 \leq \alpha \leq 1$ and consider the function

$$u(x,t) = \begin{cases} -1 & -\infty < x \leq -t \\ x/t & -t < x \leq -\alpha t \\ -\alpha & -\alpha t < x \leq 0 \\ \alpha & 0 < x \leq \alpha t \\ x/t & \alpha t < x \leq t \\ 1 & t < x \leq \infty . \end{cases} \tag{5.2}$$

Note that $u(x,0) = \operatorname{sgn} x$, for any α. For $x \neq 0$, $u(x,t)$ is Lipschitz continuous and satisfies (5.1). Across the line $x = 0$, $u(x,t)$ experiences a jump discontinuity (unless $\alpha = 0$) from $-\alpha$ to α and the Rankine-Hugoniot relation (4.2) holds. We have thus constructed a one-parameter family of solutions of the Cauchy problem for (5.1) with initial condition $u(x,0) = \operatorname{sgn} x$. For $\alpha = 0$, $u(x,t)$ is Lipschitz continuous on the upper half-plane, while, for $\alpha = 1$, $u(x,t)$ consists of two constant states joined by a shock.

The above example indicates that a criterion is needed in order to single out admissible solutions of hyperbolic systems of balance laws. A number of such criteria have been proposed, motivated by mathematical or physical considerations. In the following sections we will introduce these admissibility criteria and will explore their interconnections.

Bibliographic Remarks: For a thorough study of Hopf's equation see [29] and Whitham [63].

6. SHOCK CURVES AND SHOCK ADMISSIBILITY CRITERIA

We describe here criteria which attempt to characterize admissible solutions of a system of hyperbolic balance laws (1.6) by imposing restrictions on jump discontinuities. In preparation for the discussion we fix $\nu \in R^m$ with $|\nu| = 1$ and $\overline{U} \in R^n$ and proceed to determine the geometric structure of the set of points $U \in R^n$ which can be joined to $\overline{U}$ with a shock propagating in the direction ν. By virtue of

$$\begin{aligned} &-s\{F^0(U)-F^0(\overline{U})\} + \sum_{\alpha=1}^{m} \nu_\alpha\{F^\alpha(U)-F^\alpha(\overline{U})\} \\ &=\{-s\int_0^1 \nabla F^0(\tau U+(1-\tau)\overline{U})d\tau + \sum_{\alpha=1}^{m} \nu_\alpha \int_0^1 \nabla F^\alpha(\tau U+(1-\tau)\overline{U})d\tau\}(U-\overline{U}), \end{aligned} \tag{6.1}$$

(4.2) and (1.7), we deduce that the speed of propagation of a shock of small amplitude that joins U with $\bar{U}$ and propagates in the direction ν must be close to a characteristic speed $\lambda(\bar{U};\nu)$ in the direction ν. By the Crandall-Rabinowitz bifurcation theorem, if $\lambda(\bar{U};\nu)$ is a simple eigenvalue of (1.7), then there is a smooth function $s(\tau;\bar{U},\nu)$ and a smooth curve $V(\tau;\bar{U},\nu)$, defined for $-\varepsilon < \tau < \varepsilon$, with $s(0;\bar{U},\nu) = \lambda(\bar{U};\nu)$, $V(0;\bar{U},\nu) = \bar{U}$, such that

$$-s(\tau;\bar{U},\nu)\{F^0(V(\tau;\bar{U},\nu))-F^0(\bar{U})\}+\sum_{\alpha=1}^{m}\nu_\alpha\{F^\alpha(V(\tau;\bar{U},\nu))-F^\alpha(\bar{U})\}=0. \tag{6.2}$$

Furthermore, any U near $\bar{U}$ which can be joined to $\bar{U}$ with a shock propagating in the direction ν with speed near $\lambda(\bar{U};\nu)$ must lie on the curve $V(\tau;\bar{U},\nu)$. Thus $V(\tau;\bar{U},\nu)$ is called the *shock curve* associated with the characteristic field of $\lambda(\dot{U};\nu)$. In particular, if the system is strictly hyperbolic, n distinct shock curves emanate from $\bar{U}$.

For systems with multiple characteristic speeds, the situation is more complicated. As an illustration we consider (2.5), which, we recall, has zero as a characteristic speed of multiplicity $m^2 - m$, and investigate whether there are any shocks propagating with speed at or near zero. The Rankine-Hugoniot relation $(4.3)_2$ yields

$$\sum_{j,\alpha,\beta=1}^{m}\{\nu_\alpha\int_0^1\frac{\partial T_{i\alpha}}{\partial F_{j\beta}}(\tau F^+ + (1-\tau)F^-)d\tau\}[F_{j\beta}] = -s[v_i]. \tag{6.3}$$

If $s \neq 0$, using $(4.3)_1$ and (2.7),

$$\sum_{j=1}^{m}\{\int_0^1 Q_{ij}(\tau F^+ + (1-\tau)F^-;\nu)d\tau\}[v_j] = s^2[v_i] \tag{6.4}$$

which cannot hold for s near zero since Q is positive definite. On the other hand, when $s = 0$ (4.4) implies

$$[F_{i\beta}] = \sum_{\alpha=1}^{m}\nu_\alpha\nu_\beta[F_{i\alpha}] \overset{\text{def}}{=} w_i\nu_\beta \tag{6.5}$$

so that (6.3) yields

$$\sum_{j=1}^{m}\{\int_0^1 Q_{ij}(\tau F^+ + (1-\tau)F^-;\nu)d\tau\}w_j = 0 \tag{6.6}$$

which is impossible unless $w = 0$. So no shocks may propagate with speed at or near zero.

We now return to (6.2) and investigate the local geometry of the shock curve $V(\tau;\bar{U},\nu)$. Differentiating (6.2) with respect to τ,

$$\dot{s}(\tau;\bar{U},\nu)\{F^0(V(\tau;\bar{U},\nu))-F^0(\bar{U})\} \tag{6.7}$$

$$= \{\sum_{\alpha=1}^{m} \nu_\alpha \nabla F^\alpha(V(\tau;\bar{U},\nu))-s(\tau;\bar{U},\nu)\nabla F^0(V(\tau;\bar{U},\nu))\}\dot{V}(\tau;\bar{U},\nu).$$

In particular, setting $\tau = 0$ in (6.7) we deduce

$$s(0;\bar{U},\nu) = \lambda(\bar{U},\nu), \quad \dot{V}(0;\bar{U},\nu) = r(\bar{U};\nu), \tag{6.8}$$

that is, a shock curve emanating at $\bar{U}$ and associated with a particular characteristic field is tangent at $\bar{U}$ to the corresponding right characteristic eigenvector. Next we form the inner product of (6.7) with the left characteristic eigenvector $\ell(V(\tau;\bar{U},\nu);\nu)$, associated with the characteristic speed $\lambda(V(\tau;\bar{U},\nu);\nu)$, thus obtaining

$$\dot{s}(\tau;\bar{U},\nu)\ell(V(\tau;\bar{U},\nu);\nu)\cdot\{F^0(V(\tau;\bar{U},\nu))-F^0(\bar{U})\} \tag{6.9}$$

$$= \{\lambda(V(\tau;\bar{U},\nu);\nu)-s(\tau;\bar{U},\nu)\}\nabla F^0(V(\tau;\bar{U},\nu))\dot{V}(\tau;\bar{U},\nu).$$

We differentiate (6.9) with respect to τ and then set $\tau = 0$ and use (6.8), (1.8) to get

$$2\dot{s}(0;\bar{U},\nu) = \nabla\lambda(\bar{U};\nu)\cdot r(\bar{U};\nu) \tag{6.10}$$

which provides an alternative interpretation of the concept of genuine nonlinearity (3.11), introduced in Section 3.

For simple systems, such as (2.8) or (2.10), one can determine the global geometry of shock curves. For instance, using the Rankine-Hugoniot relations (4.5) is is easily seen that for (2.10) the shock curves emanating from $(\bar{u},\bar{v})$ associated with the two characteristic fields are globally defined and given by

$$v = \bar{v} \pm\sqrt{-(p(u)-p(\bar{u}))(u-\bar{u})}\,. \tag{6.11}$$

We have now prepared the ground for introducing the shock admissibility criteria. A shock that propagates in the direction with speed s_ρ, associated with the characteristic speed $\lambda(U;\nu)$ and joining the states U^- and U^+, is said to satisfy the *Lax shock admissibility criterion* if

$$\lambda(U^-;\nu) \geq s_0 \geq \lambda(U^+;\nu), \tag{6.12}$$

that is, characteristics overtake the shock of the same family from the left and are, in turn, overtaken by the shock from the right. In particular, if the left (and/or right) of (6.12) is satisfied as an equality, the shock is called a left (and/or right) *contact discontinuity*. In view of (6.8),(6.10), weak shocks associated with genuinely nonlinear characteristic fields are never contact discontinuities.

For our model system (2.10) and by virtue of (2.11) and (4.6), Lax's shock admissibility criterion is equivalent to the following geometric condition: In a shock joining the states (u^-,v^-) and (u^+,v^+) and propagating with speed s, when $s(u^+-u^-) > 0$ (or < 0) the chord that connects, on the p-u plane, the points $(u^-,p(u^-))$ and $(u^+,p(u^+))$ lies above (or below) the tangents to the graph of $p(u)$ at $u = u^-$ and $u = u^+$.

The Lax shock admissibility criterion has been motivated and justified in different ways. My interpretation of the evidence is that this criterion ensures that the interaction of the shock, as an entity, with the states on its left and right is stable. Moreover, it also induces internal stability of weak shocks, especially those associated with genuinely nonlinear fields. On the other hand, the Lax condition has no effect on the internal stability of strong shocks. For this reason we need a more discriminating admissibility criterion which we now proceed to introduce.

Consider a shock which propagates with speed s_0 in the direction ν and joins the states U^- and U^+. Assume U^+ lies on a shock curve through U^- associated with a particular characteristic speed $\lambda(U;\nu)$, say $U^+ = V(\tau^+;U^-,\nu)$. Let $s(\tau;U^-,\nu)$ denote the speed of the shock that joins $V(\tau;U^-,\nu)$ with U^-. In particular, $s(\tau^+;U^-,\nu) = s_0$. We say that the given shock satisfies the *Liu shock admissibility criterion* if

$$s_0 = s(\tau^+;U^-,\nu) \leq s(\tau;U^-,\nu), \quad 0 \leq \tau \leq \tau^+. \tag{6.13}$$

It is also true that $U^- = V(\tau^-;U^+,\nu)$, $s_0 = s(\tau^-;U^+,\nu)$ and (6.13) is equivalent to

$$s_0 = s(\tau^-;U^+,\nu) \geq s(\tau;U^+,\nu), \quad \tau^- \leq \tau \leq 0\ . \tag{6.14}$$

In particular, setting $\tau = 0$ in (6.13),(6.14) and recalling (6.8) we recover (6.12), namely, Liu's admissibility criterion implies Lax's admissibility criterion. The converse is also true, provided the shock speed satisfies the nondegeneracy condition $s^{(k)}(0;U^-,\nu) \neq 0$, for some positive integer k, and the shock is weak. Indeed, the nondegeneracy condition implies $\dot{s}(\tau;U^-,\nu) \neq 0$ on a small interval $0 < \tau < \varepsilon$ and then Lax's criterion (6.12) together with (6.8) yield $\dot{s}(\tau;U^-,\nu) < 0$, $0 < \tau < \varepsilon$. Thus Liu's admissibility criterion (6.13) holds providing $0 < \tau^+ < \varepsilon$. By (6.10), the nondegeneracy condition is satisfied, in particular, when the characteristic field of $\lambda(U;\nu)$ is genuinely nonlinear at U^-.

For the model system (2.10) and in virtue of (4.6), Liu's admissibility criterion is equivalent to the following geometric condition: In a shock joining the states (u^-,v^-) and (u^+,v^+) and propagating with speed s, when $s(u^+-u^-) > 0$ (or < 0) the chord that connects on the p-u plane, the points $(u^-,p(u^-))$ and $(u^+,p(u^+))$ lies above (or below) the graph of $p(u)$ between the points u^- and u^+. Comparing this statement with the geometric condition of Lax's criterion for the same system, stated earlier in this section, it becomes clear that Liu's criterion implies Lax's criterion but the converse is not generally true. However, the two criteria become equivalent for (2.10) under the genuine nonlinearity condition (3.14).

Liu's admissibility criterion was motivated by the detailed study of the single equation and the model system (2.10). As we will see in the following sections, this criterion can be justified, a posteriori, through comparisons with other criteria.

Bibliographic Remarks: The study of shock curves for genuinely nonlinear systems and Lax's admissibility criterion were introduced in [33]. For motivation and justification of Lax's criterion see Conway and Smoller [10] and Majda [48]. Liu's admissibility criterion was proposed in [39]. For earlier work on the single equation and system (2.10) which motivated Liu's criterion see Oleinik [53], Wendroff [62], and Zhang and Hsiao [64]. For insight on the relationship between positive definiteness of the acoustic tensor and the existence of stationary shocks for (2.5) see Ball [1].

7. ENTROPY ADMISSIBILITY CRITERION

In continuum physics the list of balance laws is supplemented with the Clausius-Duhem inequality which expresses the second law of thermodynamics. In the theory of thermoelastic nonconductors of heat, whose set of balance laws (2.1) is, the Clausius-Duhem inequality takes the form

$$\partial_t \eta \geq \frac{r}{\theta} \, . \tag{7.1}$$

For the isentropic approximation, with balance laws (2.5), we may combine $(2.1)_3$ with (7.1) to get the inequality

$$-\partial_t\{e(F)+\frac{1}{2}\sum_{i=1}^{m}v_i^2\}+\sum_{i,\alpha=1}^{m}\partial_\alpha\{v_iT_{i\alpha}(F)\}\geq-\sum_{i=1}^{m}f_iv_i\ . \tag{7.2}$$

In particular, in one space dimension, for system (2.10), (7.2) reduces to

$$-\partial_t\{-\int^u p(\xi)d\xi+\frac{1}{2}v^2\}-\partial_x\{vp(u)\}\geq 0. \tag{7.3}$$

Physically admissible processes are required to satisfy the Clausius-Duhem inequality, in addition to the balance laws of mass, momentum, energy, etc. At the same time it is generally assumed that all smooth physical processes that satisfy the balance laws are admissible, that is, they conform with the Clausius-Duhem inequality automatically. In fact, combining (2.1) with (2.2) one deduces easily that any Lipschitz continuous solution of (2.1) satisfies identically (7.1), as an equality. Similarly, any Lipschitz continuous solution of (2.5) or (2.10) satisfies automatically (7.2) or (7.3), as equalities. However, in the context of solutions of class BV, inequalities such as (7.1),(7.2) or (7.3) are used to rule out certain solutions that otherwise comply with the balance laws (2.1),(2.5) or (2.10).

In order to incorporate this structure into the general theory of balance laws, we seek conditions under which any Lipschitz continuous solution of (1.6) satisfies automatically an additional *entropy balance law*

$$\sum_{\alpha=0}^{m}\partial_\alpha H^\alpha(U(x)) = D(U(x),x) \tag{7.4}$$

where $H^\alpha(U)$, $\alpha=0,1,\dots,m$, and $D(U,x)$ are smooth functions. Of course we are only interested in the case where (7.4) is non-trivial, in the sense that the row (m+1)-vector field $\{H^0(U),H^1(U),\dots,H^m(U)\}$ is not a linear combination of the rows of the $(m+1)\times n$-matrix field $\{F^0(U),F^1(U),\dots,F^m(U)\}$. These requirements will be satisfied if and only if there is a non-constant integrating factor $B(U)$, taking values in R^n, such that

$$B(U)^T\nabla F^\alpha(U) = \nabla H^\alpha(U),\ u\in R^n,\ \alpha = 0,1,\dots,m, \tag{7.5}$$

$$B(U)^T G(U,x) = D(U,x),\ U\in R^n,\ x\in\mathscr{B}. \tag{7.6}$$

The integrability condition for (7.5) reads

$$\nabla B(U)^T\nabla F^\alpha(U) = \nabla F^\alpha(U)^T\nabla B(U)\ ,\ U\in R^n,\ \alpha=0,1,\dots,m. \tag{7.7}$$

System (7.7) involves $\frac{1}{2}n(n-1)(m+1)$ equations and n unknowns.

For a single balance law in any number of space dimensions ($n=1$, m arbitrary) $B(U)$ may be chosen arbitrarily. For $n=2$, $m=1$, (7.7) is a hyperbolic system of two equations in two unknowns which generally has solutions satisfying prescribed initial conditions. However, in all other cases the number of equations in (7.7) exceeds the number of unknowns so it is only special systems (1.6) that are endowed with nontrivial entropy balance laws (7.4). Note that if (1.6) is symmetric, (7.7) admits the solution $B(U) = U$. Conversely, if (7.7) has a solution $B(U)$ with $\nabla B(U)$ nonsingular, the change of variables $\bar{U} = B(U)$ transforms (1.6), at least locally, into a symmetric system of balance laws in $\bar{U}$. It is thus symmetrizable systems of balance laws that are endowed with a nontrivial entropy balance law.

In systems resulting from the balance laws of continuum physics, restrictions are imposed on the fluxes $F^{\alpha}(U)$ so as to render the equations in (7.7) compatible. This is precisely the role of thermodynamic relations (2.2) in connection with (2.1).

In the archetypical example (2.1), $H^0 = \eta$ so we will generally call H^0 the *entropy* and $\{H^1,\dots,H^m\}$ the *entropy flux*, even though, in other examples H^0 may have a different physical interpretation. For instance, in (7.2) and (7.3), for systems (2.5) and (2.10), H^0 stands for minus mechanical energy.

Assuming now that our system (1.6) of balance laws is endowed with an entropy balance law (7.4), we will say that a solution $U(x)$ of (1.6), of class BV, satisfies the *entropy admissibility criterion* if

$$\sum_{\alpha=0}^{m} \partial_{\alpha} H^{\alpha}(U(x)) \geq D(U(x),x) \quad , \tag{7.8}$$

in the sense of measures or distributions. We have selected the direction of the inequality (7.8) so as to comply with the physical notion of entropy. Of course, since changing the signs of $\{H^0,H^1,\dots,H^m\}$ and D does not affect (7.4), some normalization condition has to be imposed in order to render (7.8) meaningful. Motivated both by the physical models and the discussion to follow, we conclude that the natural normalization condition is that the symmetric matrix

$$P(U) = \nabla B(U)^T \nabla F^0(U) \tag{7.9}$$

be "negative", in an appropriate sense. Note that when $F^0(U) = U$, $P(U)$ is the Hessian of the entropy.

At this point we should alert the reader that the majority of authors, following an influential paper by Lax, postulate an entropy inequality like (7.8) but having the opposite direction. Consequently, what these authors call "entropy" is minus entropy in our sense and, for compatibility, their assumption on the sign of $P(U)$ has to be opposite to ours.

For future reference we record here the definition

$$Q^\alpha(U,V) = H^\alpha(V)-H^\alpha(U)-B(V)^T\{F^\alpha(V)-F^\alpha(U)\} \tag{7.10}$$

and note that, by virtue of (7.5),(7.9),

$$Q^\alpha(V,V) = 0, \quad \nabla_V Q^\alpha(U,V)_{U=V}=0, \quad \alpha = 0,1,\dots,m, \tag{7.11}$$

$$\nabla_V(\nabla_V Q^0(U,V))^T_{U=V} = -P(V). \tag{7.12}$$

We now state and prove an important implication of the entropy admissibility criterion when $P(U)$ is negative definite.

PROPOSITION 7.1. Assume (1.6) is endowed with an entropy balance law (7.4) and $P(U)$, defined by (7.9), is negative definite. Suppose $\overline{U}(x)$ is a bounded locally Lipschitz continuous solution of (1.6) in the strip $S_T = \{x \in R^{m+1}: 0 \le x_0 \le T\}$. If $U(x)$ is any bounded BV solution of (1.6) in S_T which satisfies the entropy admissibility criterion (7.8) and assumes the same initial conditions as $\overline{U}(x)$, i.e., $U(x) = \overline{U}(x)$ on the hyperplane $x_0 = 0$, then $U(x) = \overline{U}(x)$ for every x in S_T.

Proof. To simplify the computation, we write the proof for the case $G(U,x) \equiv 0$. We have

$$\sum_{\alpha=0}^m \partial_\alpha F^\alpha(\overline{U}(x)) = 0, \qquad \sum_{\alpha=0}^m \partial_\alpha H^\alpha(\overline{U}(x)) = 0, \tag{7.13}$$

$$\sum_{\alpha=0}^m \partial_\alpha F^\alpha(U(x)) = 0, \qquad \sum_{\alpha=0}^m \partial_\alpha H^\alpha(U(x)) \ge 0. \tag{7.14}$$

Combining (7.10),(7.13),(7.14) and (7.7) we obtain

$$\sum_{\alpha=0}^m \partial_\alpha Q^\alpha(U(x),\overline{U}(x)) \le \sum_{\alpha=0}^m (\partial_\alpha \overline{U}(x))^T \nabla B(\overline{U}(x))^T \{F^\alpha(U(x))-F^\alpha(\overline{U}(x))\} \tag{7.15}$$

$$= \sum_{\alpha=0}^m (\partial_\alpha \overline{U}(x))^T \nabla B(\overline{U}(x))^T \{F^\alpha(U(x))-F^\alpha(\overline{U}(x))-\nabla F^\alpha(\overline{U}(x))(U(x)-\overline{U}(x))\}.$$

The right-hand side of (7.15) is of quadratic order in $U-\overline{U}$ and so is $Q^{\alpha}(U,\overline{U})$, $\alpha = 0,1,\ldots,m$, by virtue of (7.11). Moreover, on account of (7.12) and since $P(U)$ has been assumed negative definite, $Q^{0}(U,\overline{U})$ is positive definite.

We fix $\beta > 0$ sufficiently large that $\beta Q^{0}(U,\overline{U}) \geq \sum_{\alpha=1}^{m} \nu_{\alpha} Q^{\alpha}(U,\overline{U})$, for any $\nu \in R^{m}$ with $|\nu| = 1$, and consider the cone $\mathscr{K} = \{x \in R^{m+1} : 0 \leq x_0 \leq T, \sum_{\alpha=1}^{m} x_{\alpha}^{2} \leq \beta^{2}(T-x_0)^{2}\}$. Integrating the inequality (7.15) over the frustum $\mathscr{K}_t = \{x \in \mathscr{K} : 0 \leq x_0 \leq t\}$, $0 \leq t \leq T$, and applying Gronwall's inequality we deduce $U(x) - \overline{U}(x) = 0$ in $\mathscr{K}$. Letting $\beta \to \infty$ we obtain $U(x) = \overline{U}(x)$ for every x in S_T. ∎

We have thus shown that when $P(U)$ is negative definite the entropy admissibility criterion ensures the uniqueness of Lipschitz continuous solutions to the Cauchy problem for a system of balance laws (1.6), within the broader class of BV solutions. In many examples of physical interest, for instance (2.8) and (2.10), $P(U)$ is indeed negative definite. On the other hand, in the case of (2.1) or (2.5), $P(U)$ would be negative definite only if the internal energy $e(F,\eta)$ or $e(F)$ were a strictly convex function of F and this last condition is incompatible with the principle of frame indifference which requires that $e(F,\eta)$ and $e(F)$ be invariant under rigid rotations, i.e., $e(OF,\eta) = e(F,\eta)$ and $e(OF) = e(F)$, for any orthogonal matrix O. Nevertheless, a uniqueness theorem, analogous to Proposition 7.1, holds for (2.1) and (2.5) because the extra balance law (2.3) compensates for the nonconvexity of $e(F,\eta)$ and $e(F)$.

Even when $P(U)$ is negative definite, the entropy admissibility criterion is not sufficiently discriminating to single out a unique BV solution of the Cauchy problem for (1.6). In fact, for reasons to be seen shortly, it is plausible to expect that the entropy criterion with $P(U)$ negative definite guarantees the uniqueness of BV solutions with weak shocks. However, to the present this conjecture has only been verified for piecewise smooth solutions of genuinely nonlinear systems of two equations in one space dimension ($n=2$, $m=1$).

In the rare occurrence that a system of balance laws is endowed with several (not trivially related) entropy balance laws, it is common practice to adopt the admissibility criterion induced by every one of them in order to enhance the prospects for uniqueness of solutions. My opinion is, however, that no entropy admissibility criterion should be imposed unless it is either motivated by the second law of thermodynamics or justified by means of another legitimate admissibility criterion (shock, viscosity, etc.).

In view of the construction of $H^\alpha(U)$, $\alpha=0,1,\ldots,m$, and $D(U,x)$ through (7.5) and (7.6), (7.4) is automatically satisfied, in the sense of measures, on the set $\mathscr{C}$ of points of approximate continuity of any BV solution $U(x)$ of (1.6). Therefore, the entropy admissibility criterion (7.8) need only be tested on the set $\mathscr{J}$ of points of approximate jump discontinuity of $U(x)$. Restricted on $\mathscr{J}$, (7.8) assumes the form of an *entropy shock admissibility criterion*

$$\sum_{\alpha=0}^{m} N_\alpha\{H^\alpha(U^+)-H^\alpha(U^-)\} \geq 0 \tag{7.16}$$

or, equivalently, setting $N = (-s_0,\nu)$ with $\nu \in R^m$, $|\nu| = 1$,

$$-s_0\{H^0(U^+)-H^0(U^-)\} + \sum_{\alpha=1}^{m} \nu_\alpha\{H^\alpha(U^+)-H^\alpha(U^-)\} \geq 0. \tag{7.17}$$

We proceed to compare the entropy shock criterion with the admissibility criteria of Lax and Liu discussed in Section 6. Let us assume U^+ lies on the shock curve $V(\tau;U^-,\nu)$ emanating from U^- and associated with the characteristic field of $\lambda(U;\nu)$, say $U^+ = V(\tau^+;U^-,\nu)$, $s_0 = s(\tau^+;U^-,\nu)$, $\tau^+ > 0$. Assume, further, that $P(U^-)$ is negative definite in the direction of the right characteristic eigenvector $r(U^-;\nu)$, associated with $\lambda(U^-;\nu)$, i.e.,

$$r(U^-;\nu)^T P(U^-) r(U^-;\nu) < 0. \tag{7.18}$$

We define

$$\Phi(\tau) = -s(\tau;U^-,\nu)\{H^0(V(\tau;U^-,\nu))-H^0(U^-)\} \tag{7.19}$$

$$+\sum_{\alpha=1}^{m} \nu_\alpha\{H^\alpha(V(\tau;U^-,\nu))-H^\alpha(U^-)\}.$$

Differentiating (7.19) with respect to τ and using (7.5), (6.7) and (7.10), we obtain

$$\dot\Phi(\tau) = -\dot s(\tau;U^-,\nu)Q^0(U^-,V(\tau;U^-,\nu)). \tag{7.20}$$

By virtue of (7.11), (7.12), (6.8) and (7.18), there is $\delta > 0$ such that

$$Q^0(U^-,V(\tau;U^-,\nu)) > 0,\quad \dot Q^0(U^-,V(\tau;U^-,\nu)) > 0,\quad 0<\tau<\delta. \tag{7.21}$$

We now claim that when $0 < \tau^+ < \delta$, (6.13) implies (7.17). Indeed, we integrate (7.20) over $(0,\tau^+)$, integrate by parts and use (6.13), (7.21) to get

$$\Phi(\tau^+) = -s_0 Q^0(U^-,U^+) + \int_0^{\tau^+} s(\tau;U^-,\nu)\dot{Q}^0(U^-,V(\tau;U^-,\nu))d\tau \tag{7.22}$$

$$\geq -s_0 Q^0(U^-,U^+) + s_0 \int_0^{\tau^+} \dot{Q}^0(U^-,V(\tau;U^-,\nu))d\tau = 0$$

which is precisely (7.17). Thus we have shown that for shocks of moderate strength Liu's shock admissibility criterion implies the entropy shock admissibility criterion. We now prove that the converse is also true, provided the shock speed satisfies the nondegeneracy condition $\overset{(k)}{s}(0;U^-,\nu) \neq 0$, for some positive integer k, and the shock is weak. Indeed, the nondegeneracy condition implies $\dot{s}(\tau;U^-,\nu) \neq 0$ on a small interval $0 < \tau < \varepsilon$ and then, if $0 < \tau^+ < \min\{\delta, \varepsilon\}$, (7.17), i.e., $\Phi(\tau^+) \geq 0$, together with (7.20) and (7.21) yield $\dot{s}(\tau;U^-,\nu) < 0$, $0 < \tau < \tau^+$, whence (6.13) follows. We recall that, as discussed in Section 6, nondegeneracy follows from genuine nonlinearity of the field of $\lambda(U;\nu)$ at U^-.

Using (7.7) and after a simple calculation one shows that right characteristic eigenvectors $r(U;\nu)$, $\bar{r}(U;\nu)$ associated with distinct characteristic speeds $\lambda(U;\nu)$, $\bar{\lambda}(U;\nu)$ are orthogonal relative to $P(U)$, i.e.,

$$r(U;\nu)^T P(U)\bar{r}(U;\nu) = 0. \tag{7.23}$$

It follows that if the system is strictly hyperbolic and (7.18) is to hold for every right characteristic eigenvector, then $P(U)$ is negative definite. The situation is different in systems that are not strictly hyperbolic. As shown in Section 6, for system (2.5) only the fields with nonzero characteristic speed generate shock curves. For this system it turns out that, since the acoustic tensor (2.7) is positive definite, (7.18) is satisfied for all right characteristic eigenvectors associated with nonzero characteristic speed, even though, as mentioned earlier in this section, $P(U)$ is not generally negative definite.

In order to see how the three shock admissibility criteria compare in the case of strong shocks, we consider the model system (2.10) with entropy inequality (7.3). The entropy shock admissibility criterion (7.17) here takes the form

$$s\{-\int_{u^-}^{u^+} p(\xi)d\xi + \tfrac{1}{2}(v^+)^2 - \tfrac{1}{2}(v^-)^2\} - \{v^+p(u^+) - v^-p(u^-)\} \geq 0. \tag{7.24}$$

On using the Rankine-Hugoniot jump conditions (4.5) and after a short computation we conclude that (7.24) is equivalent to

$$s\{\frac{1}{2}(p(u^+)+p(u^-))(u^+-u^-)-\int_{u^-}^{u^+} p(\xi)d\xi\} \geq 0. \qquad (7.25)$$

It is clear that (7.25) expresses the following geometric condition: In a shock joining the states (u^-,v^-) and (u^+,v^+) and propagating with speed s, when $s(u^+-u^-) > 0$ (or < 0) the (signed) area between the chord that connects, on the p-u plane, the points $(u^-,p(u^-))$ and $(u^+,p(u^+))$ and the graph of $p(u)$ between u^- and u^+ is nonnegative (or nonpositive). Comparing this with the corresponding statements for the Lax and Liu admissibility criteria, recorded in Section 6, we conclude that Lax's criterion does not generally imply the entropy criterion nor does the entropy criterion always imply Lax's or Liu's criterion. On the other hand, Liu's criterion implies the entropy criterion in (2.10) for shocks of arbitrary strength. When (2.10) is genuinely nonlinear, Lax's, Liu's and the entropy criterion are all equivalent.

Bibliographic Remarks: For information on the Clausius-Duhem inequality in continuum thermodynamics see, e.g., Truesdell and Noll [59]. The entropy admissibility criterion is discussed in Friedrichs and Lax [25], Lax [36] and Ruggeri and Strumia [56]. Uniqueness of piecewise smooth solutions is established in DiPerna [20]. For uniqueness results in adiabatic thermoelasticity see Dafermos [13]. For comparisons between the entropy criterion and shock admissibility criteria see Conlon and Liu [8] and Dafermos [14].

8. VISCOSITY ADMISSIBILITY CRITERION

The balance laws of mass, momentum and energy for a one-dimensional, heat conducting, viscous gas, with Newtonian viscosity and heat conductivity proportional to density, take the form

$$\begin{cases} \partial_t u - \partial_x v = 0 \\ \partial_t v + \partial_x p(u,\eta) = \mu\partial_{xx}^2 v \\ \partial_t\{e(u,\eta)+\frac{1}{2}v^2\} + \partial_x\{vp(u,\eta)\} = \mu\partial_x(v\partial_x v) + \kappa\partial_{xx}^2\theta. \end{cases} \qquad (8.1)$$

The symbols u,v,p,e,η,θ have the same physical meaning as in system (2.8) while the conductivity κ and the viscosity μ are positive constants. The thermodynamic relations (2.9) are in effect here as well.

Similarly, the balance laws of mass and momentum for a one-dimensional viscous gas with Newtonian viscosity proportional to density read

$$\begin{cases} \partial_t u - \partial_x v = 0 \\ \partial_t v + \partial_x p(u) = \mu \partial_{xx}^2 v. \end{cases} \tag{8.2}$$

Due to the presence of diffusion, (8.1),(8.2) are "tamer" them (2.8),(2.10).

Virtually every real material is viscous and conducts heat. However, such materials are often treated as elastic nonconductors of heat when it is judged that under the conditions of motion the effects of viscosity and heat diffusion are negligible. With this in mind, it is natural to dub admissible those solutions of (2.8) or (2.10) that are limits, as $\mu \downarrow 0$, $\kappa \downarrow 0$, of solutions of (8.1) or (8.2).

Motivated by the above discussion, given any system of balance laws (1.6) we consider the second order system

$$\sum_{\alpha=0}^{m} \partial_\alpha F^\alpha(U(x)) - G(U(x),x) = \mu \sum_{\alpha,\beta=0}^{m} \partial_\alpha \partial_\beta A^{\alpha\beta}(U(x)) \tag{8.3}$$

where $\mu > 0$ and $A^{\alpha\beta}(U)$, $\alpha,\beta = 0,1,\dots,m$, are given smooth vector fields taking values in R^n, and say that a solution $U(x)$ of (1.6) on $\mathscr{B}$ satisfies the *viscosity admissibility criterion* relative to (8.3) if there is a set $U_\mu(x)$ of solutions of (8.3) such that $U_\mu(x) \to U(x)$, boundedly almost everywhere in $\mathscr{B}$.

It is not obvious how the fields $A^{\alpha\beta}(U)$ are to be selected in order to induce a reasonable admissibility criterion. As a minimal requirement, the Cauchy problem for (8.3) should be well-posed and should possess a unique globally defined smooth solution. For that purpose, the effect of the second order term must be "dissipative". The hope that the $\mu \downarrow 0$ limit of solutions of (8.3) is independent of the choice of $A^{\alpha\beta}(U)$ has been dispelled by means of counterexamples. A program to classify the $A^{\alpha\beta}(U)$ into equivalence classes whose members yield the same $\mu \downarrow 0$ limit has been carried out partially for systems of two equations in one space variable ($n=2$, $m=1$), in the context of the viscosity shock admissibility criterion which will be discussed below. Whenever our system (1.6) expresses the balance laws of continuum physics, one may be guided by physics to motivate the choice of the $A^{\alpha\beta}(U)$. However, even in those cases questions may arise. For instance, in considering (2.8) as a limiting case of (8.1), one has to specify the asymptotic behavior of the ratio

μ/κ as both μ and κ tend to zero.

When the system (1.6) is endowed with an entropy balance law (7.4) and the induced entropy admissibility criterion has been adopted, the compatibility of this criterion with the viscosity admissibility criterion will have to be ascertained. We claim that any solution of (1.6) complying with the viscosity admissibility criterion relative to (8.3) satisfies automatically the entropy admissibility criterion (7.8), provided that (i) there exist smooth functions $q^{\alpha\beta}(U), \alpha, \beta=0,1,\dots,m$, defined on R^n such that

$$B(U)^T \nabla A^{\alpha\beta}(U) = \nabla q^{\alpha\beta}(U) \ , \ U \in R^n, \ \alpha,\beta=0,1,\dots,m \tag{8.4}$$

and (ii) for any collection $\{V_0, V_1, \dots, V_m\}$ of vectors in R^n,

$$\sum_{\alpha,\beta=0}^{m} V_\alpha^T \nabla B(U)^T \nabla A^{\alpha\beta}(U) V_\beta \leq 0, \qquad U \in R^n. \tag{8.5}$$

In (8.4),(8.5), $B(U)$ stands for the integrating factor associated with the entropy balance law (see (7.5),(7.6),(7.7)). To prove our assertion, we assume $U_\mu(x)$ is a solution of (8.3) on $\mathscr{B}$ such that $U_\mu(x) \to U(x)$, $\mu \downarrow 0$, boundedly almost everywhere. We form the inner product of (8.3) with $B(U_\mu)$ and use (7.5), (7.6),(8.4) to deduce the identity

$$\sum_{\alpha=0}^{m} \partial_\alpha H^\alpha(U_\mu(x)) - D(U_\mu(x),x) = \mu \sum_{\alpha,\beta=0}^{m} \partial_\alpha \partial_\beta q^{\alpha\beta}(U_\mu(x)) \tag{8.6}$$

$$- \mu \sum_{\alpha,\beta=0}^{m} (\partial_\alpha U_\mu(x))^T \nabla B(U_\mu(x))^T \nabla A^{\alpha\beta}(U_\mu(x)) \partial_\beta U_\mu(x).$$

On the right-hand side of (8.6), the first term tends to zero, as $\mu \downarrow 0$, in the sense of distributions, while the second term is nonnegative, by virtue of (8.5). It follows that the limit $U(x)$ of $U_\mu(x)$ satisfies (7.8), as it was asserted.

By account of (7.5), the choice $A^{\alpha\beta}(U) = I^{\alpha\beta} F^0(U)$, where $I^{\alpha\beta}$ is 1 or 0, satisfies (8.4) with $q^{\alpha\beta}(U) = I^{\alpha\beta} H^0(U)$. With this choice of $A^{\alpha\beta}(U)$, (8.5) will hold provided the matrix $P(U)$, defined by (7.9), is negative semidefinite, a condition that we have already encountered in our discussion of the entropy admissibility criterion.

There are other choices of $A^{\alpha\beta}(U)$ that satisfy (8.4),(8.5). For instance, the viscosity criterion for (2.10) relative to

(8.2) is compatible with the entropy inequality (7.3) since (8.6) takes the form

$$-\partial_t\{-\int^u p(\xi)d\xi + \frac{1}{2}v^2\} - \partial_x\{vp(u)\} = -\frac{1}{2}\mu\,\partial^2_{xx}v^2 + \mu(\partial_x v)^2. \tag{8.7}$$

There is no straightforward procedure to test whether a given solution complies with the viscosity criterion. It seems that the main purpose of this criterion is to suggest a program for constructing admissible discontinuous solutions of hyperbolic systems (1.6) as limits of smooth solutions of "parabolic" systems (8.3). The success of such a program hinges on the availability of a priori estimates for solutions of (8.3) which are independent of $\mu > 0$ and are sufficiently strong to ensure compactness. Effective estimates of this type are known for the case of a single equation in any number of space dimensions (n=1, m arbitrary). Recently, the method of "vanishing viscosity" has been applied to systems of two equations in one space dimension (n=2, m=1) by employing the idea of "compensated compactness" to compensate for the lack of strong estimates. The results are reported in the article by L. Tartar in this volume.

We now proceed to motivate a variant of the viscosity criterion which can be tested effectively on solutions. We note that, when $G \equiv 0$, (8.3) admits solutions $U_\mu(x) = V(\tau)$, with $\tau = \mu^{-1}\sum_{\alpha=0}^{m} N_\alpha x_\alpha$, $N \in R^{m+1}$ where $V(\tau)$ satisfies the ordinary differential equation

$$\sum_{\alpha,\beta=0}^{m} N_\alpha N_\beta \ddot{A}^{\alpha\beta}(V(\tau)) = \sum_{\alpha=0}^{m} N_\alpha \dot{F}^\alpha(V(\tau)). \tag{8.8}$$

It is then clear that a solution of (1.6), with $G \equiv 0$, on R^{m+1} which consists of two constant states joined by a plane shock, say $U(x) = U^-$ for $\sum_{\alpha=0}^{m} N_\alpha x_\alpha < 0$, $U(x) = U^+$ for $\sum_{\alpha=0}^{m} N_\alpha x_\alpha > 0$, complies with the viscosity criterion relative to (8.3), provided there is a solution $V(\tau)$ of (8.8) on $-\infty < \tau < \infty$ which satisfies the boundary conditions

$$V(-\infty) = U^-,\; V(\infty) = U^+. \tag{8.9}$$

Extrapolating from the above special situation, we now consider any BV solution $U(x)$ of (1.6) and say that it satisfies the *viscosity shock admissibility criterion*, relative to (8.3), if for every $x \in \mathcal{J}$ at which the normal on the "tangent" hyperplane is N and the left and right approximate limits are U^- and U^+, there is a solution $V(\tau)$ of (8.8) on $(-\infty,\infty)$ with boundary conditions (8.9).

In Section 7 we saw that the entropy criterion and the entropy shock criterion are equivalent. However, the equivalence between the viscosity criterion and the viscosity shock criterion has not been established to the present day.

Integrating (8.8) once, we reduce it to the first order system

$$\sum_{\alpha,\beta=0}^{m} N_\alpha N_\beta \dot{A}^{\alpha\beta}(V(\tau)) = \sum_{\alpha=0}^{m} N_\alpha \{F^\alpha(V(\tau)) - F^\alpha(U^-)\}. \tag{8.10}$$

We note that U^- and U^+ are critical points of (8.10), by virtue of the Rankine-Hugoniot relations (4.1). Therefore, admissibility means that there is an orbit connecting U^- with U^+. Actually, in order to encompass degenerate cases, we call the shock admissible when U^- can be connected with U^+ by means of any invariant *monodromic* curve which consists of critical points and orbits that connect pairs of these critical points, provided the direction of the flow along the curve is not reversed as we traverse it from U^- to U^+.

When (1.6) is endowed with an entropy balance law (7.4) and the corresponding integrating factor $B(U)$ satisfies (8.4) and (8.5), the viscosity shock admissibility criterion, relative to (8.3), implies the entropy shock criterion. To see this, we form the inner product of (8.8) with $B(V(\tau))$ and use (7.5) and (8.4) to get the identity

$$\frac{d}{d\tau} \sum_{\alpha=0}^{m} N_\alpha H^\alpha(V(\tau)) = \frac{d^2}{d\tau^2} \sum_{\alpha,\beta=0}^{m} N_\alpha N_\beta q^{\alpha\beta}(V(\tau)) \tag{8.11}$$
$$- \sum_{\alpha=0}^{m} N_\alpha N_\beta \dot{V}(\tau)^T \nabla B(V(\tau))^T \nabla A^{\alpha\beta}(V(\tau))\dot{V}(\tau).$$

Integrating (8.11) over $(-\infty,\infty)$ and using (8.9) and (8.5) we arrive at (7.16).

Extensive research has been conducted to compare the viscosity shock admissibility criterion with the shock admissibility criteria of Lax and Liu, discussed in Section 6. It has been established that, under reasonable assumptions on $A^{\alpha\beta}(U)$, the viscosity shock admissibility criterion is equivalent to Lax's criterion for weak shocks in systems with genuinely nonlinear characteristic fields and to Liu's criterion for shocks of moderate strength in fairly general systems. For special systems, the viscosity shock criterion is equivalent to Liu's criterion for shocks of arbitrary strength.

As an illustration we discuss the viscosity shock admissibility criterion, relative to (8.2), for system (2.10). Assuming that the shock joins the states (u^-,v^-) and (u^+,v^+) and propagates with speed s, (8.8) takes the form

$$\begin{cases} \ddot{v}(\tau) = -s\dot{v}(\tau) + \dot{p}(u(\tau)) \\ 0 = s\dot{u}(\tau) + \dot{v}(\tau) \end{cases} \tag{8.12}$$

with boundary conditions $u(-\infty) = u^-$, $v(-\infty) = v^-$, $u(\infty) = u^+$, $v(\infty) = v^+$. Eliminating $v(\tau)$ between the $(8.12)_1$ and $(8.12)_2$ and integrating the resulting equation once, we obtain

$$-s\dot{u}(\tau) = p(u(\tau))-p(u^-) + s^2(u(\tau)-u^-). \tag{8.13}$$

Thus, in order to get an orbit (or at least a monodromic invariant curve) connecting u^- with u^+, between u^- and u^+ the right-hand side of (8.13) should be nonpositive when $s(u^+-u^-) > 0$ and nonnegative when $s(u^+-u^-) < 0$. Thus, in view of (4.6), the viscosity shock admissibility criterion for (2.10) coincides with Liu's criterion, recorded at the end of Section 6.

Bibliographic Remarks: For general information on the constitutive theory of materials with viscosity, see, e.g., Truesdell and Noll [59]. Existence theorems for (8.1) and (8.2) are given in Dafermos and Hsiao [17] and Greenberg, MacCamy and Mizel [28]. For applications of the method of "vanishing viscosity" to a single conservation law see Volpert [61]. For comparisons between the viscosity shock criterion and Lax's criterion see Foy [24], Conley and Smoller [6] and Mock [49]. The relationship between the viscosity shock criterion and Liu's criterion is discussed in Liu [40], Conlon [7] and Pego [54].

9. ENTROPY RATE ADMISSIBILITY CRITERION

From the examples discussed in the last three sections it has become clear that the entropy admissibility criterion is generally less discriminating than either the Liu shock criterion or the viscosity criterion. In physical terms, the second law of thermodynamics, which states that entropy increases, along thermodynamical processes, at a faster rate than entropy production, does not appear sufficiently powerful to rule out all undesirable solutions. Under the circumstances, it is tempting to experiment with a strengthening of the traditional formulation of the second law and require that not only is entropy increasing but in fact it is increasing as rapidly as the balance laws of mass, momentum and energy would allow. This motivates still another criterion which we now proceed to formulate in the framework of the general theory.

We assume that the system (1.6) of hyperbolic balance laws is endowed with an entropy balance law (7.4). Let $U(x)$ be a BV solution of (1.6) in $\mathscr{B}$. The entropy rate of $U(x)$ at $x \in \mathscr{B}$ is defined by

$$\mathscr{D}\mathscr{A}\{U\}(\bar{x}) = \limsup_{\rho\downarrow 0} \frac{1}{b_{m-1}\rho^{m+1}} \{\sum_{\alpha=0}^{m} \partial_\alpha H^\alpha(U)\}(\mathscr{K}_\rho(\bar{x})) \tag{9.1}$$

where b_{m-1} denotes the measure of the unit ball in R^{m-1} $(b_0=1)$ and $\mathscr{K}_\rho$ is the cylinder

$$\mathscr{K}_\rho(\bar{x}) = \{x \in R^{m+1} : 0\leq x_0-\bar{x}_0\leq \rho^2, \sum_{\alpha=1}^{m} |x_\alpha-\bar{x}_\alpha|^2 \leq \rho^2\}. \tag{9.2}$$

In particular, if $\bar{x}$ lies on the set $\mathscr{C}$ of points of approximate continuity of $U(x)$, then $\mathscr{D}\mathscr{A}\{U\}(\bar{x}) = 0$ while if $\bar{x}$ belongs to the set $\mathscr{J}$ of points of approximate jump discontinuity of $U(x)$,

$$\mathscr{D}\mathscr{A}\{U\}(\bar{x})=-s\{H^0(U^+)-H^0(U^-)\}+ \sum_{\alpha=1}^{m} \nu_\alpha\{H^\alpha(U^+)-H^\alpha(U^-)\}, \tag{9.3}$$

where U^- and U^+ are the one-sided, left and right, approximate limits of $U(x)$ at $\bar{x}$; s is the speed and ν the direction of propagation of the shock at $\bar{x}$. Another noteworthy case is when $\bar{x}$ is a point on the residual set $\mathscr{R}$ such that a fan of k shocks emanate from it, all propagating in the same direction ν but with individual speeds $s_1,\ldots,s_k$. Then

$$\mathscr{D}\mathscr{A}\{U\}(\bar{x}) = \sum_{i=1}^{k} \{-s_i\{H^0(U_i^+)-H^0(U_i^-)\}+ \sum_{\alpha=1}^{m} \nu_\alpha\{H^\alpha(U_i^+)-H^\alpha(U_i^-)\}\}. \tag{9.4}$$

A solution $U(x)$ of class BV on $\mathscr{B}$ is said to satisfy the *entropy rate admissibility criterion* if for any $\bar{x} \in \mathscr{B}$ and every test solution $\bar{U}(x)$, which is defined on some cylinder $\mathscr{K}_\rho(\bar{x})$ and coincides with $U(x)$ on the lower base, i.e., $\bar{U}(\bar{x}_0,x_1,\ldots,x_m) = U(\bar{x}_0,x_1,\ldots,x_m)$, we have

$$\mathscr{D}\mathscr{A}\{U\}(\bar{x}) \geq \mathscr{D}\mathscr{A}\{\bar{U}\}(\bar{x}). \tag{9.5}$$

Unfortunately, there is no straightforward procedure for testing whether a given BV solution complies with the above criterion, the reason being that it is not easy to generate the test solutions $\overline{U}(x)$ used in the definition. However, there is a situation in which testing is feasible. Assume $G \equiv 0$ and suppose $U(x)$ consists of two constant states U^- and U^+ joined by a plane shock that propagates in the direction ν with speed s. It is then possible to construct test solutions $\overline{U}(x)$ by resolving the jump discontinuity (U^-,U^+) into a fan of plane waves, all propagating in the same direction ν but with individual speeds (in Section 10 we will discuss how such resolutions can be attained). The entropy rate of $\overline{U}(x)$ may then be computed via (9.4) and compared with the entropy rate of $U(x)$, given by (9.3).

Extrapolating from the above special situation, in analogy with what was done in Section 8 with the viscosity criterion, we now postulate that a BV solution $U(x)$ of (1.6) satisfies the *entropy rate shock admissibility criterion* if, relative to any $x_+ \in \mathcal{J}$ at which the left and right approximate limits are U^- and U^+ and the shock propagates in the direction ν with speed s, there is no resolution of the jump discontinuity (U^-,U^+) into a fan of plane waves, propagating in the same direction ν, which has higher entropy rate than the original shock.

In order to compare the entropy rate shock criterion with the Liu shock criterion, let us consider a shock of moderate strength which propagates in the direction ν with speed s_0, joins the states U^- and U^+ and is associated with the characteristic speed $\lambda(U^-;\nu)$. Thus $U^+ = V(\tau^+;U^-,\nu)$, $s_0 = s(\tau^+;U^-,\nu)$, where $V(\tau;U^-,\nu)$ denotes the shock curve emanating at U^- associated with the characteristic speed $\lambda(U^-;\nu)$ and $s(\tau;U^-,\nu)$ is the corresponding shock speed function. Suppose Liu's shock criterion (6.13) fails in the sense that there is τ^0, $0 < \tau^0 < \tau^+$, such that $s(\tau^0;U^-,\nu) = s_0$ and $\dot{s}(\tau^0;U^-,\nu) \neq 0$. One may then visualize the given shock as the composition of two shocks, one that joins U^- with $U^0 = V(\tau^0;U^-,\nu)$ and one that joins U^0 with U^+, both propagating in the direction ν with speed s_0. Assuming that the system (1.6) is strictly hyperbolic and employing elementary bifurcation theory one shows that our shock may resolve into a wave fan consisting of two shocks associated with $\lambda(U;\nu)$, one propagating with speed slightly less than s_0 the other with speed slightly greater than s_0, together with weak shocks associated with the other characteristic speeds. Furthermore, it turns out that if the matrix $P(U^-)$, defined by (7.9) is negative definite in the direction of the right characteristic eigenvector $r(U^-;\nu)$ associated with $\lambda(U^-;\nu)$, i.e., when (7.18) holds, then this wave fan has higher entropy rate than the given shock. It follows that the entropy rate shock admissibility criterion implies the Liu shock admissibility criterion, for shocks of moderate strength.

For the model system (2.10) it has been established that the entropy rate shock criterion is equivalent to Liu's criterion for shocks of arbitrary strength. On the other hand, for system (2.8) with constitutive relation $e = pu/(\gamma-1)$ (polytropic gas) it turns out that the entropy rate shock criterion and Liu's criterion are equivalent, for shocks of arbitrary strength, when $\gamma \geq 5/3$ (monatonic gases) but not equivalent when $\gamma < 5/3$ and the shocks are very strong.

In concluding our discussion on the different shock admissibility criteria, we may summarize our findings as follows: The Lax, Liu, entropy, viscosity and entropy rate criterion are all equivalent when applied to weak shocks with nondegenerate shock speed (thus, in particular, to shocks associated with genuinely nonlinear fields). For shocks of moderate strength, the Liu, viscosity and entropy rate criterion are equivalent and generally more discriminating than either the Lax or the entropy criterion. Finally, when applied to shocks of great strength the above criteria are not generally equivalent.

Bibliographic Remarks: See Dafermos [15] and Hsiao [30].

10. SIMPLE WAVES AND THE RIEMANN PROBLEM

For the remainder of this article we will focus attention on strictly hyperbolic systems of n conservation laws in one space dimension, $m = 1$. Since $U \mapsto F(U)$ is locally a diffeomorphism, we assume, without considerable loss of generality, that $F^0(U) = U$. Moreover, it is here convenient to switch to the standard notation $x_0 = t$, $x_1 = x$, $F^1(U) = F(U)$ so our system reads

$$\partial_t U(x,t) + \partial_x F(U(x,t)) = 0. \tag{10.1}$$

The strict hyperbolicity assumption implies that, for any $U \in R^n$, the $n \times n$ matrix $\nabla F(U)$ has n real, distinct, eigenvalues $\lambda_1(U) < \cdots < \lambda_n(U)$ and corresponding linearly independent right characteristic eigenvectors $r_1(U), \ldots, r_n(U)$.

In this section we consider the *Riemann Problem*, namely the Cauchy problem for (10.1) with initial conditions

$$U(x,0) = \begin{cases} U^- & , \quad x < 0 \\ U^+ & , \quad x > 0 \end{cases} \tag{10.2}$$

where U^- and U^+ are given constant vectors in R^n.

It turns out that the solution of this problem provides a

magnification of the local structure of any BV solution of (10.1). Furthermore, as we will see in Section 11, the explicit solution of the Riemann problem can be used as a building block for constructing the solution to the Cauchy problem for (10.1) by the random choice method.

Since, both, Equation (10.1) and the initial data (10.2) remain invariant under a uniform expansion of coordinates $x \mapsto \alpha x$, $t \mapsto \alpha t$, $\alpha > 0$, the solution of (10.1),(10.2) must depend on x/t alone, that is,

$$U = U(\xi), \quad \xi = \frac{x}{t}, \quad U(-\infty) = U^-, \quad U(\infty) = U^+. \tag{10.3}$$

Consequently, the solution of the Riemann Problem describes how the jump discontinuity (U^-,U^+) at the origin is resolved, by (10.1), into a fan of waves.

We have already encountered two types of solutions of (10.1) that are functions of x/t, namely, constant states (U = const.) and shocks emanating from the origin, joining two constant states and thus propagating with constant speed. By the discussion in Section 6, associated with each characteristic speed λ_k, there is a k-*shock curve* $V_k(\tau;\bar{U})$, through $\bar{U} \in R^n$, which contains the states near $\bar{U}$ that can be joined to $\bar{U}$ by a k-*shock* whose speed $s_k(\tau;\bar{U})$ is close to $\lambda_k(\bar{U})$. Furthermore, cf. (6.8),

$$V_k(0;\bar{U}) = \bar{U}, \quad \dot{V}_k(0;\bar{U}) = r_k(\bar{U}), \quad s_k(0;\bar{U}) = \lambda_k(\bar{U}). \tag{10.4}$$

We now consider the existence of nonconstant, continuously differentiable, solutions of (10.1) which depend on x/t alone. Upon substituting in (10.1) U = U(x/t), we deduce that $U(\xi)$ must satisfy the ordinary differential equation

$$\{\nabla F(U(\xi)) - \xi I\}\dot{U}(\xi) = 0, \tag{10.5}$$

whence

$$\dot{U}(\xi) = a(\xi)r_k(U(\xi)), \quad \lambda_k(U(\xi)) = \xi, \tag{10.6}$$

for some k=1,...,n and some scalar-valued function $a(\xi)$. Any solution U(x/t) of (10.6) is called a *centered* k-*simple wave*.

Differentiating $(10.6)_2$ with respect to ξ and using $(10.6)_1$ we obtain

$$a(\xi) = \{\nabla\lambda_k(U(\xi))\cdot r_k(U(\xi))\}^{-1} \tag{10.7}$$

which shows that the k-characteristic field must remain genuinely nonlinear on the range of a k-simple wave. We have thus deduced still another interpretation of the concept of genuine nonlinearity.

Given now $\bar{U} \in R^n$ with $\nabla\lambda_k(\bar{U})\cdot r_k(\bar{U}) \neq 0$, we set $\bar{\xi} = \lambda_k(\bar{U})$, substitute $a(\xi)$ from (10.7) into $(10.6)_1$ and solve the resulting ordinary differential equation under the initial condition $U(\bar{\xi}) = \bar{U}$ on an interval, say, $\bar{\xi} \leq \xi \leq \xi_1$. We denote this solution with $U_k(\xi;\bar{U})$ and, in analogy with the shock curve, we visualize it as a k-*simple wave curve* in R^n, parametrized by ξ, and containing the states that can be joined to $\bar{U}$ by a k-simple wave. By virtue of (10.6), (10.7),

$$U_k(\bar{\xi};\bar{U}) = \bar{U}, \dot{U}_k(\bar{\xi};\bar{U}) = \{\nabla\lambda_k(\bar{U})\cdot r_k(\bar{U})\}^{-1} r_k(\bar{U}). \tag{10.8}$$

Combining k-shocks and k-simple waves we obtain k-wave fans. An admissible k-*wave fan* is a finite (or in degenerate cases denumerable sequence of alternating k-simple waves and k-shocks that satisfy Liu's admissibility criterion. Thus a k-wave fan is characterized by a finite (or denumerable) ordering $U_1, U_2, \ldots, U_\ell$ of states, where U_{i+1} is joined to U_i by either a k-simple wave or an admissible k-shock. Note that if U_i, U_{i+1} are joined by a k-shock with speed s, which is adjacent on the left (and/or right) to a k-simple wave, then kinematic compatibility demands $\lambda_k(U_i) \leq s$ (and/or $\lambda_k(U_{i+1}) \geq s$) while Lax's condition requires $\lambda_k(U_i) \geq s$ (and/or $\lambda_k(U_{i+1}) \leq s$). It follows $\lambda_k(U_i) = s$ (and/or $\lambda_k(U_{i+1}) = s$), namely, in an admissible k-wave fan any shock adjacent to a simple wave on one of its sides must be a contact discontinuity from that side. In particular, if the k-characteristic field is genuinely nonlinear, every k-wave fan of small amplitude consists of either a single k-shock or a single k-simple wave.

With any k-wave fan, characterized by the ordering $U_1, \ldots, U_\ell$ of states, we associate a k-*wave curve* in R^n which is constructed by connecting U_i to U_{i+1}, $i=1,\ldots,\ell-1$, with a k-shock curve, when U_{i+1} is joined to U_i by a k-shock, or with a k-simple wave curve, when U_{i+1} is joined to U_i by a k-simple wave. By virtue of (10.4) and (10.8), a proper parametrization renders any k-wave curve C^1 smooth (in fact it can be shown that wave curves are C^2 smooth). It has been established that through any $\bar{U} \in R^n$ there is a unique k-wave curve $W_k(\tau;\bar{U})$ such that

$$W_k(0;\bar{U}) = \bar{U},\ \dot{W}_k(0;\bar{U}) = r_k(\bar{U}). \tag{10.9}$$

As a consequence of the above, when $|U^+ - U^-|$ is sufficiently small, the Riemann problem (10.1),(10.2) has a unique admissible solution which consists of n+1 constant states $U^- = U_0, U_1, \ldots, U_n = U^+$ and n wave fans and U_k is joined to U_{k-1}, $k=1,\ldots,n$, by a k-wave fan. To see this, consider the function

$$Z(\tau_1,\ldots,\tau_n;U^-) = W_n(\tau_n;W_{n-1}(\tau_{n-1};\ldots W_2(\tau_2;W_1(\tau_1;U^-))\ldots)) \tag{10.10}$$

which maps a neighborhood of 0 in R^n to R^n. By virtue of (10.9),

$$Z(0,\ldots,0;U^-)=U^-,\quad \nabla Z(0,\ldots,0;U^-)=\{r_1(U^-),\ldots,r_n(U^-)\} \tag{10.11}$$

hence by the implicit function theorem there is unique $(\tau_1,\ldots,\tau_n)$ near 0 such that $Z(\tau_1,\ldots,\tau_n;U^-) = U^+$. The intermediate constant states are then determined inductively by $U_0 = U^-$, $U_k = W_k(\tau_k;U_{k-1})$, $k=1,\ldots,n$.

Extensive research has been conducted for solving the Riemann problem when $|U^+-U^-|$ is large and the solution has been constructed for simple systems of physical interest, such as (2.8) and (2.10), by studying the global geometric structure of shock and simple wave curves. Another approach uses the method of "vanishing viscosity".

Bibliographic Remarks: For a study of shock and wave curves and the solution of the Riemann problem in genuinely nonlinear systems, see Lax [33]. The corresponding results for general systems are obtained in Liu [41]. Solutions in the large for various particular systems are given in Smoller [58], Chang and Hsiao [3], Leibovich [37], Dafermos and DiPerna [16] and Liu [42]. For the connection between the Riemann problem and the structure of general solutions see DiPerna [19] and Liu [38].

11. THE RANDOM CHOICE METHOD

In this final section we draw a rough sketch of the random choice method which uses the solution of the Riemann problem to construct an admissible BV solution on $(-\infty,\infty) \times [0,\infty)$ of the Cauchy problem for system (10.1) under initial conditions

$$U(x,0) = U_0(x), \quad -\infty < x < \infty, \tag{11.1}$$

provided $U_0(x)$ has small total variation on $(-\infty,\infty)$.

The solution is obtained as the $h \downarrow 0$ limit of a family of functions $U_h(x,t)$ constructed by the following procedure:

We fix $h > 0$, as a space step length, and determine the corresponding time step length $s = \Lambda^{-1}h$ where Λ is a fixed upper bound of the moduli of the n characteristic speeds. Then we partition the upper halfplane of the x-t plane into strips $S_\ell = \{(x,t)\colon -\infty < x < \infty, \ell s \le t < (\ell+1)s\}$, $\ell=0,1,2,\ldots$, and identify the *mesh-points* $[kh,\ell s]$, with $k + \ell$ even.

Assuming that $U_h(x,t)$ has been defined on $\bigcup_{i=0}^{\ell-1} S_i$, we construct $U_h(x,\ell s)$ as a step function approximation of

$$U_h(x,\ell s^-) \overset{\text{def}}{=} \lim_{t\uparrow \ell s} U_h(x,t) \tag{11.2}$$

which is constant on each interval $kh < x < (k+2)h$, $k + \ell$ even, and hence experiences its jump discontinuities at the mesh points. The approximation will have to be done in a special fashion to be discussed at length below. To start out, at $\ell = 0$, we use the above procedure with $U_h(x,0-) \stackrel{\text{def}}{=} U_0(x)$.

Once $U_h(x,\ell s)$ has been constructed, we define $U_h(x,t)$ on S_ℓ as the solution of (10.1) with initial conditions $U(x,\ell s) = U_h(x,\ell s)$. Note that $U_h(x,t)$ on S_ℓ can be constructed by resolving the jump discontinuities of $U_h(x,\ell s)$ at the mesh points $[kh,\ell s]$, $k+\ell$ even, into fans of waves, solving a Riemann problem, as discussed in Section 10. The waves emanating from the mesh points propagate at speeds not exceeding Λ and thus cannot collide over the time interval $\ell s < t < (\ell+1)s$.

So long as the oscillation of $U_h(x,t)$ remains small, the Riemann problems can be solved and the above procedure may be continued until $U_h(x,t)$ has been determined on $(-\infty,\infty) \times [0,\infty)$. By its construction, $U_h(x,t)$ satisfies (10.1) inside every strip S_ℓ, $\ell=0,1,2,\ldots$, and experiences jump discontinuities across the lines $t = 0,s,2s,\ldots$.

In order to show that $U_h(x,t)$ is an "approximate solution" of (10.1), (11.1), we will have to establish (i) the relative *compactness* of the family $\{U_h(x,t):h > 0\}$ so as to ensure that a sequence, denoted again by $\{U_h\}$, may be selected such that

$$U_h(x,t) \xrightarrow{\text{a.e.}} U(x,t),\ h \to 0,\ -\infty < x < \infty,\ 0 \le t < \infty; \tag{11.3}$$

and (ii) the *consistency* of the algorithm so as to guarantee that the limit $U(x,t)$ in (11.3) is indeed a solution of (10.1),(11.1). We will show that a special and ingenious way of approximating $U_h(x,\ell s^-)$ by the step function $U_h(x,\ell s)$ is necessary in order to achieve these two objectives.

The issue of compactness will be resolved by establishing estimates

$$TV_x U_h(x,t) \le \text{const.}\ TV_x U_0(x),\quad 0 \le t < \infty, \tag{11.4}$$

$$\int_{-\infty}^{\infty} |U_h(x,t)-U_h(x,\tau)|dx \le \text{const.}\,|t-\tau|\,TV_x U_0(x),\ 0 \le \tau < t < \infty, \tag{11.5}$$

with constants that are independent of $h > 0$. Such estimates would imply that the family $\{U_h(x,t)\}$ has uniformly bounded variation and therefore a sequence may be selected which converges, as in (11.3), to a function $U(x,t)$ of class BV.

Assuming for the time being that we already have (11.4),

(11.5),(11.3), we turn to the question of consistency. We fix a smooth test function $\phi(x,t)$ of compact support in $(-\infty,\infty) \times [0,\infty)$ and use that $\partial_t U_h + \partial_x F(U_h) = 0$ inside each strip S_ℓ to obtain

$$\int_0^\infty\int_{-\infty}^\infty \{U\partial_t\phi + F(U)\partial_x\phi\}dxdt + \int_{-\infty}^\infty U_0(x)\phi(x,0)dx$$

$$= \lim_{h\downarrow 0} \sum_{\ell=0}^\infty \int\int_{S_\ell} \{U_h\partial_t\phi + F(U_h)\partial_x\phi\}dxdt + \int_{-\infty}^\infty U_h(x,0^-)\phi(x,0)dx$$

$$= -\lim_{h\downarrow 0} \sum_{\ell=0}^\infty \int_{-\infty}^\infty \{U_h(x,\ell s) - U_h(x,\ell s^-)\}\phi(x,\ell s)dx. \tag{11.6}$$

Noting that $U(x,t)$ is a solution of (10.1),(11.1) if and only if the left-hand side of (11.6) is zero, we conclude that, for consistency, we will have to approximate $U_h(x,\ell s^-)$ by the step function $U_h(x,\ell s)$ in such a way that the right-hand side of (11.6) vanishes.

Vanishing of the right-hand side of (11.6) may be achieved easily by assigning as value of $U_h(x,\ell s)$ on the interval $(k-1)h < x < (k+1)h$, $k + \ell$ odd, the mean value of $U_h(x,\ell s^-)$ over the same interval (Godunov's scheme). Unfortunately, easily as it settles consistency, Godunov's scheme stumbles on the issue of compactness.

The reverse difficulty arises when one approximates by assigning as value of $U_h(x,\ell s)$ on the interval $(k-1)h<x<(k+1)h, k+\ell$ odd, the value of $U_h(x,\ell s^-)$ at a predetermined point of this interval, namely, compactness may be achieved but the resulting algorithm is not necessarily consistent. In this approximating procedure one starts with a sequence $\alpha_0,\alpha_1,\ldots$, taking values in the interval $(-1,1)$, and then defines

$$U_h(x,\ell s)=U_h(y_{k,\ell},\ell s^-), (k-1)h<x<(k+1)h, \ell=0,1,2,\ldots,k+\ell \text{ odd}, \tag{11.7}$$

where

$$y_{k,\ell} = (k+\alpha_\ell)h, \quad \ell=0,1,2,\ldots, \quad k+\ell \text{ odd}. \tag{11.8}$$

It is not hard to identify sequences $\alpha_0,\alpha_1,\ldots$ for which the right-hand side of (11.6) remains bounded away from zero and hence consistency fails. However, this difficulty may be overcome by selecting $\{\alpha_\ell\}$ as a sequence of randomly equidistributed numbers on $(-1,1)$. Then on the right-hand side of (11.6) we have the sum of independent random variables with zero mean. Therefore, for almost all choices of the sequence $\{\alpha_\ell\}$, the right-hand side of (11.6) vanishes and consistency is achieved. This is the *random choice method* of Glimm.

In order to get a different perspective of why the random choice of $\{\alpha_\ell\}$ induces consistency, let us consider the trivial hyperbolic equation $\partial_t u = 0$ with initial condition $u_0(x) = 0$ for $x < 0$ and $u_0(x) = 1$ for $x > 0$. The solution consists of a stationary shock, $x = 0$, which joins the constant states 0, from the left, and 1, from the right. We determine $u_h(x,t)$ for this problem using the appoximating policy (11.7),(11.8). In each strip S_ℓ, $u_h(x,t)$ is 0 for $x < K_\ell h$ and 1 for $x > K_\ell h$, where $K_\ell = K_{\ell-1}+1$ when $\alpha_\ell < 0$ and $K_\ell = K_{\ell-1}-1$ when $\alpha_\ell > 0$. In other words, in passing from $S_{\ell-1}$ to S_ℓ the stationary shock is shifted by h to the right, when $\alpha_\ell < 0$, or to the left, when $\alpha_\ell > 0$. To see that consistency generally fails, note that if $\alpha_0, \alpha_1, \ldots$ are all negative, then as $h \downarrow 0$, $u_h(x,t)$ converges to the function $u(x,t)$, with $u(x,t) = 0$ for $x < (h/s)t$, $u(x,t) = 1$ for $x > (h/s)t$, which is not a solution of $\partial_t u = 0$. It is clear that $u_h(x,t)$ will converge to a solution of $\partial_t u = 0$ only if $\{\alpha_\ell\}$ has an "equal number of terms" in the intervals (-1,0) and (0,1) because it is only then that the shock will stay, in the average, on course. By the same argument one easily sees that the algorithm will be consistent for every linear equation $\partial_t u + c\partial_x u$, c an arbitrary constant, if and only if the sequence $\{\alpha_\ell\}$ is *equidistributed* on (-1,1) in the sense

$$\lim_{m\to\infty} \frac{1}{m} B(\alpha_\ell, m, I) = \frac{1}{2}\mu(I), \tag{11.9}$$

where I is any subinterval of (-1,1) with length $\mu(I)$ and $B(\alpha_\ell, m, I)$ denotes the number of ℓ, $0 \leq \ell \leq m$, such that $\alpha_\ell \in I$. Almost all random sequences $\{\alpha_\ell\}$ will satisfy (11.9). A study based on the idea of wave partitioning, which will be discussed later, has revealed that (11.9) is a necessary and sufficient condition for the consistency of the algorithm on any system (10.1). This is a *deterministic* version of the random choice method.

We now turn to the question of compactness. We have to establish estimates (11.4),(11.5). However, since the waves emanating from each mesh-point propagate with speed not exceeding Λ, (11.5) is an immediate consequence of (11.4).

We note that for t on any interval $\ell s<t<(\ell+1)s$, $TV_x U_h(x,t)$ is constant and may be measured by the sum of the strength of waves that emanate from the mesh-points $[kh,\ell s]$, $k+\ell$ even. We proceed to estimate how the sum of the strengths of waves changes from the strip S_ℓ to the strip $S_{\ell+1}$. To this end we consider the family of diamond shpaed domains $D_{k,\ell}$, $k+\ell$ odd, with vertices the points $[y_{k,\ell}, \ell s]$, $[y_{k+1,\ell+1}, (\ell+1)s]$, $[y_{k,\ell+2}, (\ell+2)s]$, $[y_{k-1,\ell+1}, (\ell+1)s]$. Inside $D_{k,\ell}$ lies the mesh-point $[kh, (\ell+1)s]$ from which emanates a wave fan of n waves $\{\varepsilon_1, \ldots, \varepsilon_n\}$. Through the edge of $D_{k,\ell}$ which connects the vertices $[y_{k,\ell},\ell s]$ and $[y_{k-1,\ell+1},(\ell+1)s]$ there penetrates $D_{k,\ell}$ a fan of waves $\{\gamma_1,\dots,\gamma_n\}$ which is part (possibly none or all) of the wave fan that emanates from the mesh-point $[(k-1)h,\ell s]$. Similarly, through the edge of $D_{k,\ell}$ which connects the vertices $[y_{k,\ell},\ell s]$ and $[y_{k+1,\ell+1},(\ell+1)s]$ there penetrates $D_{k,\ell}$ a fan of waves $\{\delta_1,\dots,\delta_n\}$ which is part of the wave fan that emanates from the mesh-point $[(k+1)h,\ell s]$. It turns out that $\{\varepsilon_1,\dots,\varepsilon_n\}$ approximates the wave pattern that would have resulted had the wave fans $\{\delta_1,\dots,\delta_n\}$ and $\{\gamma_1,\dots,\gamma_n\}$ been allowed to propagate beyond $t = (\ell+1)s$ and thus interact. It can be shown that the strengths (denoted with vertical bars) of incoming and outgoing waves are related by

$$\sum_{i=1}^{n} |\varepsilon_i| = \sum_{i=1}^{n} (|\gamma_i|+|\delta_i|) + O(Q(D_{k,\ell})), \tag{11.10}$$

where

$$Q(D_{k,\ell}) = \sum_{i,j} \{|\gamma_i||\delta_j| \ :\ \gamma_i \text{ and } \delta_j \text{ "approach"}\} . \tag{11.11}$$

If the quadratic term were not present on the right-hand side of (11.10), the total variation of $U_h(x,t)$, as measured by the sum of the strengths of waves, would not increase from S_ℓ to $S_{\ell+1}$. The effect of the quadratic term in (11.10) can be controlled in the following way. We consider the polygonal curve J_ℓ whose arcs connect nodes $[y_{k,\ell},\ell s]$ and $[y_{k\pm1,\ell+1},(\ell+1)s]$, $k+\ell$ even. With J_ℓ we associate the *Glimm Functional*

$$\mathscr{F}(J_\ell) = \mathscr{L}(J_\ell) + C_0\mathscr{M}(J_\ell) \tag{11.12}$$

where

$$\mathscr{L}(J_\ell) = \sum\{|\beta| \ :\ \beta \text{ crosses } J_\ell\}, \tag{11.13}$$

$$\mathscr{M}(J_\ell) = \sum\{|\alpha||\beta| \ :\ \alpha,\beta \text{ cross } J_\ell \text{ and "approach"}\}, \tag{11.14}$$

and C_0 is a large positive constant. Using (11.10),(11.11) it is easy to see that, so long as the total variation remains small, $\mathscr{F}(J_\ell)$ is nonincreasing in ℓ. At the same time $\mathscr{F}(J_\ell)$ is essentially equivalent to $TV_x U_h(x,t)$, for $\ell s<t<(\ell+1)s$, and this establishes estimate (11.4).

It follows from the above discussion that the random choice

method, as well as its deterministic version, satisfy the consistency and compactness requirements and thus establish the existence of solutions of class BV to the Cauchy problem (10.1), (11.1) for a general, strictly hyperbolic, system of n conservation laws in one space dimension, under initial data of small total variation.

Returning to the wave interactions in the diamond $D_{k,\ell}$, it is instructive to view γ_i and δ_i as the "parents" of ε_i. Retracing our steps backwards in time, it is possible to identify all the "progenitors" of ε_i and thus trace its family tree all the way back to $t = 0$. This induces a *wave partitioning* which monitors the global wave pattern and thus provides detailed information on solutions. Using this information one establishes the consistency of the deterministic version of the random choice algorithm, under assumption (11.9), derives the results on local structure of solutions, reported in Section 4, and shows that solutions constructed by the random choice method satisfy Liu's admissibility criterion.

The wave partitioning approach provides also an effective tool for studying the asymptotic behavior of solutions, as $t \to \infty$. When the initial data have compact support, waves associated with distinct characteristic families are eventually separated and their asymptotic behavior is governed by a single conservation law. In particular, waves associated with genuinely nonlinear characteristic fields propagate asymptotically as N-*waves*.

The remaining major open problem is to establish the convergence of the random choice algorithm when the total variation of the initial data is not small. In this direction it has been shown that the algorithm converges, for genuinely nonlinear systems of two equations, provided the initial data have small oscillation, but not necessarily small total variation. For very special systems, such as (2.10) with $p(u) = 1/u$, it is known that the random choice method converges for arbitrary initial data.

The random choice method is strictly one-space dimensional but within this limitation it is quite versatile and robust so it can be modified to solve boundary value problems, it may be adapted to systems with special structure and can be combined with the method of fractional steps to construct BV solutions for systems of nonhomogeneous balance laws

$$\partial_t U(x,t) + \partial_x F(U(x,t),x,t) + G(U(x,t),x,t) = 0. \tag{11.15}$$

Finally, the method is an effective tool for the numerical computation of solutions yielding particularly sharp resolution of shock profiles.

Bibliographic Remarks: The random choice method for genuinely nonlinear systems was developed by Glimm [26]. The deterministic version is due to Liu [38,43]. Examples of extensions, adaptations and applications can be found in Glimm and Lax [27], Nishida [50], Nishida and Smoller [51] and Liu [44]. The asymptotic behavior of solutions is described in DiPerna [21] and Liu [45,46]. For extensions to nonhomogeneous balance laws see Liu [47] and Dafermos and Hsiao [18]. For numerical applications see Chorin [5] and DiPerna [22].

REFERENCES

[1] Ball, J.M., Strict convexity, strong ellipticity and regularity in the calculus of variations. 1980. Math. Proc. Camb. Phil. Soc. 87, pp. 501-513.

[2] Boilat, G., *La Propagation des Ondes*. Paris: Gauthier-Villars. 1965.

[3] Chang, T., and Hsiao, L., Riemann problem for the system of conservation laws of aerodynamics. 1979. Acta Math. Sinica, Dec. 1979.

[4] Chen, P.J., Growth and decay of waves in solids. *Handbuch der Physik* VIa/3 ed. C. Truesdell. Berlin: Springer-Verlag 1973, pp.303-402.

[5] Chorin, A., Random choice methods with applications to reacting gas flow. 1977. J. Comp. Phys. 25, pp.253-371.

[6] Conley, C.C., and Smoller, J.A., Viscosity matrices for two dimensional nonlinear hyperbolic systems. 1970. Comm. Pure Appl. Math. 23, pp.867-884.

[7] Conlon, J., A theorem in ordinary differential equations with an application to hyperbolic conservation laws. 1980. Adv. in Math. 35, pp.1-18.

[8] Conlon, J. and Liu, T.-P., Admissibility criteria for hyperbolic conservation laws. 1981. Indiana U. Math. J. 30, pp.641-652.

[9] Conway, E.D. and Smoller, J.A., Global solutions of the Cauchy problem for quasi-linear first order equations in several space variables. 1966. Comm. Pure Appl. Math. 19, pp.95-105.

[10] Conway, E.D. and Smoller, J.A., Shocks violating Lax's condition are unstable. 1973. Proc. Am. Math. Soc. 39,pp.353-356.

[11] Courant, R. and Friedrichs, K.O., *Supersonic Flow and Shock Waves*. New York: Interscience 1948.

[12] Dafermos, C.M., Quasilinear hyperbolic systems that result from conservation laws. *Nonlinear Waves* ed. S. Leibovich and A.R. Seebass. Ithaca: Cornell U. Press 1974, pp.82-102.

[13] Dafermos, C.M., The second law of thermodynamics and stability. 1979. Arch Rat. Mech. Anal. 70, pp.167-179.

[14] Dafermos, C.M., The equations of elasticity are special. *Trends in Applications of Pure Mathematics to Mechanics*. Vol. III ed.R.J. Knops. London: Pitman 1981.

[15] Dafermos, C.M., The entropy rate admissibility criterion for solutions of hyperbolic conservation laws. 1973. J. Diff. Eqs. 14, pp.202-212.

[16] Dafermos, C.M. and DiPerna, R.J., The Riemann problem for certain classes of hyperbolic systems of conservation laws. 1976. J. Diff. Eqs. 20, pp.90-114.

[17] Dafermos, C.M. and Hsiao, L., Global smooth thermomechanical processes in one-dimensional nonlinear thermoviscoelasticity. 1982. Nonlinear Analysis 6, pp.435-454.

[18] Dafermos, C.M. and Hsiao, L., Hyperbolic systems of balance laws with inhomogeneity and dissipation. 1982. Indiana U. Math. J. 31, pp.471-491.

[19] DiPerna, R.J., Singularities of solutions of nonlinear hyperbolic systems of conservation laws. 1975. Arch. Rat. Mech. Anal. 60, pp.75-100.

[20] DiPerna, R.J., Uniqueness of solutions to hyperbolic conservation laws. 1979. Indiana U. Math. J. 28, pp.137-188.

[21] DiPerna, R.J., Decay and asymptotic behavior of solutions to nonlinear hyperbolic systems of conservation laws. 1975. Indiana U. Math. J. 24, pp.1047-1071.

[22] DiPerna, R.J., Finite difference schemes for conservation laws. 1982. Comm. Pure Appl. Math. 35, pp.379-450.

[23] Federer, H., *Geometric Measure Theory*. New York: Springer-Verlag. 1969.

[24] Foy, R.L., Steady state solutions of hyperbolic systems of conservation laws with viscosity terms. 1964. Comm. Pure Appl. Math. 17, pp.177-188.

[25] Friedrichs, K.O., and Lax, P.D., Systems of conservation equations with a convex extension. 1971. Proc. Nat. Acad. Sci. USA 68, pp.1686-1688.

[26] Glimm, J., Solutions in the large for nonlinear hyperbolic systems of equations. 1965. Comm. Pure Appl. Math. 18, pp.697-715.

[27] Glimm, J., and Lax, P.D., Decay of solutions of systems of nonlinear hyperbolic conservation laws. Mem. Am. Math. Soc. No. 101. Providence: Am. Math. Soc. 1970.

[28] Greenberg, J.M., MacCamy, R.C., and Mizel, V.J., On the existence, uniqueness and stability of solutions of the equation $\sigma'(u_x)u_{xx}+\lambda u_{xtx}=\rho u_{tt}$. 1968. J. Math. Mech. 17, pp.707-728.

[29] Hopf, E., The partial differential equation $u_t+uu_x=\mu u_{xx}$. 1950. Comm. Pure Appl. Math. 3, pp.201-230.

[30] Hsiao, L., The entropy rate admissibility criterion in gas dynamics. 1980. J. Diff. Eqs. 38, pp.226-238.

[31] Jeffrey, A., *Quasilinear Hyperbolic Systems and Waves*. London: Pitman.1976.

[32] Klainerman, S., and Majda, A., Formation of singularities for wave equations including the nonlinear vibrating string. 1980. Comm. Pure Appl. Math. 33, pp.241-263.

[33] Lax, P.D., Hyperbolic systems of conservation laws. 1957. Comm. Pure Appl. Math. 10, pp.537-566.

[34] Lax, P.D., *Hyperbolic Systems of Conservation Laws and the Mathematical Theory of Shock Waves*. Philadelphia: SIAM 1973.

[35] Lax, P.D., Development of singularities of solutions of nonlinear hyperbolic partial differential equations. 1964. J. Math. Phys. 5, pp.611-613.

[36] Lax, P.D., Shock waves and entropy. Contributions to Functional Analysis. ed. E.A. Zarantonelo. New York: Academic Press 1971, pp.603-634.

[37] Leibovich, L., Solutions of the Riemann problem for hyperbolic systems of quasilinear equations without convexity conditions. 1974. J. Math. Anal. Appl. 45, pp.81-90.

[38] Liu, T.-P., Admissible solutions of hyperbolic conservation laws. Mem. Am. Math. Soc. No. 240. Providence: Am. Math.Soc. 1981.

[39] Liu, T.-P., The Riemann problem for general 2×2 conservation laws. 1974. Trans. Am.Math. Soc. 199, pp.89-112.

[40] Liu, T.-P., The entropy condition and the admissibility of shocks. 1976. J. Math. Anal. Appl. 53, pp.78-88.

[41] Liu, T.-P., The Riemann problem for general system of conservation laws. 1975. J. Diff. Eqs. 18, pp.218-234.

[42] Liu, T.-P., Shock waves in the nonisentropic gas flow. 1976. J. Diff. Eqs. 22, pp.442-452.

[43] Liu, T.-P., The deterministic version of the Glimm scheme. 1977. Comm. Math. Phys. 57, pp.135-148.

[44] Liu, T.-P., Initial-boundary value problems for gas dynamics. 1977. Arch. Rat. Mech. Anal. 64, pp.137-168.

[45] Liu, T.-P., Decay to N-waves of solutions of general systems of nonlinear hyperbolic conservation laws. 1977. Comm. Pure Appl. Math. 30, pp.585-610.

[46] Liu, T.-P., Linear and nonlinear large-time behavior of solutions of hyperbolic conservation laws. 1977. Comm. Pure Appl. Math. 30, pp.767-796.

[47] Liu, T.-P., Quasilinear hyperbolic systems. 1979. Comm. Math. Phys. 68, pp.141-172.

[48] Majda, A., The existence of multi-dimensional shock fronts. Mem. Am. Math. Soc. No. 275. Providence: Am. Math. Soc. 1983.

[49] Mock, M.S., A topological degree for orbits connecting critical points of autonomous systems. 1980. J. Diff. Eqs. 38, pp.176-191.

[50] Nishida, T., Global solution for an initial boundary value problem of a quasilinear hyperbolic system. 1968. Proc. Japan Acad. 44, pp.642-646.

[51] Nishida, T. and Smoller, J.A., Solutions in the large for some nonlinear hyperbolic conservation laws. 1973. Comm. Pure Appl. Math. 26, pp.183-200.

[52] Oleinik, O.A., Discontinuous solutions of non-linear differential equations. 1957. Usp. Mat. Nauk 12(3), pp.3-73.

[53] Oleinik, O.A., Uniqueness and stability of the generalized solution of the Cauchy problem for a quasi-linear equation. 1959. Usp. Mat. Nauk 14, pp.165-170.

[54] Pego, R., Viscosity matrices for systems of conservation laws. 1982. Report No. 2, Center for Pure and Appl. Math., U. Calif. Berkeley.

[55] Rozhdestvensky, B.L. and Yanenko, N.N., *Quasilinear Systems and their Application to the Dynamics of Gases* (in Russian). Moscow: Nauk 1968.

[56] Ruggeri, T. and Strumia, A., Main field and convex covariant density for quasi-linear hyperbolic systems. 1981. Ann. Inst. Henri Poincaré, 34, pp.65-84.

[57] Smoller, J.A., *Shock Waves and Reaction-Diffusion Equations*. New York: Springer-Verlag 1982. (In press)

[58] Smoller, J.A., On the solution of the Riemann problem with general step data for an extended class of hyperbolic systems. 1969. Mich. Math. J. 16, pp.201-210.

[59] Truesdell, C.A., and Noll, W., The nonlinear field theories of mechanics. *Handbuch der Physik* III/3, ed. S. Flügge. Berlin: Springer-Verlag 1965.

[60] Truesdell, C.A., and Toupin, R.A., The classical field theories. *Handbuch der Physik* III/1, ed. S. Flügge. Berlin: Springer-Verlag 1960, pp.226-790.

[61] Volpert, A.I., The spaces BV and quasilinear equations. 1967. English transl. Math. USSR Sbornik 2, pp.225-267.

[62] Wendroff, B., The Riemann problem for materials with non-convex equation of state. 1972. J. Math. Anal. Appl. 38, pp.454-466 and 640-658.

[63] Whitham, G.B., *Linear and Nonlinear Waves*. New York: Wiley 1974.

[64] Zhang, T. and Hsiao, L., Riemann problem and discontinuous initial value problem of typical quasilinear hyperbolic system without convexity. 1977. Acta Math. Sinica, 20, pp.229-231.

ILL-POSED PROBLEMS IN THERMOELASTICITY THEORY

J.L. Ericksen

Department of Aerospace Engineering and Mechanics
and School of Mathematics,
University of Minnesota,
Minneapolis, Minnesota, U.S.A.

LECTURE 1 TWINNING OF THERMOELASTIC MATERIALS

In crystals, it is not unusual to observe the phenomena of twinning or trilling. The nonlinear theories of elasticity and thermoelasticity seem to be capable of describing some special types of twinning, for example. To be discussed is some of the elementary theory which has evolved. Such theory suggests some rather deep mathematical problems which need attention.

INTRODUCTION

Twinning is one of those subjects which are discussed in many works on crystal behaviour, but which tend to be ignored in expositions written by experts in mechanics. As the term is used, it covers a variety of phenomena. There is a select subset to which thermoelasticity theory might reasonably be applied, if one gives up the common assumption that the equilibrium equations always satisfy the strong ellipticity condition. Considering the mathematical complications attending this, it could be useful to have physical examples to indicate how nature behaves, in such cases. Actually, I will abstract features which seem to be of import, and are involved in physical examples.

One kind of example arises in studies of the so-called memory alloys. As is rather clear from, say, the discussion of Müller and Wilmanski [1], workers are making some use of one-dimensional versions of thermoelasticity theory, and twinning plays a significant role. Similar remarks apply to cubic-tetragonal transitions

J. M. Ball (ed.), Systems of Nonlinear Partial Differential Equations, 71–93.

encountered in some of the superconducting crystals, discussed by Ericksen [2], for example. Basically, these are the prototypes from which I abstract. This is not to say that these are the only cases for which the format to be described might be applicable. The likely prospects are among those for which the adjective "Martensitic" is appropriate, but only some such.

BASIC EQUATIONS

For homogeneous materials, thermoelasticity theory employs constitutive equations of the form

$$\phi = \phi(C,\theta) \quad , \tag{1}$$

where ϕ is the Helmholtz free energy per unit mass, $\theta > 0$ is the absolute temperature, and C is a symmetric, positive definite tensor, given in terms of the usual deformation gradient F by

$$C = F^T F, \ \det. \ F > 0 \quad . \tag{2}$$

For our purposes, it is important that ϕ be invariant under some group G,

$$\phi(H^T CH,\theta) = \phi(C,\theta), \ H \in G \quad . \tag{3}$$

Mathematically, it is not very important what G is, although twinning is commonly associated with crystal symmetries. Mathematically, at least, the study by Varley and Day [3] is interesting, and can be included by identifying G with the orthogonal group.

We will be concerned only with cases where θ is independent of position and C is either constant or piecewise constant. The values of C of concern are those which, for fixed values of θ, correspond to absolute or relative minima of ϕ, what are commonly considered as natural states. For unconstrained materials, the usual first derivative test gives

$$\partial\phi/\partial C = 0 \quad , \tag{4}$$

which means that the stress vanishes. Similarly, constrained materials can be unstressed, or subject to stresses which, for these, are trivial. We won't make use of this or other criteria for minima, but this is one of the factors involved, in common thinking. The idea is that we might have more than one such minimiser. Generally speaking, different ones would be expected to give different values of ϕ, unless some condition of symmetry implies that they must be energetically equivalent. In the discussion of Müller and Wilmanski [1], the Austenitic and Martensitic phases generally correspond to different values of ϕ, but ϕ must be the same for the two twin-related Martensitic phases which they consider. In the cases discussed by Ericksen [1], the tetragonal phase involves twins, for which ϕ must have the same value, by symmetry. Of course, it is such relations which motivate using the words "twin" or "twinning".

Suppose

$$C = C_1(\theta) \tag{5}$$

is one such minimiser. Then, because of the assumed invariance, this generates an orbit of minimisers, with the same value of ϕ. These will all be the same if

$$H^T C_1 H = C_1 \quad , \quad H \in G \quad , \tag{6}$$

as is commonly assumed, at least tacitly, in mathematical work on thermoelasticity theory. Our concern is with cases where this does not hold, so that, for at least one $H = \bar{H} \in G$, and some temperature,

$$C_2 = \bar{H}^T C_1 \bar{H} \neq C_1 \tag{7}$$

gives a second minimiser. Reasonably, one can expect (7) to apply over temperature intervals, at least. In practice, the word "twin" does not imply that there are only two such possibilities, as will be made clear. Clearly, (7) cannot hold, if G consists of just the identity, or the two elements ± 1, but any other choice of G is possible, if C_1 has a suitable form. As is discussed by Gurtin and Williams [4], there are good reasons to take G as a subgroup of the unimodular group. Since it is automatic that (3) holds when $H = -1$, we can then assume that

$$\det. \bar{H} = 1 \quad . \tag{8}$$

Physically, the two configurations (phases) mentioned in (7) then have the same specific volume. In turn, this means that, by the standards of classical thermodynamics, they are indistinguishable. For our purpose, this is not of much import, except to indicate that discussions of the phenomena are not likely to be found in a thermodynamics text.

Now, suppose that F_1 is some deformation gradient corresponding to C_1. It then follows from (2), (7) and (8) that any deformation gradient F_2 corresponding to C_2 must be of the form

$$F_2 = RF_1\bar{H} \quad , \tag{9}$$

where R represents some rigid rotation,

$$R^{-1} = R^T, \text{ det. } R = 1 \quad . \tag{10}$$

We need another assumption, which can be motivated as follows. Suppose that the body has undergone some deformation F_1. Rigidly rotating it would leave it in a physically equivalent configuration, so there is no loss in generality in assuming that

$$F_1 = C_1^{\frac{1}{2}} \Rightarrow F_1 = F_1^T \quad , \tag{11}$$

and that F_1 is positive definite. Now suppose that part of the body undergoes a deformation which leaves the displacement continuous, but converts C_1 to C_2 in this part. Where the parts join, we will have a surface of discontinuity, a finite discontinuity in F. For simplicity, assume it is a plane, the twin plane, with unit normal n. Application of the usual kinematic conditions of compatibility then gives the condition that the value of F_2 which occurs must satisfy an equation of the type

$$F_2 = (1 + a \otimes n)\, F_1 \quad , \quad n \cdot n = 1 \quad , \tag{12}$$

where a is some vector, the amplitude vector. Combining (9) and (12) to eliminate F_2, we get

$$RF_1\bar{H} = (1 + a \otimes n)\, F_1 \quad . \tag{13}$$

In part, the elementary theory involves determining the possible choices of R, $\bar{H}$, a and n which can satisfy (13), considering F_1 and G as given. Using (8), it is easy to show that

$$\text{det. } (1 + a \otimes n) = 1 \Rightarrow a \cdot n = 0 \quad , \tag{14}$$

or, equivalently,

$$(1 + a \otimes n)^{-1} = 1 - a \otimes n \quad . \tag{15}$$

Equations of this kind are commonly used in analysing various kinds of Martensitic transformations, in the discussion of Nishiyama [5,§6.3], for example. The possibility of coexistence of two such symmetry-related phases is commonly implied by the word "twinning". If you like, our twins are Siamese.

For some purposes, it is convenient to rearrange the equations, by using F_1 to map the twin plane back to the fixed reference configuration. This yields a plane with unit normal

$$N = F_1^T n / \| F_1^T n \| \quad . \tag{16}$$

Then, setting

$$A = \| F_1^T n \| \; F_1^{-1} a \quad , \tag{17}$$

we can rewrite (13) as

$$R F_1 \bar{H} = F_1 (1 + A \otimes N) \quad , \tag{18}$$

with

$$(1 + A \otimes N)^{-1} = 1 - A \otimes N, \quad A \cdot N = 0 \quad . \tag{19}$$

A crystallographer might balk at the idea that N could vary continuously with θ, but I see no harm in being open-minded about this. As is easily verified, the assumption of coexisting configurations implies that the equilibrium equations cannot satisfy the strong ellipticity condition for all kinematically possible deformations. One might restrict the domain so as to keep the condition satisfied, and some would prefer to do this. Using the equations given, it is rather easy to show that one can introduce any number of parallel twin planes, with C alternating between the values C_1 and C_2 in the strips between the planes. Of course, this gives infinitely many unstressed, energetically equivalent configurations. Observations confirm the notion that a great variety of such configurations can occur. The behaviour of the memory alloys involves changes in the number and location of twin planes in the Martensitic phases, as well as transitions from Austenitic to Martensitic phases, obtained by applying and removing loads. In the cases discussed by Ericksen [2], cooling the unloaded cubic phase below a critical temperature produces tetragonal phases, which tend to be twinned.

Certainly, the mathematical problems of the kind suggested are not routine, so we grope for some way to master the phenomena.

TWIN PLANES

For a particular material, it is of interest to determine the possible orientations of twin planes, and the amplitude vector a or A. Equations of the same form arise in the problem of bonding together differently oriented pieces of a body, to avoid developing stresses near the bond, as the temperature changes. Theory for this is presented by Ericksen [6]. This assumes the discontinuity surface to be a material surface, but relaxing this restriction causes no difficulty. Here, there is no good reason to prejudice the analysis, by assuming this. Crystal experts lean more on pictures of atomic configurations so, again, we abstract some of the essence. Physically, one relies on such pictures, in deciding whether the theory being described might apply, to estimate G, etc., so, in fact, the differences in reasoning are not great. To adapt the aforementioned analysis, we need to impose one limitation, that $\bar{H}$ represent a rigid rotation

$$\bar{H}^{-1} = \bar{H}^T, \quad \det. \bar{H} = 1 \quad . \tag{20}$$

For solids, most workers estimate G to be a subgroup of the orthogonal group, in which case (20) follows automatically. As is discussed by Ericksen [2], molecular theories of crystal elasticity imply that G should contain non-orthogonal elements and, in some respects, accounting for this promotes a better physical understanding of the phenomena. However, it adds mathematical complications, so it might be better to ignore this, for the present.

To adapt the aforementioned theory of stress-free joints, consider as given F_1, satisfying (11), and $\bar{H}$ satisfying (20). The first step is to determine the set of rigid rotations $\bar{R}$ defined by

$$\{\bar{R} \mid \bar{R}^T F_1 \bar{R} = F_1\} \quad . \tag{21}$$

The next step is to find one such that

$$\hat{R} = \bar{R}\,\bar{H} \quad . \tag{22}$$

represents a 180° rotation about some axis. Any one will do, and if none exists, the problem is insoluble. Any possibility can be represented in the form

$$\hat{R} = -1 + 2E \otimes E \quad , \quad E \cdot E = 1 \quad , \tag{23}$$

E being the axis. Solutions are given by one of two prescriptions. First, we can take

$$\begin{aligned} N &= E \quad , \\ R &= \tilde{R}\bar{R} \quad , \\ \tilde{R} &= -1 + 2\|F_1^{-1} E\|^{-2} (F_1^{-1} E) \otimes (F_1^{-1} E) \quad , \\ A &= 2(1 - \|F_1^{-1} E\|^{-2})(F_1^{-2} E) \quad . \end{aligned} \tag{24}$$

The second prescription is

$$\begin{aligned} A &= \alpha E \\ \alpha N &= 2(E - \|F_1 E\|^{-2} F_1^{2} E) \quad , \\ R &= \tilde{R}\bar{R} \\ \tilde{R} &= 1 + 2\|F_1 E\|^{-2} (F_1 E) \otimes (F_1 E) \quad , \end{aligned} \tag{25}$$

with α determined by the condition that $\|N\| = 1$. In the latter, it seems that N is undefined, if the prescription gives $\alpha = 0$. It can be shown that this, or having $A = 0$ (in 24) are excluded by assumptions made earlier.

The cases considered by Ericksen [2] provide one illustrative example, which I will sketch. As a reference, take one of the cubic configurations which occur at the higher temperatures, with lattice vectors proportional to an orthonormal set of vectors E_1, E_2 and E_3. Here, by conventional reasoning, the likely choice for G is the crystallographic group suggested, including pairwise interchanges of the three vectors or reversing one of them. For C_1 to be invariant under this, it is well known that C_1 must be proportional to the identity. Physically, this is in agreement with observations, as we cool the crystal, until we reach a critical temperature. As we continue to cool, the crystal changes, continuously or nearly so, to a tetragonal form, containing twins. One deformation giving the tetragonal form can be represented in the form .

$$C_1(\theta) = \alpha^2 1 + (\beta^2 - \alpha^2) E_1 \otimes E_1 \quad , \tag{26}$$

with α and β functions of θ, $\beta^2 = \alpha^2$ above, but not below the critical temperature. Application of the cubic group generates an orbit containing two more

$$C_2(\theta) = \alpha^2 1 + (\beta^2 - \alpha^2)\ E_2 \otimes E_2 \quad , \tag{27}$$

and

$$C_3(\theta) = \alpha^2 1 + (\beta^2 - \alpha^2)\ E_3 \otimes E_3 \quad . \tag{28}$$

Observations indicate that α and β vary continuously with θ, or suffer such a slight discontinuity that it is masked by experimental error. Clearly, as the basic pair, we could take C_1 and C_2, C_2 and C_3 or C_1 and C_3. With the choice fitting previous notation, we can take

$$F_1 = \alpha 1 + (\beta - \alpha)\ E_1 \otimes E_1 \quad , \tag{29}$$

with α and β positive. Clearly,

$$C_2 = \bar{H}^T C_1 \bar{H} \tag{30}$$

holds, with $\bar{H}$ the rotation defined by

$$\bar{H}E_1 = -E_2, \quad \bar{H}E_2 = E_1, \quad \bar{H}E_3 = E_3 \quad . \tag{31}$$

It is easy to see that the rotations $\bar{R}$ defined by (21) are those such that

$$\bar{R}E_1 = \pm\ E_1 \quad . \tag{32}$$

By routine calculations, to sort out those which are consistent with (22) and (23), one finds that

$$\pm\ 2\ E = E_1 \pm E_2 \quad . \tag{33}$$

Applying the prescriptions, one finds that N must take one of these directions, A the other. Bearing in mind the third tetragonal configuration, one gets other possibilities, by replacing E_1 or E_2 by E_3, giving, altogether, six possible orientations of the twin planes. An expert on crystals might do the calculation in another way, but would consider it to be elementary. Descriptions such as are given by Batterman and Barrett [7] suggest that only one occurs, in a particular sample. The reader might like to try out for himself the analysis of the possibility of intersecting twin planes, using algorithms presented by James [8].

Perhaps this is enough to indicate what is involved, mathematically, at the most elementary level. I will close with one of the queries which come to mind. Is there a mathematically acceptable criterion, according to which one of our twinned configurations can be judged to be metastable? Generally, it seems to require some force, to get twin planes to move through a material, and there seems to be no generally accepted criterion for this.

This material is based on work supported by the National Science Foundation under grant CME 1112.

REFERENCES

[1] Müller, I., and Wilmanski, K.: 1981, *'Continuum models of discrete systems'*, (ed. O. Brulin and R.K. Hsieh), North Holland, 4, pp.495-509.

[2] Ericksen, J.L.: 1980, Arch. Rational Mech. Analysis, 73, pp.99-124.

[3] Varley, E., and Day, A.: 1966, Arch. Rational Mech. Analysis, 22, pp.253-269.

[4] Gurtin, M.E., and Williams, W.O.: 1966, Arch. Rational Mech. Analysis, pp.163-172.

[5] Nishiyama, Z.: 1978, *'Martensitic Transformation'*, Academic Press.

[6] Ericksen, J.L.: *Theory of stress-free joints*, to appear in J. Elasticity.

[7] Batterman, B.W., and Barrett, C.S.: 1964, Phys. Rev. Lett., 13, pp.390-392.

[8] James, R.D.: *Finite deformation by mechanical twinning*, to appear in Arch. Rational Mech. Analysis.

LECTURE 2 PROBLEMS FOR INFINITE ELASTIC PRISMS

For elastic prisms with free sides, some of the solutions of greatest physical interest are solutions for infinitely long prisms. Plausibility arguments indicate that it is unusual for a solution to have a continuation belonging to this set, so it is a rather select set. In some way, we should seek to characterise its structure. The lecture will cover some of the evidence, and smaller problems which could shed light.

INTRODUCTION

For reasons which should become more apparent, I will take a roundabout route to get into the basic problem. Occasionally, I have pondered the rather formal derivations of cruder, more tractable mechanical theories from those less crude, not finding the efforts very convincing. Thus, I have tried to think of some different line of thought. Here, I will consider the traditional problems of the equilibrium of elastic prisms with free sides. Here, the cruder classical rod theories, associated with the names of Euler, Bernoulli and Kirchhoff have been quite successful, in the small-strain regime, for prisms which are not too short. I refer to the theories involving a system of ordinary differential equations, discussed in some detail by Love (1, Ch. XVIII). At least in this regime, many think that the solutions should give a pretty good caricature of solutions obtained from three-dimensional theory, and that it should be feasible to derive these rod equations, or something much like them, from three-dimensional theory.

An old idea, still used in such problems, involves expanding the displacement in powers of the thickness coordinates, retaining only terms of lower order. Love (1, Ch. XVIII) indicates skepticism of this, mentioning two reasons for his doubts. One is that Cauchy got the wrong answer for the torsional rigidity of prisms, with cross-sections of rectangular shape. The other, which is different, is that the implied power series might not converge over the entire cross-section. That the latter possibility is real is illustrated by simple examples given by Jenkins [2]. This refers to attempts to derive rod equations, from classical linear elasticity theory. In itself, this is a dubious business since, by accounting for large rotations, rod theory covers buckling phenomena which are outside the range of such linear theory.

By any reasonable estimate, the three-dimensional partial differential equations admit infinitely more solutions than do the ordinary differential equations. A satisfactory derivation must then somehow explain how infinitely many can be collapsed to one. Expansions quite different from the thickness expansions have been used, for example by Parker [3], to avoid difficulties mentioned, including an attempt to deal with the last issue, in a rather formal way. Here, a different issue concerns me. As is clear from the dicussion of Love (1, Ch. XVIII), many of the solutions of the classical rod equations have a periodic character. Expansions such as are used by Parker [3] seem not to be very well designed, to capture such things.

My purpose is not to look for a better kind of perturbation theory, but to try to move closer to the heart of the problem. Since rod theory is a theory applicable to long prisms, it should do best when we make them as long as we can, pushing the two ends to infinity. This motivated me to ponder what little we know or might reasonably speculate about this.

THE SETS S(L)

In terms of rectangular Cartesian coordinates, selected in an obvious way, a reference configuration of prismatic form involves a region of the form $\Delta = D \times I$, with

$$(x^1, x^2) \in D \qquad , \tag{1}$$

a finite planar domain, independent of x^3, and

$$x^3 \in I = [-L, L] \qquad , \tag{2}$$

$2L > 0$ being the length of the undeformed prism. For simplicity, I will assume domains, solutions, etc. smooth enough to avoid obvious difficulties associated with roughness. We are concerned with deformation, $y : \Delta \longrightarrow \mathbb{R}^3$, or

$$y = y(x) \qquad , \tag{3}$$

with

$$F > 0, \quad F = \nabla y \qquad . \tag{4}$$

There is good physical motivation for assuming such mappings are

1:1, for our smooth deformations. Rod theory admits solutions having the prism pass through itself, giving some motivation for relaxing this. Pragmatically, it is too hard for analysts to now cope with the condition, anyhow. Again for simplicity, we ignore theories of constrained materials.

The usual equations of equilibrium are of the Euler-Lagrange form

$$\nabla \cdot \left(\rho \frac{\partial \phi}{\partial F}\right) = 0 \quad , \tag{5}$$

with ϕ, the strain energy per unit mass, depending smoothly on F, in the combination

$$C = F^T F \quad , \tag{6}$$

a consequence of the assumption that rigid body rotations leave it unaffected. Similarly, it does not depend explicitly on y or we assume, on x^3. Commonly, ϕ is assumed to be independent of x^1 and x^2 as well, but I don't think that this is of much import, mathematically, and other possibilities are of physical interest. Here, $\rho > 0$ is the reference mass density. Mathematically, nothing would be lost be setting $\rho = 1$.

We accept the usual assumption that ϕ has a strong isolated minimum at $C = 1$ so, there,

$$\frac{\partial \phi}{\partial C} = 0 \quad , \tag{7}$$

and

$$d^2\phi = \frac{\partial^2 \phi}{\partial C_{ij} \partial C_{kl}} \quad dC_{ij} dC_{kl} \geqslant 0 \quad , \tag{8}$$

the inequality being strict. Said differently, if we set $u=y-x$, and linearise the equations in u, we get the equations of the classical linear theory. In this sense, all equations considered share some nice properties, near $C = 1$, making it reasonable to think that, in this, the small strain regime, we might get a uniform theory, by suitably interpreting the vaguely defined regime. It is this rather limited goal which we will be attempting to assess. Of course, we would like to broaden the regime.

No boundary conditions are imposed at the ends, but we are to satisfy the free boundary condition on the sides, a null boundary condition of the Neumann type,

$$\frac{\partial\phi}{\partial F^i{}_j} N_j = \frac{\partial\phi}{\partial y^i{}_{,j}} N_j = 0 \quad \text{on } (\partial D)\times I \quad . \tag{9}$$

Here, N is the unit normal, with $N_3 = 0$.

Considering the function ϕ and domain D as fixed, it clearly makes sense to consider these problems for different values of L. For any particular value of L, we denote by $S(L)$ the set of solutions which exist. Rod theory would only be used, when L is fairly large, compared to the diameter of D.

One property of the sets $S(L)$ follows from simple logic. Given $L^1 < L$, one can take restrictions of any solution in $S(L)$, and translate x^3 a bit, to get solutions in $S(L^1)$. In the sense made clear by this,

$$S(L) \subset S(L^1) \quad . \tag{10}$$

Actually, if the first solution is not too trivial, it will yield infinitely many different restrictions. To go the other way, one must take a solution in $S(L^1)$, and find a continuation, qualifying it to be in $S(L)$. It is just not plausible that every solution can be continued indefinitely. Consider setting a boundary value problem, in the classical linear theory, of the traction kind. With the corners where the ends meet the sides, and data switching from zero to non-zero values as one crosses the corners, we rather expect to find some singularity here. Experience indicates that this is the only place where singularities are to be expected. When it displays such a singularity, the solution can hardly be continued to any larger value of L. So, plausibly, we lose more and more solutions as L increases. So, it becomes of interest to think about what we might be left with in $S(\infty)$, solutions defined for $I = (-\infty, \infty)$, particularly when we insist that $C - 1$ not be very large.

One thing which gave me pause was considering the analogous question, for the classical rod theories. Here one is dealing with systems of ordinary differential equations of a nice form, so one can apply the usual theory of existence, uniqueness, and continuation. The so-called energy integral provides bounds, which can be used to prove that any solution can be continued indefinitely. Further, the measures analogous to $C - 1$ remain bounded for any particular solution. Lettings bars denote the corresponding sets, we thus have the rather different result that

$$\bar{S}(L) = \bar{S}(\infty) \quad . \tag{11}$$

At least in the naive sense, such solutions form a finite-dimensional manifold. That is, with (11), a solution can be fixed, by giving a set of initial data, and values of assignable parameters, the resultant force being in this category. This provides a means of coordinatising the set. It might be illuminating to dig into the subtleties involved in topologising the set to give it the structure of a smooth manifold but I haven't, so we have only a scheme for counting solutions. Even this is subject to interpretation, since there are obvious invariance groups which map $\bar{S}(\infty)$ onto itself.

Putting the two lines of thought together, an interesting speculation emerges. It just might be that, with some restriction on the size of $C - 1$, $S(\infty)$ is a rather small set, something that might have the structure of a finite-dimensional manifold, more or less. If so, there might well be some 1:1 map of the set to a set like $\bar{S}(\infty)$, which can be associated with a system of ordinary differential equations, more or less like that offered by the classical rod theories. If so, it would make clearer what we need to do; to get the right rod equations we need to understand the structure of $S(\infty)$. Applying the idea in reverse, we can use what we know about $\bar{S}(\infty)$, which is quite a bit, to make educated guesses about the structure of $S(\infty)$.

To me, at least, the next step is to try to estimate whether it is at all plausible that such a guess about the structure of $S(\infty)$ is correct. One way of doing so is to go back to what we know about the classical linear theory, which assumes $C - 1$ small, and more. Where linear theory makes any sense, this means that the classical infinitesimal measure of strain must be small. One might require a uniform pointwise bound, or a less stringent requirement like

$$E(x^3) = \int_D \rho\phi \leqslant \text{const.}, \tag{12}$$

with $\rho\phi$ being the approximate form used in linear theory, equivalent to an L_2 norm on the strains. Using results such as are presented by Oleinik and Yosifian [4], one can see that (12) is enough to imply that, as estimated by linear theory, $S(\infty)$ is a finite dimensional vector space. Pushing this a bit further, I find that it consists of solutions for which the infinitesimal strains are independent of x^3. These form a subset of the very well known St.-Venant solutions. For extremely long prisms, such linear theory is not considered to be very reliable, but the structure estimated is compatible with our speculation. David Kinderlehrer has shown me an analysis of single nonlinear partial differential equations, with analogous domains, boundary conditions, <u>etc</u>. With a similar L_2 bound one gradients, and a condition of ellipticity, one gets a rather trivial finite-dimensional manifold. Perhaps this better estimates what we might have, if ϕ were a convex function of F. As is now appreciated, I think, our assumptions contradict this and, probably, this lets $S(\infty)$ have a more complicated structure. Still, the idea that it might be something like a finite dimensional manifold does not seem so ludicrous.

What we know, essentially, is that $S(\infty)$ contains a variety of solutions suggested by linear theory, solutions with C independent of x^3. Kinematical characterisations of such deformations are given by Ericksen [5]. Some relevant existence theory is presented by Ball [6], yielding solutions in rather rough function spaces. Physical experience certainly suggests that they tend to be quite smooth, when the strains are not too large, at least. If we accept the idea that classical rod theory gives a reasonable estimate, we have reason to expect many solutions which are periodic, in the sense described by Ericksen [7], others with F approaching limits at infinity. There is not space here to discuss what is suggested by generalisations of classical rod theory, perturbation theory, etc. In any event, such guesswork needs to be replaced by solid theorems.

If the indicated speculation is more or less correct, we still face the problem that, physically, we are interested in prisms of finite length. Here, the only hope seems to be that, for L large, restrictions of $S(\infty)$ might serve as a reasonably good approximation of $S(L)$. As I see it, this is a second issue, and I prefer to regard it as such. Here is where we might replace infinitely many solutions by one, using something roughly like St.-Venant's Principle.

ACKNOWLEDGEMENT

This material is based on work supported by the National Science Foundation under grant CME 79 11112.

REFERENCES

[1] Love, A.E.H.: 1927, *A treatise on the mathematical theory of elasticity*, 4th ed., Cambridge.

[2] Jenkins, J.T.: 1981, Int. J. Mech. Sci., 23, pp.65-67.

[3] Parker, D.F.: 1979, J. Appl. Mech., 46, pp.861-866.

[4] Oleinik, O.A., and Yosifian, G.A.: 1982, Arch. Rational Mech. Analysis, 78, pp.29-53.

[5] Ericksen, J.L.: 1977, *Advances in applied mechanics*, (ed. C.S. Yih), Academic Press, 17, pp.189-244.

[6] Ball, J.M.: 1977, *Nonlinear analysis and mechanics: Heriot-Watt Symposium*, (ed. R.J. Knops), Pitman, 1, pp.187-238.

[7] Ericksen, J.L.: 1980, Quart. Appl. Math., 37, pp.443-446.

LECTURE 3 ST.-VENANT'S PROBLEM FOR ELASTIC PRISMS

In trying to analyse physical situations, we rarely know the data which seems, or is known to be needed to properly set a boundary value problem. For the traditional problems of elastic prisms, with free sides, St.-Venant drew attention to the basic problem, proposed a practical resolution of the difficulty, which has proved to be quite successful. The basic problem remains, when we turn to nonlinear theory, but there are numerous reasons to think that a resolution, if possible, must be quite different in nature. The lecture will cover some observations and speculations concerning this.

INTRODUCTION

Here as in my previous lecture [1], we will be concerned with the problems of elastic prisms with free sides, carrying over the equations and notations.

Often, in practice, we do not know boundary conditions in sufficient detail to set up a well-posed problem, even within the context of classical linear theory, where we have a good grasp of what these words mean. Clearly it is necessary to find some way to cope with this. This is what I mean by St.-Venant's Problem, he being the pioneer who recognised it, and took a major step toward resolving it, for linear theory. Pragmatically, the resolution he suggested, involving his Principle, has proved to be successful. Toupin [2] had a seminal influence in clearing up the mathematical status of his Principle. As is clear from his discussion, it has some faults, but there is much truth in it. For nonlinear theory, the Problem remains, but there are various reasons for thinking that the Principle can't be carried over, without serious modification.

In general terms, it was St,-Venant's idea that a resolution is feasible, if one has a small amount of information, of the right kind. The notion is to divide solutions into near-equivalence classes, the minimal information serving to pick out the class. If one has additional information, one might distinguish some members of the class from others, reducing error and, no doubt, some kinds of information are more valuable than others, for this. For linear theory, it might be time to give more consideration to such refinements but, for nonlinear theory, it seems best to ignore this. As St.-Venant pursued it, the next step is to produce a set of representative solutions, one for each class. So he devised schemes for producing these, the St.-Venant solutions, discussed at length in common textbooks. If one is going to single out one in such a class, it would be at least aesthetically pleasant to introduce some simple concept, as a basis for the selection. Also, this might enhance the prospects of extrapolating the ideas to other kinds of problems. Various workers have tried to give his selections some such status, with mixed success. It seems to be particularly hard to do this for the flexure solutions. At least partly for this reason, expositions of nonlinear theory tend to avoid flexure. Taking a different tack, Maissoneuve [3] introduced a simple idea of minimum energy, which selects a different set of representatives. Mathematically, the Principle does not really imply that the differences are very slight, so something of interest might come from taking a hard look at this. Either way, one is attempting to approximate all solutions, by a set having the structure of a finite dimensional linear vector space. Here, we will explore a different kind of concept, one which favours the St.-Venant solutions a bit more.

Classical rod theory gets by with what must be regarded as only rough averages or estimates of boundary data, getting far enough into what we consider to be the nonlinear regime to accommodate Euler buckling, at least. The averages used are not only the resultant forces and moments called for by the Principle. Actually, different kinds and combinations are used, in different problems. That such theory is successful seems to argue in favour of the notion that a resolution of the Problem is feasible, at least in the small-strain regime, and that a finite-dimensional manifold of solutions might suffice. Further, it gives a caricature of what we might expect to find, as a pretty good set of representatives. As is rather clear from the discussion of Horgan and Knowles [4], there is some mathematical evidence that something like St.-Venant's Principle applies rather broadly, to problems more or less of the kind considered. Conditions of ellipticity play an important role in the analyses and are, no doubt, of real import. For operators not always of this kind, subsets of solutions might still fit the requirements, and it seems to me only sensible to keep this in mind.

INTEGRALS

Built into the strain energy function ϕ is invariance under a Lie group of transformations. The classical theory of Noether [5] covers our Euler-Lagrange equations, implying that we will have a divergence identity for each parameter in the group or, better, integrals over closed surfaces which vanish. With the null boundary conditions assumed, this can easily imply that certain integrals over D are independent of x^3. Roughly, these correspond to what we call "integrals" for systems of ordinary differential equations, so we give them this name. Invariance giving a divergence identity need not produce an integral, and some integral might not be obtainable in this way. Roughly, one can't reasonably expect two solutions to be close to each other, unless the values of such integrals are close. Roughly, St.-Venant's Principle amounts to an assertion that if those he knew of have the same value, then the solutions must be pretty close. For such reasons, they must be of some relevance.

Let

$$\psi_i = \rho \frac{\partial \phi}{\partial F^i_3} \tag{1}$$

denote the components of the traction vector on $x^3 = \text{const.}$, oriented with its normal pointing in the direction of increasing x^3. In particular, ϕ is invariant under the translations $y \longrightarrow y + \text{const.}$, and this gives a vector integral,

$$f = \int_D \psi = \text{const.}, \tag{2}$$

interpretable as the condition that forces are in balance.

The invariance under rotations implied by equation (6), of my previous lecture (1), similarly gives the balance of moments, or

$$m = \int_D y \wedge \psi = \text{const.} \tag{3}$$

For classical linear theory, one uses for this an infinitesimal approximation for the rotations. This has the effect of replacing (3) by

$$m_\ell = \int_D x \wedge \psi = \text{const.} \tag{4}$$

Certainly, St.-Venant was aware of (1) and (4), and based his Principle on them. Finally, the invariance $x^3 \longrightarrow x^3 + \text{const.}$

gives

$$e = \int_D (\rho\ \phi - y,_3 \cdot \psi) = \text{const.} \tag{5}$$

The last came on the scene rather late, in a discussion by Ericksen [6]. The failure of St.-Venant's Principle to account for it might be what underlies some of the flaws in the Principle, such as are discussed by Toupin [2], but little has been done to clear up this point. One point seems worth bearing in mind. In dealing with infinite or semi-infinite prisms, it is rather common to assume something about the behaviour of these at infinity. The effect can be to fix the value of such integrals, perhaps unwittingly. It would then be interesting to know what would happen, if we merely fix the integrals. The discussion of Horgan and Knowles [4] treats a mathematically similar but different problem, involving a single nonlinear equation, of elliptic type. For similar group-theoretic reasons, one has such integrals, their discussion serving as an example of the situation just described, since they introduce relevant assumptions about behaviour at infinity. At least for such simpler prototypes, it should be feasible to clear up such points, to clarify the role played by invariance.

For similar group-theoretic reasons, constants analogous to f, m and e are encountered in classical rod theory. There, as is clear from the exposition of Love [7], the analogue of e, the so-called energy integral, has long played a significant role. There, it is known that one can have a variety of quite different solutions with matching values of f and m, different values of e. This is one of the indications that St.-Venant's Principle needs to be modified, to apply to finite deformations. The obvious thing to try is to amend it, to require matching of e. Various rough calculations, etc. lead me to believe that this is not good enough, that something other than these integrals must be brought into the picture. One might search for other integrals. My hunch is that something else must be done. Certainly, it can do no harm to try to look at the basic problem, from different perspectives. Really, there are two somewhat different issues. One is to pick a set of approximating solutions. The other is to decide how to group solutions into sets close to one of these. In the latter, at least, the integrals are of obvious import.

THE SET $S(\infty)$

Before, I [1] hinted that we might try to use restrictions of solutions in $S(\infty)$ as a set to be used to approximate $S(L)$, when L is not too small. As mentioned before, there are indications that suitably bounded solutions of this kind don't form a terribly large set, perhaps one having the structure of a finite-dimensional manifold, somewhat comparable to what we have in classical rod theory, or in the St.-Venant set of solutions. For $S(\infty)$ itself, it is hard to see how we could say that one solution differs from another by something that could be regarded as primarily an end effect. This seems to mean that our set of representatives should at least include $S(\infty)$ and, maybe, we need to add something else. We should keep the set to a minimum, so let's ponder $S(\infty)$.

The suggestion from classical linear theory is that we need something else. To sort out the status of classical rod theories, it would seem necessary to pick out solutions in $S(\infty)$ for which $C - 1$ is suitably bounded, and not too large. Physically, classical linear theory is only good in a part of this regime, where rotations are fairly small, where $C - 1$ being small agrees with the statement that the infinitesimal measures of strain are small, at least bounded. Before, I [1] asserted that, with such a condition, such linear theory has $S(\infty)$ containing only solutions for which the infinitesimal strains are independent of x^3. Essentially, the St,-Venant solutions cover all of these, and more, the flexure solution being excluded. It is rather easy to see why. If the infinitesimal strains are bounded, so are the infinitesimal measures of stress. Here, m_ℓ must be constant. This leads to the condition that $x^3 f_1$ and $x^3 f_2$ must be bounded. For the infinite prism, x^3 is unbounded, so we must have $f_1 = f_2 = 0$. To cover flexure, one needs solutions for which these components of f are non-zero, so St.-Venant had to find something. For classical rod theory, the argument fails. Accounting for the large rotations, it estimates the moment as being more like m than m_ℓ and, for flexure, the difference can be great. Take a steel measuring tape in its case, aligned to come out horizontally. Put a small weight on the end, and start feeding out the tape. When a reasonable part is out, you will see the part of the tape near the weighted end hanging nearly vertically. Here, x^3 might be getting large, but the relevant part of the "lever arm", y^3 stays within reasonable bounds, and you see no large stretching of the tape. Actually, classical rod theory does quite a good job of describing flexure, because it better accounts for the finite rotations. It is then likely that a properly invariant three-dimensional theory will similarly cover flexure, with bounded solutions in $S(\infty)$, although classical linear theory can't. Thus, the idea of using $S(\infty)$ seems to remain tenable.

Clearly, we can't seriously explore the indicated proposition, without having a better knowledge of what we really have, in $S(\infty)$. What is suggested by classical linear theory are solutions with C

independent of x^3. A kinematical characterisation of such deformations is given by Ericksen [8]. With some restrictions on ϕ, Ball [9] has established existence of weak solutions of this kind, using a variational formulation. Certainly, the experience in mechanics is that they tend to be pretty smooth. General experience with approximate theory suggests that $S(\infty)$ includes various other kinds of solutions, but this is more a matter of guesswork. We want theorems, not guesswork.

Studies of simplified prototype problems can be illuminating. Mathematically, the work of Horgan and Knowles [4] represents a study of the analogous problem, for a single equation of elliptic type. For simplified problems of this kind, David Kinderlehrer has shown me an analysis, characterising the analogue of $S(\infty)$, call it $\hat{S}(\infty)$. An L_2 norm of the gradient, involving integration over the cross-section, is required to be bounded. From this, one gets $\hat{S}(\infty)$ by applying a physically trivial operation to the simple shearing solutions which play a prominent role in the analysis of Horgan and Knowles. One adds an arbitrary constant to their displacement u, giving a two-dimensional manifold, for $\hat{S}(\infty)$. It seems clear from their discussion that what they intend to explore are similarly bounded, smooth solutions for the semi-infinite prisms, a set $\hat{S}^+(\infty)$. They seem to restrict their attention to a subset, by making rather strong assumptions about the limiting behaviour of solutions, at infinity. For the subset, they conclude that $\hat{S}(\infty)$ does a pretty good job of approximating the set, except for an end effect, estimated to decay exponentially, with distance from the end. By customary standards, this is a tolerable approximation. It then becomes important to know whether they are dealing with $\hat{S}^+(\infty)$, or a proper subset, involving some solutions which cannot be approximated by $\hat{S}(\infty)$. I have not had time to do more than take a quick look at the analyses mentioned, with this in mind. From this, I would guess that the analyses can be refined to settle the matter, with the conclusion that $\hat{S}(\infty)$ satisfactorily approximates $\hat{S}^+(\infty)$. If this were a proved theorem for the prototype problem, it would remain to consider the approximation of solutions for prisms of finite length, certainly a more difficult business. Here, one might be forced to resort to linearised theory, or another kind of perturbation theory. With linearised theory, one has a problem much like that considered by Toupin [1]. Hopefully, we can come a bit closer to the nonlinear problem.

The problem originally posed is tougher, but I think that some progress can be made with it.

To keep the length of discussion of a complex subject within reasonable bounds is not easy. What we need, I think, is a strategy and, in a sketchy way, I have presented mine, along with some of the reasoning which underlies it. If we can agree on a strategy, we can sensibly discuss tactics, a matter pretty well ignored here. To decide between my strategy and another, we need to know the other, then to look more closely at the available evidence. If some proposal is left unmentioned in the discussion of Horgan and Knowles [4], I am not aware of it. In a nutshell, that is where we are, as I see it.

ACKNOWLEDGEMENT

This work is based in part on work supported by the National Science Foundation under grant CME 79 11112. Also, the author gratefully acknowledges support from the Ordway Endowment of the School of Mathematics, University of Minnesota.

REFERENCES

[1] Ericksen, J.L.: *Problems for infinite elastic prisms*, in this publication, p. 80.

[2] Toupin, R.A.: 1965, Arch. Rational Mech. Anal., 18, pp.83-96.

[3] Maissoneuve, O.: 1971, Thèse (Poitiers).

[4] Horgan, C.O., and Knowles, J.K.: 1981, J. Elasticity, 11, pp.271-291.

[5] Noether, E.: 1918, Nachr. Akad. Wiss. Goettingen, Math.-Phys. Kl, 2, pp.235-257.

[6] Ericksen, J.L.: 1977, *'Nonlinear analysis and mechanics: Heriot-Watt Symposium'* (ed. R.J. Knops), 1, pp.158-186, Pitman.

[7] Love, A.E.H.: 1927, *'A treatise on the mathematical theory of elasticity'*, 4th ed., Cambridge.

[8] Ericksen, J.L.: 1977, *'Advances in applies mechanics'*, (ed. C.-S. Yih), 17, pp.189-244, Academic Press.

[9] Ball, J.M.: 1977, *'Nonlinear analysis and mechanics: Heriot-Watt Symposium'*, (ed. R.J. Knops), 1, pp.187-238, Pitman.

NONLINEAR SYSTEMS IN OPTIMAL CONTROL THEORY AND RELATED TOPICS

Lawrence C. Evans

Department of Mathematics
University of Maryland
College Park, Maryland 20742

This paper describes the formal dynamic programming derivation of certain nonlinear PDE relevant in control theory and explains some recent work regarding the solution of these problems. These PDE are either first order (deterministic control) or second order, elliptic (stochastic control), and are either a system (if it costs to switch the control) or else a single fully nonlinear equation (if switching is free).

1. INTRODUCTION

My purpose in this paper is first to explain the dynamic programming derivation of certain nonlinear PDE relevant in various control theoretic applications and then to indicate some recent techniques for studying these problems. The basic heuristic viewpoint is that whereas such PDE are often very badly nonlinear, they nevertheless always admit some kind of maximum principle: this, properly exploited, is a key for their analysis.

The basic control theoretic set up is that we are given a family of differential equations (deterministic or stochastic), whose evolution in time we influence by continuously adjusting certain parameters in the relevant DE. Our goal is to find if possible an optimal such control, so as to minimize some cost functional. This last may have several forms and it is of particular interest as to whether or not there is a cost incurred each time we switch the control settings.

Dynamic programming (cf. Bellman [3]) provides a unified approach to all sorts of such problems. The main idea is to

J. M. Ball (ed.), Systems of Nonlinear Partial Differential Equations, 95–113.

study the value function (or functions - see §2) and to show this formally satisfies some kind of PDE, the exact structure of which depending upon the particular problem at hand. Next, conversely, we try to prove that should the dynamic programming PDE have any nice enough solution, however obtained, then this in fact is the value function. Finally, further analysis of the PDE will sometimes yield a recipe for designing an optimal control (cf. [12, p. 159]). All this motivates a direct, rigorous study of the formally derived dynamic programming PDE.

This paper is organized as follows. Section 2 indicates the formal derivation of the relevant PDE in several interesting cases: for deterministic and stochastic DE, and for positive and zero switching costs. (For simplicity we will assume the set of control values to be finite.) Next, §3 analyzes those PDE for the stochastic case; here we face either an elliptic system of quasi-variational inequalities (if it costs to switch the control) or else a scalar, fully nonlinear elliptic PDE (if switching is free). We explain how to solve these problems using some fairly delicate estimates resulting from various tricky applications of the maximum principle. Finally in §4 we carry out a similar analysis for the dynamic programming PDE from deterministic control problems. The resulting system of QVI and scalar, fully nonlinear PDE are now first order, and so the good estimates available in the elliptic case fail. Nevertheless the maximum principle motivates an interpretation of solutions existing in the weak, "viscosity" sense, a notion introduced by Crandall-Lions [6].

Specific references are given later, but I should explicitly acknowledge here the large number of ideas of P. L. Lions, of M. G. Crandall, and of N. V. Krylov in what follows.

2. DYNAMIC PROGRAMMING

Control of deterministic DE

a. Zero switching costs. For our first problem consider the ordinary differential equation

$$\text{(ODE)} \quad \begin{cases} \dfrac{dy(t)}{dt} = a(y(t),\alpha(t)) & (t > 0) \\ y(0) = x \; ; \end{cases}$$

here $\alpha:[0,\infty) \to \{1,\ldots m\}$ is a given measurable mapping called the control, $a:\mathbb{R}^n \times \{1,\ldots m\} \to \mathbb{R}^n$ is bounded, Lipschitz in its first argument, and the solution $y = y_x^{\alpha(\cdot)}:[0,\infty) \to \mathbb{R}^n$ is the response of the system to the control $\alpha(\cdot)$.

In addition we define the associated <u>cost functional</u>

$$\text{(C)} \quad J_x(\alpha(\cdot)) \equiv \int_0^\infty f(y(s),\alpha(s))e^{-\lambda s}ds,$$

where $f : \mathbb{R}^n \times \{1,\ldots m\} \to \mathbb{R}$ is the <u>running cost</u> and $\lambda > 0$ the <u>discount</u> (or <u>inflation</u>) <u>rate</u>.

Let us consider also the <u>value function</u>

$$u(x) \equiv \inf_{\alpha(\cdot)} J_x(\alpha(\cdot)) \tag{2.1}$$

and, following the tenets of dynamic programming, attempt to derive (formally!) a PDE u satisfies.

For this, note first that if we choose an arbitrary control $\alpha(\cdot)$ and employ it for time $0 \le s \le t$, we incur thereby a cost

$$\int_0^t f(y(s),\alpha(s))e^{-\lambda s}ds.$$

We also change the state of the system to $y(t)$, and so thereafter the cost is $u(y(t))e^{-\lambda t}$, provided we control optimally for times $s \ge t$. As $u(x)$ is the cost should we control optimally from the start, we must therefore have

$$u(x) \le \int_0^t f(y(s),\alpha(s))e^{-\lambda s}ds + u(y(t))e^{-\lambda t} \tag{2.2}$$

for all measurable $\alpha: [0,t] \to \{1,\ldots m\}$, y solving (ODE) for $\alpha(\cdot)$.

This estimate embodies the heuristic principle that "it is better to control optimally from the very beginning, than to use an arbitrary control for time t and only thereafter to control optimally". Furthermore, supposing for simplicity that there does in fact exist an optimal control $\alpha^*(\cdot)$, then our employing $\alpha^*(\cdot)$ for $0 \le s \le t$ yields

$$u(x) = \int_0^t f(y(s),\alpha^*(s))e^{-\lambda s}ds + u(y(t))e^{-\lambda t},$$

y solving (ODE) for $\alpha^*(\cdot)$.

We combine this with (2.2) to derive the <u>optimality condition</u>

$$u(x) = \inf_{\alpha(\cdot)} \left(\int_0^t f(y(s),\alpha(s))e^{-\lambda s}ds + u(y(t))e^{-\lambda t} \right) \quad (t > 0, x \in \mathbb{R}^n). \tag{2.3}$$

This we rearrange and then divide by $t > 0$. We next send $t \searrow 0$ and so obtain formally

$$\text{(HJ)} \qquad \max_{1 \le d \le m} \{\lambda u(x) - a(x,d) \cdot Du(x) - f(x,d)\} = 0 \qquad (x \in \mathbb{R}^n),$$

the Hamilton-Jacobi equation of dynamic programming for the control problem (ODE), (C). Note carefully that its derivation, while perhaps plausible in part, is certainly suspicious: we have assumed that u is everywhere differentiable, that an optimal control exists for each x, and that the limiting procedure in passing from (2.3) to its infinitesimal version (HJ) is justified.

We temporarily ignore these objections and pass on to another problem.

b. Positive switching costs. Suppose now the preceding problem is modified in that we now incur a positive switching cost $k(i,j)$ each time we change the control from setting i to setting j $(1 \le i,j \le m)$. In this context it is appropriate to define a control α to comprise a sequence of switching times θ_i and control settings d_i:

$$\alpha = \{\theta_i, d_i\}_{i=0}^{\infty},$$

where

$$\begin{cases} \theta_i \in [0,\infty],\ 0 = \theta_0 \le \theta_1 \le \ldots, & \theta_i \to \infty \\ d_i \in \{1,\ldots m\}, d_i \ne d_{i-1} \quad \text{if} \quad \theta_i < \infty & (i = 0,\ldots). \end{cases}$$

Set

$$A^d \equiv \{\alpha \mid \alpha \text{ is a control with } d_0 = d) \qquad (1 \le d \le m).$$

For any $\alpha \in A^d$, the corresponding response $y = y_x^\alpha$ is the solution of

$$\text{(ODE)}' \begin{cases} \dfrac{d^+ y(t)}{dt} = a(y(t),d_i) & \theta_i \le t \le \theta_{i+1},\ i = 0,1,\ldots \\ y(0) = x \end{cases}$$

and the corresponding cost functional is

$$(C)' \quad J_x^d(\alpha) \equiv \sum_{i=1}^{\infty} \{\int_{\theta_{i-1}}^{\theta_i} f(y(t),d_{i-1})e^{-\lambda t}dt + k(d_{i-1},d_i)e^{-\lambda\theta_i}\}.$$

As before we define the value functions

$$u^d(x) \equiv \inf_{\alpha \in A^d} J_x^d(\alpha(\cdot)) \qquad (1 \le d \le m). \tag{2.4}$$

Note that these are in general m distinct functions: since it is now assumed to cost us each time we switch the control setting, the minimal cost obtainable from x depends also on the initial control setting $d = d_0$.

As above we next try to show the value functions $(u^1,\dots u^m)$ solve some kind of PDE. First note

$$u^d(x) \le u^{\hat{d}}(x) + k(d,\hat{d}) \qquad (x \in \mathbb{R}^n, 1 \le d,\hat{d} \le m);$$

this is because we can always if needs be immediately switch our control setting from d to $\hat{d}$ (at a switching cost $k(d,\hat{d})$), but could possibly do better by some other switching scheme. Set

$$M^d[u] \equiv \min_{\hat{d}\neq d}\{u^{\hat{d}} + k(d,\hat{d})\};$$

then

$$u^d \le M^d[u] \qquad (x \in \mathbb{R}^n,\ 1 \le d \le m). \tag{2.5}$$

In addition we argue on the same principles as before that

$$u^d(x) \le \int_0^t f(y(s),d)e^{-\lambda s}ds + u^d(y(t))e^{-\lambda t} \tag{2.6}$$

for all $t > 0$, and furthermore

$$\left.\begin{array}{l} u^d(x) = \displaystyle\int_0^t f(y(s),d)e^{-\lambda s}ds + u^d(y(t))e^{-\lambda t} \\ \text{for all sufficiently small } t>0, \text{ provided a strict} \\ \text{inequality occurs in (2.5) at } x. \end{array}\right\} \tag{2.7}$$

Statements (2.5)-(2.7) comprise the <u>optimality conditions</u> for our problem with positive switching costs.

In analogy with our previous derivation of (HJ) from (2.3) we divide (2.6) (and (2.7)) by $t > 0$ and let $t \searrow 0$. The formal consequence is that $u = (u^1,\ldots,u^d)$ solves the system of <u>quasi-variational inequalities</u>

$$\text{(QVI)}\quad \max\{\lambda u^d(x) - a(x,d)Du^d(x) - f(x,d),\, u^d(x) - M^d[u](x)\} = 0$$
$$(x \in \mathbb{R}^n,\ 1 \le d \le m).$$

This is the infinitesimal version of the optimality conditions. Note that our derivation suffers from the same objections levelled earlier against the calculation leading to (HJ).

<u>Control of stochastic DE</u>

a. <u>Zero switching costs</u>. The foregoing considerations extend to embrace certain problems concerning the optimal control of stochastic differential equations. For this let us suppose $W(t)$ is a standard d-dimensional Brownian motion, and let W_t denote the history of the process to time t. Consider now the <u>stochastic differential equation</u>

$$\text{(SDE)}\quad \begin{cases} dy(t) = a(y(t),\alpha(t))dt + b(y(t),\alpha(t))dW(t) \equiv (t > 0) \\ y(0) = x \quad \text{a.s.}, \end{cases}$$

where the control $\alpha(\cdot)$ is a progressively measurable (w.r.t. W_t) stochastic process taking values in $\{1,\ldots m\}$, a is as before, and $b : \mathbb{R}^n \times \{1,\ldots m\} \to \{n \times d \text{ matrices}\}$ is uniformly Lipschitz in its first argument. The random process $y = y_x^{\alpha(\cdot)}$ solving (SDE) in the Itô sense is the <u>response</u> to the control $\alpha(\cdot)$; this can be regarded as the solution of the ordinary differential equation (ODE), now subject to an additional random forcing term $b(y(t),\alpha(t))\xi(t)$, where – formally – $\xi(t) \equiv \dfrac{dW(t)}{dt}$ is a "white noise". (See L. Arnold [1], Gihman-Skorohod [13] and Krylov [14] for careful definitions, more explanations, etc.; in particular see Krylov [14] concerning the solvability of (SDE)).

Associate with (SDE) the <u>expected cost functional</u>

$$\text{(EC)}\quad J_x(\alpha(\cdot)) \equiv E_x\left(\int_0^\infty f(y(t),\alpha(t))e^{-\lambda t}dt\right)$$

and <u>value function</u>

$$u(x) \equiv \inf_{\alpha(\cdot)} J_x(\alpha(\cdot)).$$

Once more we seek a PDE u solves. This is accomplished by the dynamic programming heuristics discussed before, with the additional complication now that we must use the Itô calculus for computing

$$\lim_{t\to 0} E_x \frac{(u(x)-u(y(t)))}{t},$$

y solving (SDE). We omit the details (see [1]) and just record the result:

$$\text{(HJB)} \quad \max_{1\le d\le m} \{\lambda u(x) - a(x,d)\cdot Du(x) - \operatorname{tr} \sigma(x,d)D^2u(x) - f(x,d)\} = 0 \quad (x \in \mathbb{R}^n).$$

This is the <u>Hamilton-Jacobi-Bellman</u> equation. Here

$$\operatorname{tr} \sigma(x,d)D^2u(x) \equiv \sum_{i,j=1}^{n} \sigma_{ij}(x,d) \frac{\partial^2 u}{\partial x_i \partial x_j},$$

for

$$\sigma_{ij}(x,d) \equiv \frac{1}{2} \sum_{\ell=1}^{d} b_{i\ell}(x,d) b_{j\ell}(x,d);$$

these second order terms are the analytic manifestations of the random terms occuring in (SDE).

b. <u>Positive switching costs</u>. Now a control α consists of a system of (W_t-adapted) Markov <u>switching times</u> θ_i and <u>control settings</u> d_i:

$$\alpha = \{\theta_i, d_i\}_{i=0}^{\infty},$$

where

$$\begin{cases} \theta_i \in [0,\infty),\ 0 \equiv \theta_0 \le \theta_1 \ldots, \theta_i \to \infty & \text{a.s.} \\ d_i \in \{1,\ldots m\},\ d_i \ne d_{i-1} \ \text{if} \ \theta_i < \infty. \end{cases}$$

Again

$$A^d \equiv \{\alpha \mid \alpha \text{ is a control with } d_0 = d\}.$$

For $\alpha \in A^d$ the <u>response</u> $y = y_x^\alpha$ solves

$$\text{(SDE)}' \quad \begin{cases} dy(t) = a(y(t),d_i)dt + b(y(t),d_i)dW(t) \ (\theta_i \le t \theta \le_{i+1}) \\ y(0) = x, \end{cases}$$

with the <u>expected cost functional</u>

$$\text{(EC)}' \quad J_x^d(\alpha) = E_x\left(\sum_{i=1}^{\infty}\left\{\int_{\theta_{i-1}}^{\theta_i} f(y(t),d_{i-1})e^{-\lambda t}dt + k(d_{i-1},d_i)e^{-\lambda\theta_i}\right\}\right).$$

In analogy with the foregoing above it turns out that the <u>value functions</u>

$$u^d(x) \equiv \inf_{\alpha \in A^d} J_x^d(\alpha(\cdot)) \qquad (1 \le d \le m)$$

formally satisfy the system of <u>quasi-variational inequalities</u>

$$\text{(QVI)*} \quad \max\{\lambda u^d(x) - a(x,d)Du^d(x) - \operatorname{tr}\sigma(x,d)D^2u^d(x) - f(x,d), u^d(x) - M^d[u](x)\} = 0 \quad (x \in \mathbb{R}^n,\ 1 \le d \le m).$$

<u>Additional references</u>

See Fleming-Rishel [12, Chap. VI], Krylov [14], Gihman-Skorohod [13] for more explanation about dynamic programming and stochastic control.

3. PDE FROM STOCHASTIC CONTROL THEORY

First we investigate (QVI)* and (HJB), as these under the nondegeneracy hypothesis (3.1) turn out to be amenable to a direct analytic investigation.

Henceforth we will assume

$$\sum_{i,j=1}^{n} \sigma_{ij}(x,d)\xi_i\xi_j \ge \theta|\xi|^2 \qquad (3.1)$$

for some constant $\theta > 0$ and all ξ, $x \in \mathbb{R}^n$, $1 \le d \le m$; this says the linear operators

$$L^d v \equiv \lambda v - \sum_{i=1}^{n} a_i(\cdot,d)v_{x_i} - \sum_{i,j=1}^{n} \sigma_{ij}(\cdot,d)v_{x_i x_j} \qquad (1 \le d \le m)$$

are uniformly elliptic. Suppose also f, a_i, σ_{ij}, etc. are smooth and set $f^d(\cdot) \equiv f(\cdot,d)$.

a. Positive switching costs. Consider the system of quasi-variational inequalities

$$(\text{QVI})^* \quad \max(L^d u^d - f^d, u^d - M^d[u]) = 0 \qquad (1 \le d \le m),$$

and recall from §2 that the value functions $(u^1,\dots u^m)$ for the control problem (SDE)', (EC)' formally solve this.

We commence a direct study of (QVI)* by considering an approximating quasi-linear elliptic system (cf. [10], [19]):

$$\left.\begin{array}{lll} L^1 u_\varepsilon^1 + \sum_{j=1}^{m} \beta_\varepsilon(u_\varepsilon^1 - u_\varepsilon^j - k(1,j)) & = & f^1 \\ \vdots & & \vdots \\ L^m u_\varepsilon^m + \sum_{j=1}^{m} \beta_\varepsilon(u_\varepsilon^m - u_\varepsilon^j - k(m,j)) & = & f^m \end{array}\right\} \qquad (3.2)_\varepsilon$$

where $k(i,i) \equiv 0$ $(1 \le i \le m)$. Here $\varepsilon > 0$ and

$$\beta_\varepsilon(x) \equiv \beta\left(\frac{x}{\varepsilon}\right),$$

for $\beta : \mathbb{R} \to \mathbb{R}$ satisfying

$$\left.\begin{array}{l} \beta'' \ge 0, \quad 0 \le \beta' \le 1, \quad \beta(x) = 0 \quad \text{for} \quad x \le 0, \\ \beta(x) > 0 \quad \text{for} \quad x > 0. \end{array}\right\} \qquad (3.3)$$

In view of the a priori estimates described below $(3.2)_\varepsilon$ has a unique bounded, smooth solution $u_\varepsilon = (u_\varepsilon^1, \dots u_\varepsilon^m)$ for each $\varepsilon > 0$. We propose to show that as $\varepsilon \searrow 0$ u_ε converges to a function u solving (QVI)*.

We will first need

Estimate I: $$\sup_{\substack{1 \le i \le m \\ x \in \mathbb{R}^n}} |u_\varepsilon^i(x)|, |Du_\varepsilon^i(x)|, |D^2 u_\varepsilon^i(x)| \le C,$$

the constant C independent of ε and the $k(i,j)$.

These bounds are obtained from the maximum principle. To most clearly explain the ideas, let us derive the gradient bound in the case that the coefficients of the operators L^d do not depend on x. Introduce auxiliary functions

$$w^j \equiv |Du^j_\varepsilon|^2 \qquad (1 \le j \le m) \tag{3.4}$$

and suppose, for simplicity, there exists a point $x_0 \in \mathbb{R}^n$ and an index i such that

$$w^i(x_0) \ge w^j(x) \qquad (x \in \mathbb{R}^n,\ 1 \le j \le m). \tag{3.5}$$

Then

$$L^i w^i \ge \lambda w^i \qquad \text{at } x_0,$$

by the maximum principle, and so some easy calculations yield

$$\lambda w^i \le -2\theta |D^2 u^i_\varepsilon|^2 + 2Du^i_\varepsilon L^i(Du^i_\varepsilon) \qquad \text{at } x_0.$$

Next differentiate the ith row of $(3.2)_\varepsilon$ to compute

$$L^i(Du^i_\varepsilon) = Df^i - \sum_{j=1}^{m} \beta'_\varepsilon(u^i_\varepsilon - u^j_\varepsilon - k(i,j)) \cdot (Du^i_\varepsilon - Du^j_\varepsilon).$$

Substitute above and note

$$Du^i_\varepsilon \beta'_\varepsilon(\cdot)(Du^i_\varepsilon - Du^j_\varepsilon) \ge \beta'_\varepsilon(\cdot)\frac{(|Du^i_\varepsilon|^2 - |Du^j_\varepsilon|^2)}{2} \ge 0$$

at x_0, by (3.5). We thus obtain

$$\lambda w^i \le 2Du^i_\varepsilon \cdot Df^i \le 2(w^i)^{1/2}|Df| \qquad \text{at } x_0,$$

and this, with (3.5), gives the desired bound.

The situation that the a_i, a_{ij} depend on x is trickier and necessitates introduction of modified auxiliary functions

$$w^j \equiv |Du^j_\varepsilon|^2 + \mu(u^j_\varepsilon + \sigma)^2,$$

for suitably selected constants $\mu, \sigma > 0$ (cf. [10], [19]).

Derivation of second derivative estimates is harder: see Lions [15] and Evans-Lions [11].

With Estimate I in hand it is quite easy to construct a solution of (QVI)*. Indeed, there exists a subsequence $\varepsilon_\ell \to 0$ such that

$$\begin{cases} u^i_{\varepsilon_\ell} \to u^i & \text{locally uniformly} \\ Du^i_{\varepsilon_\ell} \to Du^i & \text{locally uniformly} \\ D^2u^i_{\varepsilon_\ell} \rightharpoonup D^2u^i & \text{weakly * in } L^\infty_{loc} \end{cases}$$

for each $1 \le i \le m$ and some $u = (u^1,\dots u^m)$. Since

$$|\beta_\varepsilon(u^i - u^j - k(i,j))| \le \text{Const.}$$

by $(3.2)_\varepsilon$ and Estimate I, we have

$$u^i \le u^j + k(i,j) \qquad (1 \le i,j \le m);$$

and so

$$u^i \le M^i[u] \qquad (1 \le i \le m).$$

It is also simple to prove

$$L^iu^i \le f^i \qquad \text{a.e.}$$

and furthermore

$$(L^iu^i)(x) = f^i(x) \quad \text{if} \quad u^i < M^i[u] \quad \text{at} \quad x.$$

Thus u solves (QVI)* a.e., and uniqueness follows as in [10].

b. Zero switching costs. Next we solve (HJB), following Evans-Friedman [10]. Since the control theoretic model problem here is that which leads to (QVI)*, except that now the switching costs are set equal to zero, it is reasonable heuristically to expect the solutions of (QVI)* to converge to a solution of (HJB) as the $k(i,j) \to 0$. This is true, as we will see, but is somewhat surprising analytically since the structure of (HJB) and (QVI)* differ considerably: (QVI)* is a system and (HJB) a scalar PDE, (QVI)* is quasilinear and (HJB) is fully nonlinear, etc.

Set

$$K \equiv \{k(i,j) \mid 1 \le i,j \le m\}$$

and denote by u_K the corresponding solution of (QVI)*. From Estimate I we have

<u>Estimate II</u>. $$\sup_{\substack{K\\ x\in\mathbb{R}^n}} |u_K(x)|, |Du_K(x)|, |D^2u_K(x)| \le C.$$

In light of this there exists a subsequence $K \to 0$ (i.e. $k(i,j) \to 0$, $1 \le i,j \le m$) such that

$$\begin{cases} u_K^i \to u^i & \text{locally uniformly} \\ Du_K^i \to Du^i & \text{locally uniformly} \\ D^2u_K^i \rightharpoonup D^2u^i & \text{weakly * in } L^\infty_{loc} \end{cases}$$

for each $1 \le i \le m$ and some $(u^1,\dots u^m)$.

We claim

$$u^1 = u^2 \dots = u^m \equiv u \tag{3.6}$$

and

$$u \text{ solves (HJB) a.e.} \tag{3.7}$$

First, since

$$u_K^i \le u_K^j + k(i,j) \qquad (1 \le i,j \le m),$$

we have

$$u^i \le u^j \qquad (1 \le i,j \le m).$$

This gives (3.6).

Proving (3.7) is harder, owing to the strong nonlinearity in (HJB). First we observe from (QVI)* that

$$L^iu_K^i - f^i \le 0 \quad \text{a.e.} \quad (1 \le i \le m),$$

so that in the limit

$$\max_{1\le i\le m} (L^iu - f^i) \le 0 \quad \text{a.e.} \tag{3.8}$$

To prove equality in (3.8) we argue indirectly. Choose $\phi \in C^2$

and suppose that $u - \phi$ has a strict local negative minimum at some point x_0. Then for K small enough, there exists a nearby point x_K and an index i such that

$$(u_K^i - \phi)(x_K) \le (u_K^j - \phi)(x) \tag{3.9}$$

for all nearby x and $1 \le j \le m$. This in particular implies

$$u_K^i(x_K) < M^i[u_K](x_K),$$

and so

$$L^i u_K^i = f^i \quad \text{at} \quad x_K.$$

The maximum principle gives

$$L^i(u_K^i - \phi)(x_K) \le 0.$$

Therefore

$$L^i \phi(x_K) \ge f^i(x_K),$$

and hence

$$\max_{1\le j\le m} (L^j \phi - f^j) \ge 0 \quad \text{at} \quad x_K.$$

We send $K \to 0$, $x_K \to x_0$ to discover

$$\max_{1\le i\le m} (L^i \phi - f^i) \ge 0 \quad \text{at} \quad x_0. \tag{3.10}$$

Now according to [7, Lemma 2.2] for a.e. x_0 there exists a sequence of C^2 functions $\{\phi_\ell\}$ such that

$$\begin{cases} u - \phi_\ell \text{ has a strict local negative min at } x_0, \\ \phi_\ell(x_0) \to u(x_0), D\phi_\ell(x_0) \to Du(x_0), D^2\phi_\ell(x_0) \to D^2u(x_0). \end{cases}$$

Put $\phi = \phi_\ell$ in (3.10) and let $\ell \to \infty$:

$$\max_{1\le i\le m} (L^i u - f^i) \ge 0 \quad \text{a.e.}$$

This and (3.8) establish (3.7).

Uniqueness is a consequence of the weak maximum principle (cf. [10]).

The reader should note the philosophy underlying our proof of (3.7). To pass to limits from (QVI)* to (HJB) we want to "throw" the dangerous derivatives onto a test function ϕ. However we cannot integrate by parts to accomplish this, since (HJB) is fully nonlinear. Instead we use the maximum principle at certain specific points x_K and only then pass to limits, with the nonlinear operator now acting on the test function. This yields (3.10), which itself would have been a direct consequence of the weak maximum principle had we known all along that u solved (HJB). Thus we show first that u verifies the proper kind of maximum principle, and then conclude (3.7) from this. These ideas strongly motivate our approach to the first order PDE in §4.

c. Additional references. The degenerate case that (3.1) holds only with $\theta = 0$ is of great practical interest: see Krylov [14] and P.L. Lions [16].

When (3.1) is valid the solution of (HJB) in fact belongs to $C^{2,\alpha}$ for some small $\alpha > 0$. The associated *a priori* estimate, proved in [9] (cf. also Trudinger [20]), justifies a different existence proof. P.L. Lions [17] has studied the Monge-Ampere equation in light of this theory for (HJB).

4. PDE FROM DETERMINISTIC CONTROL THEORY

Our next goal is to establish for (HJ) and (QVI) existence theorems like those obtained in §3 for (HJB) and (QVI)*. The difficulty is that these new PDE are not elliptic; they instead involve the first order linear operators

$$K^d v \equiv \lambda v - \sum_{i=j}^{n} a_i(\cdot,d) v_{x_i} \qquad (i \le d \le m).$$

Now this is really a problem as our estimates above depended very strongly on hypothesis (3.1). And this of course is not only a failure of technique: solutions of nonlinear first order PDE are generally not smooth and good estimates we may want are simply false.

Crandall and Lions [6] have recently circumvented these difficulties by introducing a new notion of weak or so-called "viscosity" solution. This concept was later reformulated and streamlined in part by Crandall-Evans-Lions [5]. Roughly, the idea is to call a "viscosity solution" any function which satisfies an appropriate kind of maximum principle. This approach is in part motivated by the remarks at the end of §3.

a. Zero switching costs. Precisely now, we say that a bounded, uniformly continuous function u is a viscosity solution of (HJ) provided for all $\phi \in C^1$

(a) if $u - \phi$ attains a local maximum at some point x_0, then

$$\max_{1 \le d \le m} \{\lambda u - \sum_{i=1}^{n} a_i(\cdot,d)\phi_{x_i} - f(\cdot,d)\} \le 0 \quad \text{at} \quad x_0, \tag{4.1}$$

and

(b) if $u - \phi$ attains a local minimum at x_1, then

$$\max_{1 \le d \le m} \{\lambda u - \sum_{i=1}^{n} a_i(\cdot,d)\phi_{x_i} - f(\cdot,d)\} \ge 0 \quad \text{at} \quad x_1. \tag{4.2}$$

A viscosity solution which is (Frechet) differentiable at some point turns out to solve (HJ) in the usual sense there. Furthermore

a viscosity solution of (HJ) is unique;

this is a special case of the principal result in [6]. The maximum principle thus provides here a uniqueness criterion for weak solutions.

We outline next the uniqueness proof for the special case that the a and f do not depend on x. If u and v are two viscosity solutions, define

$$\Phi(x,y) \equiv u(x) - v(y) + \gamma_\varepsilon(x-y) \qquad (x,y \in \mathbb{R}^n)$$

where

$$\gamma_\varepsilon(z) = \gamma(\tfrac{z}{\varepsilon}),$$

for some smooth function γ satisfying

$$\begin{cases} 0 \le \gamma \le M, \ \gamma(0) = M, \quad \gamma(z) < M \quad \text{for} \quad z \ne 0 \\ \gamma(z) = 0 \quad \text{if} \quad |z| \ge 1. \end{cases}$$

Here M is a fixed, large constant; say $M = 5\max\{\|u\|, \|v\|\}$.

Assume, for simplicity, there exists a point (x_0,y_0)

such that

$$\Phi(x_0,y_0) = \max_{x,y} \Phi(x,y).$$

Now $x \mapsto \Phi(x,y_0) = u(x) - \phi(x)$ attains its maximum at x_0, where $\phi(x) \equiv v(y_0) - \gamma_\varepsilon(x-y_0)$. Thus

$$\lambda u(x_0) + H(-D\gamma_\varepsilon(x_0-y_0)) \leq 0, \tag{4.3}$$

for

$$H(p) \equiv \max_{1\leq d\leq m} \{-a(d)\cdot p - f(d)\}.$$

On the other hand $y \mapsto -\Phi(x_0,y) = v(y) - \psi(y)$ attains its minimum at y_0, where $\psi(y) = u(x_0) + \gamma_\varepsilon(x_0-y_0)$. Thus

$$\lambda v(y_0) + H(-D\gamma_\varepsilon(x_0-y_0))) \geq 0.$$

This and (4.3) imply

$$u(x_0) - v(y_0) \leq 0.$$

But for any x

$$u(x) - v(x) + \gamma_\varepsilon(0) = \Phi(x,x) \leq \Phi(x_0,y_0) = u(x_0) - v(y_0) + \gamma_\varepsilon(x_0-y_0).$$

Consequently

$$\sup(u-v) \leq u(x_0) - v(y_0) \leq 0;$$

and, by symmetry, thus $u \equiv v$.

The general case that a_i and f depend on x follows from an elaboration of these calculations and the observation

$$|x_0 - y_0| = o(\varepsilon);$$

see [5], [6] for details.

As the name suggests we can construct a (and therefore the) viscosity solution of (HJ) as the limit as $\varepsilon \to 0$ of solutions to the regularized problems

$$\max_{1\leq d\leq m} \{K^d u_\varepsilon - f^d\} - \varepsilon\Delta u_\varepsilon = 0 \quad \text{in} \quad \mathbb{R}^n. \tag{4.4}_\varepsilon$$

Instead let us indicate a direct proof that the value function u (cf. (2.1)) is the viscosity solution of (HJ); here we follow an observation of Lions [18]. So assume $\phi \in C^1$, $u - \phi$ attaining a local maximum at some point x_0. Then

$$u(x_0) - \phi(x_0)) \geq u(y(t)) - \phi(y(t)) \tag{4.5}$$

for all sufficiently small $t > 0$, $y(\cdot)$ solving (ODE) (with initial condition x_0) for any control $\alpha(\cdot)$. The optimality condition (2.3) implies

$$\sup_{\alpha(\cdot)}\{\frac{u(x_0)-u(y(t))e^{-\lambda t}}{t} - \frac{1}{t}\int_0^t f(y(s),\alpha(s))e^{-\lambda s}ds\} = 0;$$

and thus (4.5) gives

$$\sup_{\alpha(\cdot)} \{\frac{\phi(x_0)-\phi(y(t))}{t} - u(y(t))(\frac{e^{-\lambda t}-1}{t}) - \frac{1}{t}\int_0^t f(y(s),\alpha(s))e^{-\lambda s}ds\} \leq 0.$$

If we now send $t \searrow 0$, it is quite reasonable to expect the desired estimate (4.1) to hold in the limit. This is indeed true, although a tiny bit of care is required for the analysis: see [18].

Similar considerations hold should $u - \phi$ attain a local minimum at a point x_1, and thus the value function u is the viscosity solution of (HJ). We have made rigorous the suspect, formal calculations in §2.

b. <u>Positive switching costs</u>. The system (QVI) is ammenable to an analysis quite like that for (HJ). As in Capuzzo Dolcetta-Evans [4] we define $u = (u^1,\ldots u^m)$ to be a <u>viscosity solution</u> of (QVI) provided for each $1 \leq d \leq m$ and each $\phi \in C^1$

(a) if $u^d - \phi$ attains a local maximum at x_0, then

$$\max\{\lambda u^d - \sum_{i=1}^n a_i(\cdot,d)\phi_{x_i} - f(\cdot,d), u^d - M^d[u]\} \leq 0 \quad \text{at} \quad x_0,$$

and

(b) if $u^d - \phi$ attains a local minimum at x_1, then

$$\max\{\lambda u^d - \sum_{i=1}^{n} a_i(\cdot,d)\phi_{x_i} - f(\cdot,d), u^d - M^d[u]\} \geq 0 \quad \text{at } x_1.$$

It is not very difficult to adapt the rigorous proof for (HJ) to work here. Furthermore the dynamic programming optimality conditions (2.5)-(2.7) imply that the value functions defined by (2.4) do indeed form the viscosity solution of (QVI). It is also relatively easy to show that the solutions of (QVI) converge as the switching costs tend to zero to the solution of (HJ); this is the deterministic analogue of our techniques in §3 above. Complete proofs of all these assertions may be found in [4].

c. Additional references. See P.L. Lions' book [18] for much more information on viscosity solutions, dynamic programming, etc. Consult also [8] and [2] for some earlier versions of the viscosity methods.

REFERENCES

[1] L. Arnold, Stochastic Differential Equations, J. Wiley and Sons, 1974.
[2] J. Ball and L.C. Evans, Weak convergence theorems for nonlinear PDE of first and second order, Bull. London Math. Soc. 251 (1982), 332-346.
[3] R. Bellman, Dynamic Programming, Princeton U. Press, 1957.
[4] I. Capuzzo Dolcetta and L.C. Evans, Optimal switching for ordinary differential equations, to appear.
[5] M. G. Crandall, L.C. Evans, and P.L. Lions, Some properties of viscosity solutions of Hamilton-Jacobi equations, to appear in Trans. A.M.S.
[6] M. G. Crandall and P.L. Lions, Viscosity solutions of Hamilton-Jacobi equations, to appear in Trans. A.M.S.
[7] L. C. Evans, A convergence theorem for solutions of nonlinear second order elliptic equations, Ind. U. Math. J. 27 (1978), 875-887.
[8] L. C. Evans, On solving certain nonlinear PDE by accretive operator methods, Israel J. Math. 36 (1980), 225-247.
[9] L. C. Evans, Classical solutions of the Hamilton-Jacobi-Bellman equation for uniformly elliptic operators, to appear in Trans. A.M.S.
[10] L. C. Evans and A. Friedman, Optimal stochastic switching and the Dirichlet problem for the Bellman equation, Trans. A.M.S. 253 (1979), 365-389.
[11] L. C. Evans and P.L. Lions, Resolution des equations de Hamilton-Jacobi-Bellman, C.R.A.S. (1980).

[12] W. Fleming and R. Rishel, Deterministic and Stochastic Optimal Control, Springer, 1975.
[13] I. Gihman and A.V. Skorohod, Controlled Stochastic Processes, Springer, 1979.
[14] N. V. Krylov, Controlled Diffusion Processes, Springer, 1980.
[15] P. L. Lions, Resolution analytic des problèmes de Bellman-Dirichlet, Acta. Math. 146 (1981), 151-166.
[16] P. L. Lions, Control of diffusion processes in $\mathbb{R}^n$, Comm. Pure. Appl. Math. 34 (1981),121-147.
[17] P. L. Lions, Sur les equations de Monge-Ampere I, II, III, to appear.
[18] P. L. Lions, Generalized Solutions of Hamilton-Jacobi Equations, Pitman, 1982.
[19] P. L. Lions and J. L. Menaldi, Optimal control of stochastic integrals and Hamilton-Jacobi-Bellman equations I, II, SIAM J. Control 20 (1982), 58-95.
[20] N. S. Trudinger, Fully nonlinear, uniformly elliptic equations under natural structure assumptions, to appear.

Acknowledgement: The author was partially supported by NSF Grant MCS-81-02846 and the Alfred P. Sloan Foundation.

THE REGULARITY PROBLEM OF EXTREMALS OF VARIATIONAL INTEGRALS

Mariano Giaquinta

Istituto di Matematica Applicata
Università Degli Studi di Firenze
50139 Firenze, Italy.

In these lectures I shall consider the problem of the regularity of functions minimizing a variational integral

$$\mathscr{F}(u;\Omega) = \int_\Omega F(x,u(x),Du(x))dx. \tag{0.1}$$

Here Ω is an open set in $\mathbb{R}^m$, $u:\Omega \to \mathbb{R}^N$, $Du = \{D_\alpha u^i\}$ $\alpha = 1,\dots,n$; $i = 1,\dots,N$ and $F:\Omega \times \mathbb{R}^m \times \mathbb{R}^{nN} \to$ is a Carathéodory function (i.e. measurable in x for all (u,p) and continuous in (u,p) for a.e. x) satisfying

$$\lambda|p|^m - a \leqslant F(x,u,p) \leqslant \Lambda|p|^m + a \qquad \lambda > 0,\ m > 1. \tag{0.2}$$

1. EXISTENCE

Direct methods in the calculus of variations quite easily provide existence of minimum points (or minimizers) in classes of admissible functions. For example the following semicontinuity theorem and (0.2), which ensures the weak compactness of the minimizing sequences in $H^{1,m}$, $m > 1$, easily permit us to prove the existence of a minimizer for

$$\mathscr{F}(u;\Omega) \to \min$$

$$u \in \{v \in H^{1,m}(\Omega,\mathbb{R}^N) : v - u_0 \in H_0^{1,m}(\Omega,\mathbb{R}^N)\}$$

where u_0 is given in $H^{1,m}(\Omega,\mathbb{R}^N)$ (and Ω is bounded).

J. M. Ball (ed.), Systems of Nonlinear Partial Differential Equations, 115–145.

Theorem 1.1 (of semicontinuity) ([60],[11])

Suppose that (i) $F(x,u,p) \geqslant 0$; (ii) F *is a Carathéodory function*; (iii) F *is convex in* p *for all* u *and* a.e. x. *Then the functional* $\mathscr{F}(u;\Omega)$ *is sequentially lower semicontinuous* (s.l.s.c.) *with respect to the weak convergence in* $H^{1,m}_{loc}(\Omega;\mathbb{R})$ *for* $1 \leqslant m < +\infty$.

We shall not emphasise the question of the existence of minimizers except for a few remarks. The convexity assumption in the semicontinuity theorem is natural in the scalar case $N = 1$, actually it is necessary (see e.g. [3]), but it is very far from being necessary in the vector valued case $N > 1$, a necessary condition being the quasi-convexity condition of C.B. Morrey [6] sec. 4.4. $F(x,u,p)$ *is called quasi-convex if for* a.e. $x_0 \in \Omega$ *and for all* $s_0 \in \mathbb{R}^n$, $\xi_0 \in \mathbb{R}^{nN}$, $\phi \in C_0^\infty(\Omega,\mathbb{R}^N)$:

$$\frac{1}{|\Omega|} \int F(x_0,s_0,\xi_0 + D\phi)\,dx \geqslant F(x_0,s_0,\xi_0).$$

This means that the frozen functional

$$\mathscr{F}^0(u;\Omega) = \int_\Omega F(x_0,s_0,Du)\,dx$$

has the linear functions as minimizers.

Quasi-convexity is weaker than convexity and reduces to it for $N = 1$. It arises in a natural way in elastostatics, see [2] [3][4] [5]. If $n = N$, any convex function of the invariants of the Jacobian matrix of u is a quasi-convex function.

Semicontinuity theorems under the quasi-convexity condition plus quite strong assumptions have been proved in [56][51][61] sec. 4.4 and in [2] [5]. Recently Fusco [17] and Acerbi-Fusco [1] have made a strong contribution to the question. Let me state the main theorem of [17].

Theorem 1.2 ([17])

Let $F(x,u,p)$ *be a Carathéodory function. Assume that*

$$|F(x,u,p)| \leqslant \lambda(1 + |u|^m + |p|^m) \qquad m \geqslant 1$$

and that F *be quasi-convex. Then* $\mathscr{F}(u;\Omega)$ *is weakly* s.l.s.c. in $H^{1,q}(\Omega,\mathbb{R}^N)$ *for* $q > m$.

Unfortunately the assumption $q > m$ is crucial: the result

fails if $q = m$, see [66]. But, as we shall see, theorem 1.2 still enables us to prove existence (compare section 6).

By enlarging the class of competing functions (Sobolev classes) we can 'easily' show existence. We pay for this simplicity in terms of the new problem of showing, if possible, the differentiability of the $H^{1,m}$-minimizers.

Since we want to avoid any complication due to the boundary data, from now on 'u is a minimum point or a minimizer for $\mathscr{F}$ in Ω' means that for all $\phi \in C_0^\infty(\Omega,\mathbb{R}^N)$ with $\operatorname{supp}\phi \subset\subset \Omega$

$$\mathscr{F}(u;\operatorname{supp}\phi) \leqslant \mathscr{F}(u+\phi;\operatorname{supp}\phi).$$

2. THE PROBLEM OF REGULARITY

It would be very difficult to quote all the many contributions to the regularity problem, as they start at the beginning of this century and are still developing. With respect to our aim we can distinguish two main steps.

I. Starting with the work of S. Bernstein in 1904, and through the fundamental contributions of Liechtenstein, Hopf, Petrowsky, Leray, Schauder, Caccioppoli, Friedrichs, Lewy, Ladyzhenskaya, John, Morrey, Niremberg (among others) we arrive at the following result ([57], 1954):

<u>any C^1-minimizer of an elliptic (in any reasonable sense) functional is as regular as the data permit.</u>

e.g. C^∞ or analytic if the data are C^∞ or analytic. This step is classical nowadays, and has mainly to do with the linear theory of elliptic systems; we shall not dwell on it.

II. <u>From</u> $H^{1,m}$ <u>to</u> C^1

Except for the two dimensional case ([53][54][57]) no progress was made until the celebrated result of Di Giorgi in 1957, [9], who proved that <u>any weak solution of a second order linear elliptic equation with bounded coefficients is Hölder-continuous,</u> deducing this way that any extremal of the elliptic functional

$$\int_\Omega F(Du)\,dx \qquad N = 1 \tag{2.1}$$

is as regular as the data permit.[1)] In fact, let u be a minimizer for (2.1). Then

1) Here we assume natural growths on F_p, F_{pp} i.e. $F_p \sim |p|$, $F_{pp} \sim \text{const.}$

$$\int_\Omega F_{p_\alpha}(Du) D_\alpha \phi dx = 0 \qquad \forall\ \phi \in H^1_0(\Omega).$$

Now it is an easy matter to show, by means of the quotient method, that u belongs to $H^{2,2}_{loc}$ and that the derivatives $D_s u$ satisfy the equation

$$\int_\Omega F_{p_\alpha p_\beta}(Du) D_\beta (D_s u) D_\alpha \phi dx = 0 \qquad \forall\ \phi \in H^1_0(\Omega)$$

which can be reread as a linear elliptic second order equation with L^∞ coefficients $A^{\alpha\beta}(x) = F_{p_\alpha p_\beta}(Du(x))$.

De Giorgi's paper opened a new stage, which reached its culmination in the works of Moser, Stampacchia, Morrey, Ladyzhenskaya and Uraltseva. The theory of scalar second order non linear equations has to be considered as classical; we refer to [46][61][34], see also sec. 6.

No result was obtained during the year 1957 - 1968 in the vector valued case $N > 1$. In 1968 De Giorgi [10] showed that his result for equations could not be extended to systems. He presented a quadratic functional

$$\int_\Omega A^{\alpha\beta}_{ij}(x) D_\alpha u^i D_\beta u^j dx \tag{2.2}$$

with $A^{\alpha\beta}_{ij} \in L^\infty(\Omega)$, elliptic

$$A^{\alpha\beta}_{ij} \xi^i_\alpha \xi^j_\beta \geqslant \nu |\xi|^2 \qquad \forall\ \xi \in \mathbb{R}^{nN} \quad ; \quad \nu > 0 \tag{2.3}$$

having as minimizer and unbounded $H^{1,2}$ function. More recently J. Souček, see [22], has shown that minimizers of functionals of the type (2.2)(2.3) may have discontinuities on a dense subset of Ω.

By modifying De Giorgi's example Giusti and Miranda [38] showed that functionals of the type

$$\int_\Omega A^{\alpha\beta}_{ij}(u) D_\alpha u^i D_\beta u^j dx$$

with analytic coefficients satisfying (2.3) may have singular minima for large dimension n, and the same holds for weak solutions of elliptic quasilinear systems of the type:

$$\int_\Omega A^{\alpha\beta}_{ij}(u) D_\alpha u^i D_\beta \phi^j dx = 0 \qquad \forall\ \phi \in C^\infty_0(\Omega, \mathbb{R}^N)$$

in dimension $n \geqslant 3$.

Similar examples were presented in the meantime and independently by Maz'ya [49] and now different extensions and improvements are available. We point out the example in [19] which shows that a singularity may occur at the boundary even when the boundary domain and datum are C^{∞}.

In 1975 J. Nečas presented an elliptic functional of the simplest type

$$\int_{\Omega} F(Du)\,dx$$

having as minimizer the vector valued function $(u_{ij}) = \left(\dfrac{x_i x_j}{|x|}\right)$ which is Lipschitz-continuous, but not C^1, [68][69].

We can thus conclude that, except in dimension $n = 2$, vector valued minima or extremals or weak solutions of non linear elliptic systems are in general <u>non-smooth</u> - there is hope only for <u>partial regularity</u>, i.e. regularity except possibly on a <u>closed singular set</u> Σ, hopefully <u>not too large</u>.

Partial regularity of solutions to non linear elliptic systems, essentially variational systems for multiple integrals of the kind

$$\int_{\Omega} F(x,Du)\,dx$$

was proved by Morrey [62], Giusti-Miranda [39], Giusti [36], Pepe [72] during the years 1968-'71, see [21] for an account. The method used relies on an indirect argument, very similar to the one introduced by De Giorgi and Almgren in the theory of regularity of parametric minimal surfaces. We refer to Giusti's lectures [37].

During the years 1975-79 elliptic systems of diagonal form, i.e. of the type

$$-D_{\beta}(A^{\alpha\beta}(x,u)D_{\alpha}u^{i}) = f_{i}(x,u,Du) \qquad i = 1,\dots,N$$

with

$$|f(x,u,p)| \leqslant a|p|^2 + b \qquad A^{\alpha\beta}\xi_{\alpha}\xi_{\beta} \geqslant |\xi|^2$$

have been particularly studied, mainly in connection with problems in differential geometry. We refer to Hildebrandt's lectures [40].

In 1978 Giusti and myself [23] introduced a new argument of direct type, essentially a perturbation argument, which has turned out to be very useful for studying the regularity of minimizers of a large class of general functionals

$$\int_\Omega F(x,u,Du)\,dx$$

as well as of weak solutions to non-linear elliptic systems

$$-D_\alpha A_i^\alpha(x,u,Du) = B_i(x,u,Du) \qquad i = 1,\dots,N.$$

In the next sections we shall describe some ideas and results in this connection.

3. A MODEL RESULT

Let $u \in H^{1,2}(\Omega,\mathbb{R}^N)$ be a weak solution of

$$-D_\beta(A_{ij}^{\alpha\beta}(u)D_\alpha u^i) = 0 \qquad j = 1,\dots,N \tag{3.1}$$

where $A_{ij}^{\alpha\beta}$ are (uniformly) continuous functions, bounded $|A_{ij}^{\alpha\beta}| \leq L$, satisfying the ellipticity condition

$$A_{ij}^{\alpha\beta}\xi_\alpha^i\xi_\beta^j \geq \lambda|\xi|^2 \qquad \forall\ \xi \in \mathbb{R}^{nN}\ ; \qquad \nu > 0.$$

Then the partial regularity results of Morrey, Giusti-Miranda, Giusti (see [39]) can be summarized in the simplest case as

Theorem 3.1

There exists an open set $\Omega_0 = \Omega_0(u) \subset \Omega$ such that u is locally Hölder-continuous with any exponent less than 1 in Ω_0. Moreover $\mathcal{H}^{n-s}(\Omega - \Omega_0) = 0$ for some $s > 2$. Here $\mathcal{H}^k$ denotes the k-dimensional Hausdorff measure

Remark 3.1

Note that higher regularity of u in Ω_0 then follows immediately, and that Ω_0 is connected (if Ω is); in fact an n-2 dimensional set does not disconnect.

Remark 3.2

Theorem 3.1 implies regularity, in an open subset Ω_0, for the minimizers (or extremals) of the elliptic functional

$$\int_\Omega F(Du)\,dx \qquad m = 2 \quad .$$

In fact we have, if u is an extremal,

$$\int_\Omega F_{p^i_\alpha}(Du) D_\alpha \phi^i dx = 0 \qquad \forall \quad \phi \in H^1_0(\Omega, \mathbb{R}^N).$$

Moreover $u \in H^{2,2}_{loc}$, and the vector $U = (U^j_s) = (D_s u^j)$ satisfies

$$\int_\Omega \delta_{\rho s} F_{p^i_\alpha p^j_\beta}(U) D_\beta U^j_s D_\alpha \phi^i_\rho dx = 0 \qquad \forall \quad \phi \in H^1_0(\Omega, \mathbb{R}^{nN})$$

which is a system of type (3.1). (We again assume $F_p \sim |p|$, $F_{pp} \sim$ const.)

Remark 3.3

In dimension $n = 2$ solutions are smooth everywhere.

We now give the proof of theorem 3.1 using the direct argument of [23] .

First we observe that, because of our assumptions on the A's, there exists a continuous, bounded, increasing concave function $\omega : \mathbb{R}^+ \to \mathbb{R}^+$, $\omega(0) = 0$, such that

$$|A^{\alpha\beta}_{ij}(u) - A^{\alpha\beta}_{ij}(v)| \leqslant \omega(|u-v|^2).$$

The idea of the proof consists in comparing u in a ball $B_{R/2}(x_0)$, $x_0 \in \Omega$, $R < \mathrm{dist}(x_0, \partial\Omega)$ with the solution v of the approximating problem

$$\begin{cases} -D_\beta(A^{\alpha\beta}_{ij0} D_\alpha v^i) = 0 & j = 1,\ldots,N \quad \text{in} \quad B_{R/2}(x_0) \\ v - u \in H^1_0(B_{R/2}, \mathbb{R}^N) \end{cases}$$

where $A^{\alpha\beta}_{ij0}$ are the frozen coefficients

$$A^{\alpha\beta}_{ij0} = A^{\alpha\beta}_{ij}(u_{x_0,R}) \qquad u_{x_0,R} = \fint_{B_R(x_0)} u = |B_R|^{-1} \int_{B_R} u\, dx.$$

An analogous idea has been extensively used for linear systems and appears for example in [57][7] .

Now v is a solution of a system with constant coefficients, and therefore we know a lot of estimates, in particular, see e.g. [57][7][21] , for all $\rho < R/2$

$$\int_{B_\rho(x_0)} |Dv|^2 dx \leqslant c\left(\frac{\rho}{R}\right)^n \int_{B_R(x_0)} |Dv|^2 dx .$$

Hence

$$\int_{B_\rho(x_0)} |Du|^2 dx \leqslant c\left(\frac{\rho}{R}\right)^n \int_{B_{R/2}(x_0)} |Du|^2 dx + c \int_{B_{R/2}(x_0)} |D(u-v)|^2 dx \quad (3.2).$$

The point is now to estimate the last integral in (3.2). We have

$$\int_{B_{R/2}(x_0)} A^{\alpha\beta}_{ij0} D_\alpha (u-v)^i D_\beta \phi^j dx = \int_{B_{R/2}(x_0)} [A^{\alpha\beta}_{ij0} - A^{\alpha\beta}_{ij}(u)] D_\alpha u^i D_\beta \phi^j dx$$

$$\forall\ \phi \in H^1_0(B_{R/2}, \mathbb{R}^N).$$

In particular we may take $\phi = u - v$, so that using the ellipticity and Hölder inequality we deduce

$$\int_{B_{R/2}(x_0)} |D(u-v)|^2 dx \leqslant c \int_{B_{R/2}(x_0)} \omega(|u - u_{x_0,R}|^2)^2 |Du|^2 dx. \quad (3.3)$$

Now we claim that there exists a $p > 2$ such that for any $B_\rho(y) \subset \Omega$ the following "reverse-Hölder inequality with different supports" holds

$$\left(⨍_{B_{\rho/2}(y)} |Du|^p dx\right)^{\frac{1}{p}} \leqslant c\left(⨍_{B_\rho(y)} |Du|^2 dx\right)^{\frac{1}{2}}. \quad (3.4)$$

In fact testing (3.1) with $(u - u_{y,r})\eta^2$ where $\eta \in C^\infty_0(B_r)$, $0 \leqslant \eta \leqslant 1$, $\eta \equiv 1$ on $B_{r/2}$, $|D\eta| \leqslant 2/r$ one easily gets the Caccioppoli inequality

$$\int_{B_{r/2}(y)} |Du|^2 dx \leqslant \frac{c}{r^2} \int_{B_r(y)} |u - u_{y,r}|^2 dx \quad (3.5)$$

and, using the Sobolev-Poincaré inequality

$$\int_{B_{r/2}} |Du|^2 dx \leqslant \frac{c}{r^2}\left(\int_{B_r} |Du|^q dx\right)^{\frac{2}{q}} \qquad q = \frac{2n}{n+2} < 2$$

i.e.

$$\left(⨍_{B_{r/2}} |Du|^2 dx\right)^{\frac{1}{2}} \leqslant c\left(⨍_{B_r} |Du|^q dx\right)^{\frac{1}{q}} \quad (3.6)$$

which is a "reverse-Hölder inequality with different supports" between 2 and $q < 2$. The claim is, and we shall prove it in the next section, that reverse-Hölder inequalities propagate in the exponent; therefore (3.6) implies (3.4).

Using (3.4), we then get, taking also into account the boundedness of ω,

$$\int_{B_{R/2}(x_0)} \omega^2 |Du|^2 dx \leq \Big(\int_{B_{R/2}(x_0)} |Du|^p dx\Big)^{\frac{2}{p}} \left(\int_{B_{R/2}(x_0)} \omega^{\frac{2p}{p-2}} dx \right)^{\frac{p-2}{p}} \leq$$

$$\leq \int_{B_R(x_0)} |Du|^2 dx \left(\fint_{B_R(x_0)} \omega(|u - u_{x_0,R}|^2) dx \right)^{\frac{p-2}{p}}$$

and since ω is concave

$$\fint \omega = \omega(\fint).$$

From (3.2) and (3.3) we therefore conclude

$$\int_{B_\rho(x_0)} |Du|^2 dx \leq c\left[\left(\frac{\rho}{R}\right)^n + \omega\Big(cR^{2-n} \int_{B_R(x_0)} |Du|^2 dx\Big)^{\frac{p-2}{2}} \right] \int_{B_R(x_0)} |Du|^2 dx \tag{3.7}$$

for all $x_0 \in \Omega$, $\rho < R/2$, $R < \operatorname{dist}(x_0, \partial\Omega)$, and hence for all $\rho < R$, (3.7) being trivial for $R/2 \leq \rho < R$.

Let

$$\phi(x_0,R) = R^{2-n} \int_{B_R(x_0)} |Du|^2$$

$$\chi(x_0,R) = \omega\Big(cR^{2-m} \int_{B_R(x_0)} |Du|^2 dx\Big)^{\frac{p-2}{2}}.$$

From (3.7) we deduce that for $0 < \tau < 1$

$$\phi(x_0, \tau R) \leq K[1 + \chi(x_0,R)\tau^{-n}]\tau^2 \phi(x_0,R). \tag{3.8}$$

Let now $0 < \gamma < 1$ and choose τ in such a way that $2K\tau^{2-2\gamma} = 1$. Let $x_0 \in \Omega$ and $R < \operatorname{dist}(x_0, \partial\Omega)$ be such that

$$\chi(x_0,R) < \tau^n. \tag{3.9}$$

Then we have from (3.8)

$$\phi(x_0,\tau R) \leqslant \tau^{2\gamma}\phi(x_0,R)$$

and hence

$$\chi(x_0,\tau R) \leqslant \chi(x_0,R) \leqslant \tau^n.$$

By induction we get for every k

$$\phi(x_0,\tau^k R) \leqslant \tau^{2\gamma k}\phi(x_0,R)$$

and hence for every $\rho < R$

$$\phi(x_0,\rho) \leqslant \tau^{-n+2+2\gamma}\left(\frac{\rho}{R}\right)^{2\gamma}\phi(x_0,R). \tag{3.10}$$

Now, since χ is a continuous function of x_0, if (3.9) holds for a point $x_0 \in \Omega$, then there exists a ball $B(x_0,r)$ such that for every $x \in B(x_0,r)$ we have $\chi(x,R) < \tau^n$. We conclude then that (3.10) holds uniformly for all $x \in B(x_0,r)$. It follows from a well-known lemma of Morrey [61], see also [6], that u is Hölder-continuous in $B(x_0,r)$ with exponent γ. In conclusion there is an open set Ω_0 where u is locally Hölder-continuous

$$\Omega_0 = \left\{x : \liminf_{R\to 0} R^{2-n}\int_{B_R(x)}|Du|^2dx = 0\right\} =$$

$$= \left\{x : \liminf_{R\to 0} R^{-n}\int_{B_R(x)}|u-u_{x,R}|^2dy = 0\right\}.$$

Obviously Ω_0 is non-empty and $\mathrm{meas}(\Omega-\Omega_0) = 0$.
The second part of the theorem has to do with the problem of the pointwise definition of $H^{1,p}$ functions, see e.g. [35].

4. REVERSE HÖLDER INEQUALITIES

For technical reasons we shall work in this section with n-cubes with sides parallel to the axes, instead of balls.

Let $f \in L^q_{loc}(\mathbb{R}^n)$ $q > 1$, $f \geqslant 0$. We say that f satisfies a reverse Hölder inequality if there exists a constant $b > 1$ such that for any n-cube Q with sides parallel to the axes we have

$$\left(⨍_Q f^q dx\right)^{\frac{1}{q}} \leqslant b ⨍_Q f dx. \tag{4.1}$$

Note that if $b = 1$ then $f = \text{const.}$

Reverse Hölder inequalities appear in different fields, for example in studying the boundedness of the Hilbert transform in

$L^p(fdx)$ (the Muckenhaupt classes); when studying the characterization of BMO functions; in the theory of quasi-conformal mapping; in the theory of P.D.E.: the Harnack inequality for positive solutions of second order elliptic equations can be viewed as a reverse Hölder inequality.

Here we are mainly interested in a fundamental property of functions satisfying a reverse Hölder inequality, namely that they are integrable to a higher power, as is stated in the following theorem due to Gehring [18].

Theorem 4.1

Let Q_0 *be an* n-*cube*. *Suppose that for any subcube* Q *with parallel sides we have*

$$\left(\fint_Q f^q dx\right)^{\frac{1}{q}} \leqslant b \fint_Q f dx .$$

Then $u \in L^{q+\varepsilon}(Q_0)$ *for some positive* ε *and*

$$\left(\fint_{Q_0} f^{q+\varepsilon} dx\right)^{\frac{1}{q+\varepsilon}} \leqslant c \left(\fint_{Q_0} f^q dx\right)^{\frac{1}{q}} .$$

An improvement of theorem 4.1, in terms of maximal functions of f^q and f, can be found in [18], see also [8]. Gehring's result has a global character; actually it allows to prove the integrability of higher powers of the gradient of solutions to Dirichlet boundary value problems for elliptic systems. But in order to study the local behaviour of solutions of elliptic systems we need a local version, which involves reverse Hölder inequalities with different supports, instead. The following theorem due to Giaquinta and Modica [29], supplies our needs.

Theorem 4.2

Let Q *be an* n-*cube*. *Let* $f \in L^q_{loc}(Q)$, $q > 1$, $f \geqslant 0$, *and* $g \in L^{q+\sigma}_{loc}(Q)$, $\sigma > 0$. *Suppose that*

$$\fint_{Q_{R/2}(x_0)} f^q dx \leqslant b\left(\fint_{Q_R(x_0)} f dx\right)^q + \fint_{Q_R} g^q dx$$

for each $x_0 \in Q$ *and each* $R < \mathrm{dist}(x_0, \partial\Omega)$. *Here* $Q_R(x_0)$ *denotes the cube, with sides parallel to the axes, of radius* R *around* x_0. *Then* $f \in L^p_{loc}(Q)$ *for* $p \in [p, p+\varepsilon)$, *and*

$$\left(\fint_{Q_{r/2}} f^p dx\right)^{\frac{1}{p}} \leq c\left(\fint_{Q_r} f^q dx\right)^{\frac{1}{q}} + c\left(\fint_{Q_r} g^p dx\right)^{\frac{1}{p}}$$

<u>for any</u> $Q_r \subset\subset Q$, <u>where</u> c <u>and</u> ε <u>are positive constants depending on</u> b,q,n,σ.

The inequality (3.4) now follows immediately from (3.6) by means of theorem 4.2; see section 6 for other applications.

Improvements of theorem 4.2 are due to [29][75][21]; we refer to [29][21] for its proof. Here we shall present a proof under the assumption that $g \equiv 0$.

<u>Proof</u>

We need the well known
<u>Calderon-Zygmund covering argument</u>: <u>Let</u> $f \geq 0$, $f \in L^1(Q_0)$, $\alpha \geq \fint_{Q_0} f dx$. <u>Then there exists a sequence of parallel cubes</u> $Q_j \subset Q_0$ <u>such that</u>

(i) $\alpha < \fint_{Q_j} f dx \leq 2^n \alpha$

(ii) $f \leq \alpha$ a.e. <u>on</u> $Q_0 - \bigcup_{j=1}^{\infty} Q_j$.

We also need the following lemma, which is a simple modification of a lemma due to Gehring [18].

<u>Lemma 4.3</u>

<u>Let</u> $h(t):[0,+\infty) \to [0,+\infty)$ <u>be a nonincreasing function such that</u> $\lim_{t\to\infty} h(t) = 0$ <u>and</u>

$$\forall\ t \geq t_0 \qquad -\int_t^{+\infty} s^{q-1} dh(s) \leq \alpha t^{q-1} h(\beta t)$$

<u>where</u> $\alpha > 1$ <u>and</u> $0 < \beta < 1$.
<u>Then there exists</u> $p > q$ <u>and</u> γ (p,γ <u>depending only on</u> q,α,β) <u>such that</u>

$$-\int_0^{\infty} s^{p-1} dh(s) \leq \gamma\, t_0^{p-q}\left(-\int_0^{\infty} s^{q-1} dh(s)\right).$$

We shall apply lemma 4.3 with $h(t)$ = a 'distribution function' of f. Note that if

$$h(t) = \int_{\{x \in Q : f(x) > t\}} f dx$$

then

$$\int_{\{x \in Q : f(x) > t\}} f^q dx = -\int_t^\infty t^{q-1} dh(t).$$

Now we need a few notations. Since everything is invariant by dilatations, we fix $Q = Q_1$

$$Q_1 = \left\{x \in \mathbb{R}^n : |x| < \frac{3}{2}\right\}$$

and we set

$$C_0 = \left\{x \in \mathbb{R}^n : |x| < \frac{1}{2}\right\}$$

$$C_k = \left\{x \in Q_1 : 2^{-k} \leqslant \operatorname{dist}(x, \partial Q_1) < 2^{-k+1}\right\} \qquad k \geqslant 1.$$

Obviously

$$Q_1 = C_0 \cup (\bigcup_{k \geqslant 1} C_k).$$

Moreover we denote by $P_{k,j}$ the maximal cubes contained in C_k; for any cube Q, $\overline{Q}$ denotes the cube with the same centre as Q and double the radius. Finally

$$\phi(x) = \phi_{Q_1}(x) = \operatorname{dist}(x, \partial Q_1)^n \qquad \forall \ x \in Q_1.$$

Note that on C_k, $\phi(x) \sim |P_{k,j}|$ and the same holds on $C_{k-1} \cup C_k \cup C_{k+1}$ and that if $Q \subset C_k$, then $\overline{Q}$ meets at most $C_{k-1} \cup C_k \cup C_{k+1}$.

Now we are ready for the proof. Let

$$\lambda \geqslant \gamma_0 \int_{Q_1} f dx ,$$

γ_0 to be fixed later. For fixed k,j, we have

$$\lambda \geq \gamma_0 \underset{P_{k,j}}{\bar{\int}} f dx \geq \gamma_0 2^n |P_{k,j}| \underset{P_{k,j}}{\bar{\int}} f dx \geq \frac{2^n \gamma_0}{b} \left(\underset{P_{k,j}}{\bar{\int}} f^q dx \right)^{\frac{1}{q}} |P_{k,j}| \geq$$

$$\geq \frac{2^n \gamma_0}{b} c_1 \left(\underset{P_{k,j}}{\bar{\int}} (f\phi)^q dx \right)^{\frac{1}{q}}$$

and

$$\lambda \geq \left(\underset{P_{k,j}}{\bar{\int}} (f\phi)^q dx \right)^{\frac{1}{q}}$$

if $\gamma_0 \geq b/2^n c_1$.

Then we can apply the Calderon-Zygmund argument on each $P_{k,j}$, concluding that there exists a sequence of cubes Q_k^j with disjoint interiors such that:

(i) $\lambda^q < \underset{Q_k^j}{\bar{\int}} (f\phi)^q dx \leq 2^n \lambda^q$

(ii) $f\phi \leq \lambda$ a.e. on $Q_1 - \bigcup_{j,k} Q_k^j$

(iii) $Q_k^j \subset C_k$.

We have

$$\int_{\{x \in Q_1 : f\phi > \lambda\}} (f\phi)^q dx = \sum_{j,k} |Q_k^j| \underset{Q_k^j}{\bar{\int}} (f\phi)^q dx \leq 2^n \lambda^q \sum_{j,k} |Q_k^j|. \tag{4.2}$$

The next step is to obtain an estimate for $|Q_k^j|$. Let Q be one of the Q_k^j; we have from (i)

$$\lambda \leq (\underset{Q}{\bar{\int}} (f\phi)^q dx)^{\frac{1}{q}} \leq c|Q| (\underset{Q}{\bar{\int}} f^q dx)^{\frac{1}{q}} \leq cb|Q| \underset{\bar{Q}}{\bar{\int}} f dx \leq c \underset{\bar{Q}}{\bar{\int}} f\phi dx$$

i.e.

$$\lambda|\bar{Q}| \leq c \int_{\bar{Q}} f\phi dx \leq c \int_{\bar{Q} \cap \{f\phi > \beta\lambda\}} f\phi dx + c\beta\lambda|\bar{Q}|.$$

Hence

$$\lambda|\bar{Q}| \leq \frac{c}{1 - c\beta} \int_{\bar{Q} \cap \{f\phi > \beta\lambda\}} f\phi dx$$

provided β is chosen in such a way that $1 - c\beta > 0$.

The $\overline{Q}_k^j$'s may now intersect too many times. In order to avoid this inconvenience we use a well known covering theorem (of the type of Vitali's theorem) which tells us: from the covering $\{\overline{Q}_j^k\}$ we can extract a sequence of disjoint cubes $\{R_k^j\}$ such that

$$|\bigcup_{j,k} Q_j^k| \leqslant 5^n \sum_{j,k} |R_k^j|.$$

From (4.2) we therefore conclude:

$$\int_{\{x\in Q_1:f\phi>\lambda\}} (f\phi)^q\, dx \leqslant c\, \lambda^{q-1} \int_{\{x\in Q_1:f\phi>\beta\lambda\}} f\phi\, dx \tag{4.3}$$

for any $\lambda \geqslant \gamma_0 \int_{Q_1} f\, dx$.

We now set

$$h(t) = \int_{\{x\in Q_1:f\phi>t\}} f\phi\, dx$$

then (4.3) can be rewritten as

$$-\int_t^{+\infty} s^{q-1} dh(s) \leqslant ct^{q-1} h(\beta t) \qquad \forall\ t \geqslant \gamma_0 \int_{Q_1} f\, dx$$

and the conclusion follows immediately from lemma 4.3.

q.e.d.

We remark that there is a big difference between 'reverse Hölder inequalities' (r.H.i.) and 'reverse Hölder inequalities with different supports' (r.H.i.d.s.). For example if f satisfies a r.H.i. with constant b, then f may have zeros of finite order (function of b), see [21]; while if f satisfies only a r.H.i.d.s. it can have zeros of any order. Reverse Hölder inequalities are backward propagating, more precisely r.H.i. are propagating up to a negative exponent, i.e. if f satisfies (4.1) then there exists an exponent $\sigma > 0$ such that

$$⨍_Q f\, dx \leqslant c \left(⨍_Q f^{-\sigma}\right)^{\frac{1}{\sigma}}$$

see [8]; while if f satisfies a r.H.i.d.s. then we only have

$$\fint_{Q_{R/2}} f\,dx \leqslant c(\varepsilon)\left(\fint_{Q_R} f^{\varepsilon}dx\right)^{\frac{1}{\varepsilon}}$$

for any $\varepsilon > 0$, with $c(\varepsilon) \to \infty$ as $\varepsilon \to 0$.

5. SYSTEMS OF NONLINEAR ELLIPTIC EQUATIONS: PARTIAL REGULARITY

The method in section 3 may be easily extended to prove partial regularity for elliptic systems of the type

$$-D_\beta(A^{\alpha\beta}_{ij}(u)D_\alpha u^i) = f_j(x,u,Du) \qquad j = 1,\dots,N \tag{5.1}$$

where the A's are bounded and (uniformly) continuous functions satisfying the ellipticity condition

$$A^{\alpha\beta}_{ij}\,\xi^i_\alpha\xi^j_\beta \geqslant \lambda|\xi|^2 \qquad \forall\ \xi \in \mathbb{R}^{nN};\quad \lambda > 0 \tag{5.2}$$

and

$$|f(x,u,p)| \leqslant a|p|^2 + b. \tag{5.3}$$

Formally (5.1) with f_j given by

$$f_\ell(x,u,Du) = \frac{1}{2}\,\frac{\partial A^{\alpha\beta}_{ij}(u)}{\partial u^\ell}\,D_\alpha u^i D_\beta u^j$$

is the Euler equation of the functional

$$\int_\Omega A^{\alpha\beta}_{ij}(u)D_\alpha u^i D_\beta u^j dx\ . \tag{5.4}$$

Actually functional (5.3) is not differentiable in the direction of H^1_0 functions, but only in $H^1_0 \cap L^\infty$ directions. Moreover even in the scalar case (5.4) may have unbounded extremals and (5.1) unbounded $H^{1,2}$ solutions, [15][46]. So one is led to consider $H^1 \cap L^\infty$ as the natural class to start with in order to prove regularity (we refer to [40] for more information; we only remark that, in the scalar case, bounded solutions of (5.1) are smooth [46].) But even considering $H^{1,2} \cap L^\infty$ solutions of (5.1), in the vector valued case, we are in trouble, since they can be singular in dimension $n = 2$, [14]: one needs a smallness condition (see [40]). More precisely we have:

<u>Theorem 5.1</u> ([23])

<u>Let</u> $u \in H^1 \cap L^\infty(\Omega,\mathbb{R}^N)$, $|u| \leqslant M$, <u>be a weak solution to</u> (5.1)

i.e.

$$\int_\Omega A_{ij}^{\alpha\beta}(u) D_\alpha u^i D_\beta \phi^j dx = \int_\Omega f_j \phi^j dx \qquad \forall \quad \phi \in H_0^1 \cap L^\infty(\Omega, \mathbb{R}^N)$$

<u>suppose that</u> (5.2)(5.3) <u>hold and moreover that</u>

$$2aM < \lambda \tag{5.5}$$

<u>then</u> u <u>is Hölder-continuous (and as smooth as the data permit) in an open set</u> Ω_0 <u>and</u> $\Omega - \Omega_0$ <u>has Hausdorff dimension less than</u> $n - 2$.

The theorem is optimal except maybe for the factor 2 in (5.9). We note that (5.5) is needed only in order to get the Caccioppoli inequality (3.5) and therefore (3.4); i.e. in order to prove an a-priori L^p-estimate for Du.

<u>Remark 5.1</u>

Although optimal from the point of view of solutions of systems, theorem 5.1 does not give any information on the regularity of $H^{1,2}$ minimizers of the functional (5.4).

Some improvement of the method of [23], which appears in [29] [30], also permits one to study the regularity of 'solutions' to general non-linear elliptic systems

$$-D_\alpha A_i^\alpha(x,u,Du) = B_i(x,u,Du). \tag{5.6}$$

Let me state one of the results. We assume

(i) $|A(x,u,p)| \leqslant L(1 + |p|)$

(ii) $(1 + |p|)^{-1} A_i^\alpha(x,u,p)$ are Hölder-continuous in (x,u) uniformly with respect to p.

(iii) A_i^α of class C^1 in p, and $|A^\alpha_{ip^j_\beta}(x,u,p)| \leqslant L$

(iv) $A^\alpha_{ip^j_\beta} \xi^i_\alpha \xi^j_\beta \geqslant \lambda |\xi|^2 \qquad \forall \xi \in \mathbb{R}^{nN} \quad ; \quad \lambda > 0$

(v) $|B(x,u,p)| \leqslant a|p|^2 + b$.

We remark that under these assumptions one cannot linearize system (5.6); actually, even if $B \equiv 0$, weak solutions $u \in H^{1,2}$ do not have square integrable second derivatives.

We have

Theorem 5.2 ([30])

(a) *Let* $u \in H^{1,2}_{loc}$ *be a weak solution to*

$$-D_\alpha A^\alpha_i(x,u,Du) = 0.$$

Then the first derivatives of u *are Hölder-continuous in an open set* Ω_0 *and* $\text{meas}(\Omega - \Omega_0) = 0$.

(b) *Let* $u \in H^{1,2}_{loc} \cap L^\infty(\Omega,\mathbb{R}^N)$, $|u| \leq M$, *be a weak solution to* (5.6) and $2aM < \lambda$. *Then the same conclusion of* (a) *holds.*

We refer to [30] and especially to the introduction for more information. Anyway we have to note again that theorem 5.2 gives no information on the minimizers of

$$\int_\Omega F(x,u,p)\,dx \qquad\qquad |p|^2 \leq F(x,u,p) \leq \mu|p|^2$$

which we are able to find only in $H^{1,2}$.

6. DIRECT METHODS FOR THE REGULARITY

The regularity results we have described do not cover the case of minima of regular functionals

$$\mathscr{F}(u;\Omega) = \int_\Omega F(x,u,Du)\,dx. \tag{6.1}$$

In fact they have as starting point the Euler equation or the variational equation of the functional under consideration, and therefore they require at least:

(a) some smoothness of F, moreover suitable growth conditions, not only on F, but also on its partial derivatives F_u and F_p: otherwise the functional (6.1) need not be Gateaux differentiable. Moreover

(b) under 'natural' growth conditions one needs to start with bounded minimizers, and also assume some smallness condition

(c) there is no distinction between true minima and simple extremals.

In [24][26][27] Giaquinta and Giusti have tried to develop a theory of the regularity for minimizers, working directly with the functional $\mathscr{F}$ instead of working with its Euler equation, providing, in analogy with the existence theory, some sort of direct method for regularity.

Of course the results have a different character in the scalar and in the vector valued case. We can distinguish two main steps. The first basic step consists in proving Hölder-continuity, if $N = 1$, or an L^p-estimate for the gradient, in the general case $N \geq 1$, of minimizers. It does not even need any condition of convexity or ellipticity, which are of course necessary for higher regularity. It depends on a 'minimality property', condition (6.3) below, of minimizers and, more generally, of solutions of elliptic systems. The second step deals with the Hölder-continuity of the first derivatives of minimizers.

Quasi-minima

Let us consider the functional (6.1) where $F(x,u,p)$: $\Omega \times \mathbb{R}^N \times \mathbb{R}^{nN} \to \mathbb{R}$ is a Carathéodory function satisfying

$$|p|^m - b|u|^\gamma - g(x) \leq F(x,u,p) \leq a|p|^m + b|u|^\gamma + g(x) \qquad (6.2)$$

with $1 < m \leq n$, $\gamma < m^* = \dfrac{mn}{n-m}$ $(\gamma \geq m)$

Definition

We say that $u \in H^{1,m}_{loc}(\Omega,\mathbb{R}^N)$ *is a quasi-minimum* (*Q-minimum*) *for* $\mathscr{F}$ *in* Ω (*with constant* Q) *if*

$$\mathscr{F}(u;\, \mathrm{supp}\,\phi) \leq Q\, \mathscr{F}(u+\phi;\, \mathrm{supp}\,\phi) \qquad (6.3)$$

for all $\phi \in H^{1,m}_{loc}(\Omega,\mathbb{R}^N)$ *with* $\mathrm{supp}\,\phi \subset\subset \Omega$.

The relevance of the notion of Q-minimum is due to the result below and to its unifying character shown by the examples below.

Theorem 6.1 ([24],[26]).

Let u *be a Q-minimum for* $\mathscr{F}$ *in* Ω, *and assume that* (6.2) *holds. Then we have*

(i) *if* $N = 1$ *and* $g \in L^s(\Omega)$ *for some* $s > n/m$, *then* u *is locally Hölder-continuous, in particular locally bounded*

(ii) *if* $N \geq 1$ *and* $g \in L^s(\Omega)$ *for some* $s > 1$, *then there exists* an exponent $r > m$ such that $u \in H^{1,r}_{loc}(\Omega,\mathbb{R}^N)$.

1. Of course any minimizer of $\mathscr{F}$ is a Q-minimum (we note that the above result is new even for minimizers). It is also a Q-minimum for

$$\int\{|Du|^m + b|u|^\gamma + (b+g)\}dx$$

in particular for $m = 2$, $b = 0$, $g = 0$, it is a Q-minimum for the Dirichlet integral. Any weak solution to the linear elliptic system with L^∞ coefficients

$$-D_\beta(A^{\alpha\beta}_{ij}(x)D_\alpha u^i) = 0 \quad j = 1,\dots,N \quad , \quad A^{\alpha\beta}_{ij}\xi^i_\alpha\xi^j_\beta \geq |\xi|^2 \quad \forall\ \xi$$

is a Q-minimum for the Dirichlet integral; in particular for $N = 1$ we obtain De Giorgi's result.

2. More generally, 'weak solutions' of non-linear elliptic systems under natural and usual hypotheses are Q-minima. Thus the Hölder regularity of weak solutions to non-linear elliptic equations (compare with [46]) and the L^p-estimates for the gradient of general elliptic systems [50][29][52][21] can be obtained as consequences of theorem 6.1. We refer to [24][21] for more information.

3. weak solutions of the obstacle problem, i.e. for example

$$u \geq \psi \quad \text{in} \quad \Omega: \int_\Omega Du\ D(u-v)dx \leq 0 \quad \forall\ v,\ v \geq \psi,\ \text{supp}(u-v) \subset\subset \Omega$$

is a Q-minimum for

$$\int_\Omega(|Du|^2 + |D\psi|^2)dx.$$

4. quasi-conformal (or quasi-regular) mappings are Q-minima for

$$\int_\Omega |Du|^n dx.$$

For further properties of Q-minima and for the proof of theorem 6.1 we refer to [26]. Here we confine ourselves to sketch the proof of (ii), assuming $b = g = 0$ for the sake of simplicity.

Proof

Let $x_0 \in \Omega$, $B_R = B_R(x_0) \subset \Omega$ and let $\eta \in C_0^\infty(B_s)$, $0 \leq \eta \leq 1$, $\eta \equiv 1$ on B_t, $|D\eta| \leq 2(s-t)^{-1}$, $t < s < R$. Using (6.3) we have

$$\mathscr{F}(u;\ \text{supp}\,\eta(u-u_{x_0,R})) \leq \mathscr{F}(u - \eta(u-u_{x_0,R}),\ \text{supp}\,\eta(u-u_{x_0,R}));$$

hence

$$\int_{B_t}|Du|^m dx \leq \gamma_1\Big\{\int_{B_s\setminus B_t}|Du|^m dx + (s-t)^{-m}\int_{B_s}|u-u_R|^m dx\Big\}.$$

Now we fill in the hole (compare [77]), i.e. we add to both sides

γ_1 times the left hand side and we get

$$\int_{B_t} |Du|^m dx \leqslant \Theta \int_{B_s} |Du|^m dx + \gamma_2 (s-t)^{-m} \int_{B_R} |u - u_R|^m dx, \quad \Theta = \frac{\gamma_1}{1+\gamma_1} < 1.$$

A simple algebraic argument permits us now to deduce

$$\int_{B_{R/2}} |Du|^m dx \leqslant \gamma_3 R^{-m} \int_{B_R} |u - u_R|^m dx$$

which, by means of the Sobolev-Poincaré inequality gives

$$\left(\fint_{B_{R/2}} |Du|^m dx \right)^{\frac{1}{m}} \leqslant \gamma_4 \left(\fint_{B_R} |Du|^q dx \right)^{\frac{1}{q}}, \quad q = \frac{mn}{n+m} < m.$$

The result then follows at once by applying theorem 4.2. Moreover we have

$$\left(\fint_{B_{R/2}} |Du|^r dx \right)^{\frac{1}{r}} \leqslant \gamma_5 \left(\fint_{B_R} |Du|^m dx \right)^{\frac{1}{m}} \tag{6.4}$$

for any $B_R \subset\subset \Omega$.

q.e.d.

Remark 6.1

Theorem 6.1 (ii) gives an improvement to the well known Morrey type results, i.e.: *let* u *be a Q-minimum for* (6.1), (6.2) *hold with* $m = n$; *then* u *is Hölder-continuous*.

Remark 6.2

No (partial) Hölder regularity can be shown for Q-minima (of the Dirichlet integral, for example), in the vector valued case, as they can be singular on a dense set; it is an open question whether partial Hölder-regularity can be shown for minima of (convex or elliptic) functionals, under the same assumptions (compare with [16]).

Quasi-minima and quasi-convexity

Let us recall the following variational principle in Ekeland [12]:

Theorem 6.2

Let (V,d) *be a complete metric space,* $\mathcal{F}: V \to [0,+\infty]$ *a lower semicontinuous functional,* $F \not\equiv +\infty$. *Let* $\eta > 0$ *and* $w \in V$ *verify*

$$\mathscr{F}(w) \leqslant \inf_V \mathscr{F} + \eta \, ;$$

then there exists $v \in V$ such that $\mathscr{F}(v) \leqslant \mathscr{F}(w)$, $d(v,w) \leqslant 1$ and v is the (only) minimum point for the functional

$$\mathscr{F}(u) + \eta d(u,v) .$$

Consider the functional (6.1) and assume

$$|p|^m \leqslant F(x,u,p) \leqslant Q|p|^m \tag{6.5}$$

then obviously it is lower semicontinuous in $\{u \in H^{1,1}(\Omega) : u = \tilde{u}$ on $\partial\Omega\}$. Hence we may apply theorem 6.2, and the function v we obtain is a Q-minimum for a functional of the same type, which is independent of η for η small. In particular there exists a minimizing sequence $\{u_k\}$ of Q-minima with constant Q uniform. Theorem 6.1 (ii) implies then

$$\|u_k\|_{H^{1,r}(\tilde{\Omega},\mathbb{R}^N)} \leqslant \text{const.} ,$$

where the constant depends on $\tilde{\Omega}$, but is independent of k, for any $\tilde{\Omega} \subset\subset \Omega$, with $r > m$.

Let us now assume that F be quasi-convex; then, by means of theorem 1.2, we can conclude

Theorem 6.3

Let $\tilde{u} \in H^{1,m}(\Omega,\mathbb{R}^N)$. Suppose that F be a quasi-convex function satisfying (6.5). Then there exists a minimum point for the functional (6.1) in $H^{1,m}(\Omega,\mathbb{R}^N) \cap \{u : u = \tilde{u}$ on $\partial\Omega\}$. Moreover $u \in H^{1,r}_{loc}(\Omega,\mathbb{R}^N)$ for some $r > m$

The proof given above is a rereading of the proof in [48].

Unfortunately no regularity result is known for the minimizer u in theorem 6.3 (except obviously for $m \geqslant n$).

Partial and everywhere regularity of the first derivatives of minimizers

Let us start with the simplest case of quadratic functionals, i.e.

$$\int_\Omega A^{\alpha\beta}_{ij}(x,u) D_\alpha u^i D_\beta u^j dx \tag{6.6}$$

where the A's are (uniformly) continuous, bounded and satisfy the ellipticity condition

$$A_{ij}^{\alpha\beta}\xi_\alpha^i\xi_\beta^j \geq \lambda|\xi|^2 \qquad \forall\ \xi \in \mathbb{R}^{nN} \ ; \qquad \lambda > 0.$$

Then we have

<u>Theorem 6.4</u> ([24])

<u>Let</u> $u \in H^{1,2}_{loc}(\Omega,\mathbb{R}^N)$ <u>be a minimizer for the functional</u> (6.6). <u>Then there exists an open set</u> $\Omega_0 \subset \Omega$ <u>such that</u> $u \in C^{0,\gamma}_{loc}(\Omega_0,\mathbb{R}^N)$ $\forall\ \gamma < 1;$ $\mathcal{H}^{n-q}(\Omega - \Omega_0) = 0$ <u>for some</u> $q > 2$.

Moreover, see for example [23][28], u is as regular as the data permit in Ω_0. Let us sketch the proof.

<u>Proof</u>

Let $x_0 \in \Omega$, $R < \frac{1}{2}\operatorname{dist}(x_0,\partial\Omega)$ and let v be the solution of

$$\begin{cases} \int_{B_R(x_0)} A_{ij}^{\alpha\beta}(x_0,u_{x_0,R})D_\alpha v^i D_\beta v^j dx \rightarrow \min \\ v - u \in H^1_0(B_R(x_0),\mathbb{R}^N) . \end{cases}$$

Then we have, as in section 3, for any $\rho < R$

$$\int_{B_\rho} |Dv|^2 dx \leq c_1\left(\frac{\rho}{R}\right)^n \int_{B_R} |Dv|^2 dx$$

and moreover, since $u \in H^{1,r}_{loc}(\Omega,\mathbb{R}^N)$ for some $r > 2$, theorem 6.1 gives

$$\int_{B_R} |Dv|^r dx \leq c_2 \int_{B_R} |Du|^r dx .$$

Let $w = u - v$; we have $w \in H^1_0(B_R,\mathbb{R}^N)$ and

$$c_3 \int_{B_R} |Dw|^2 dx \leq \int_{B_R} A_{ij}^{\alpha\beta}(x_0,u_{x_0,R})D_\alpha w^i D_\beta w^j dx .$$

On the other hand

$$\int_{B_R} A_{ij}^{\alpha\beta}(x_0,u_{x_0,R})D_\alpha v^i D_\beta w^j dx = 0$$

and therefore, with the obvious meaning

$$\int_{B_R} A(x_0,u_{x_0,R})DwDwdx = \int_{B_R} A(x_0,u_{x_0,R})DuDwdx =$$

$$= \int_{B_R} [A(x_0,u_{x_0,R}) - A(x,u)]D(u+v)Dwdx +$$

$$+ \int_{B_R} [A(x,v) - A(x,u)]DvDvdx +$$

$$+ \int_{B_R} A(x,u)DuDudx - \int_{B_R} A(x,v)DvDvdx .$$

Since u minimizes $\mathscr{F}$ and $v = u$ on ∂B_R, the sum of the last two terms is non positive. Hence

$$\int_{B_R} |Dw|^2 dx \leqslant c_4 \int_{B_R} (|Du|^2 + |Dv|^2)[\omega(R^2 + |u-u_R|^2 + \omega^2(R^2 + |u-v|^2)]dx$$

where ω is a continuous, bounded, increasing, concave function with $\omega(0) = 0$ and

$$|A(x,u) - A(x,v)| \leqslant \omega(|x-y|^2 + |u-v|^2).$$

Now using the L^r-estimate (6.4) for Du, the boundedness of ω and Poincaré inequality in the same way as we did in section 3 it is not difficult to conclude the proof.

q.e.d.

Let us now consider the general functional in (6.1), where we assume

a_1) $\lambda|p|^2 \leqslant F(x,u,p) \leqslant \Lambda|p|^2$

a_2) $(1+|p|^2)^{-1}F(x,u,p)$ is Hölder-continuous in (x,u) uniformly with respect to p

a_3) F is twice differentiable in p and

$$|F_{pp}(x,u,p)| \leqslant L$$

$$F_{p^i_\alpha p^j_\beta}\xi^i_\alpha \xi^j_\beta \geqslant \nu|\xi|^2 \qquad \forall\, \xi \in \mathbb{R}^{nN} \;;\quad \nu > 0.$$

Then we have

<u>Theorem 6.5</u> ([27])

<u>Let</u> $a_1)a_2)a_3)$ <u>hold and let</u> $u \in H^{1,2}_{loc}(\Omega,\mathbb{R}^N)$ <u>be a minimizer for the functional</u> (6.1).

(A) If $N = 1$, then the first derivatives of u are Hölder-continuous in Ω

(B) If $N > 1$, then the first derivatives of u are Hölder-continuous in an open set Ω_0 and $\operatorname{meas}(\Omega - \Omega_0) = 0$.

In general $\Omega - \Omega_0$ is nonempty; however in some special cases, for example when

$$\mathscr{F}(u;\Omega) = \int_\Omega [A_{ij}^{\alpha\beta}(x) D_\alpha u^i D_\beta u^j + g(x,u)]\,dx$$

when, more generally, the A's satisfy the Legendre-Hadamard condition

$$A_{ij}^{\alpha\beta}\xi_\alpha\xi_\beta\eta_i\eta_j \geqslant |\xi|^2|\eta|^2 \qquad \forall\ \eta,\xi$$

we have regularity everywhere [27].

We refer to [27] for the proof. A different proof of (B) has also been given in [45].

I did not have a chance to discuss many other interesting results, and for the most part point out open problems. But I must at least say that almost nothing is known on the structure of the singular set and on the structural conditions ensuring the regularity of the minimizers (in this respect refer to [76][44][25]).

REFERENCES

[1] Acerbi, E., and Fusco, N.: *Semicontinuity Problems in the Calculus of variations*, to appear in Arch. Rat. Mech. Anal.

[2] Ball, J.M.: 1977, *Convexity conditions and existence theorems in nonlinear elasticity*, Arch. Rat. Mech. Anal. 66, pp. 337-403.

[3] Ball, J.M.: 1976, *On the Calculus of Variations and Sequentially weakly continuous maps*, in 'Ordinary and partial differential equations' Dundee, Springer Lecture Notes in Math. 564, pp.13-25.

[4] Ball, J.M.: 1980, *Strict convexity, strong ellipticity, and regularity in the calculus of variations*, Proc. Cambridge Phil. Soc. 87, pp.501-513.

[5] Ball, J.M., Currie, J.C., and Olver, P.J.: 1981, *Null Lagrangians, weak continuity, and variational problems of arbitrary order*, J. Functional Analysis 41, pp.135-174.

[6] Campanato, S.: 1963, *Proprietà di Holderianità di alcune classi di funzioni*, Ann. S.N.S. Pisa 17, pp.175-188.

[7] Campanato, S.: 1965, *Equazioni ellittiche del secondo ordine e spazi* $L^{2,\lambda}$, Ann. Mat. Pura et Appl. 69, pp.321-380.

[8] Coiffman, R.R., and Fefferman, C.: 1974, *Weighted norm inequalities for maximal functions and singular integrals*, Studia Math. pp.241-250.

[9] De Giorgi, E.: 1957, *Sulla differenziabilità e l'analiticità delle estremali degli integrali multipli regolari*, Mem. Accad. Sci. Torino cl. Sci. Fis. Mat. Nat. (3) 3,pp.25-43.

[10] De Giorgi, E.: 1968, *Un esempio di estremali discontinue per un problem variazionale di tipo ellittico*, Boll. UMI (4) 1, pp.135-137.

[11] De Giorgi, E.: 1968-69, *Teoremi di semicontinuità nel calcolo delle variazioni*, Lezioni tenute all'Ist. Naz. Alta Mat. Roma.

[12] Ekeland, I.: 1979, *Nonconvex minimization problems*, Bull. Amer. Math. Soc. 1 (3) pp.443-474.

[13] Evans, L.C.: 1981, *A new proof of local* $C^{1,\alpha}$ *regularity for solutions of certain degenerate elliptic P.D.E.*, preprint.

[14] Frehse, J.: 1973, *A discontinuous solution of a mildly nonlinear elliptic system*, Math. Z. 134, pp.229-230.

[15] Frehse, J.: 1975, *A note on the Hölder continuity of solutions of variational problems*, Abhandlung. Math. Sem. Hamburg 43, pp.59-63.

[16] Frehse, J.: 1979, *Un problème variational bidimensional possédant des extremales bornées et discontinuées*, C.R. Acad. Sc. Paris 289 ser. A., pp.751-753.

[17] Fusco, N.: *Quasi-convessità e semicontinuità per integlrali multipli di ordine superiore*, Ricerche di Mat.

[18] Gehring, F.W.: 1973, *The* L^p*-integrability of the partial derivatives of a quasi conformal mapping*, Acta Math. 130, pp.265-277.

[19] Giaquinta, M.: 1978, *A counter-example to the boundary regularity of solutions to elliptic quasilinear systems*, Manuscripta math. 14, pp.217-220.

[20] Giaquinta, M.: 1979, *Sistemi ellittici non lineari*, Teoria della regolarità, Boll. UMI (5) 16-A, pp.259-283.

[21] Giaquinta, M.: 1981, *Multiple integrals in the Calculus of variations and nonlinear elliptic systems*, Vorlesungsreihe SFB 72 no.6, Bonn, pp.1-266.

[22] Giaquinta, M.: 1982, *On the differentiability of the extremals of variational integrals*, Proc. of the Spring School on 'Nonlinear Analysis, Function spaces and Applications 2' Pisek, May 24-28, 1982, to appear.

[23] Giaquinta, M., and Giusti, E.: 1978, *Non linear elliptic systems with quadratic growth*, Manuscripta Math. 24, pp.323-349.

[24] Giaquinta, M., and Giusti, E.: *On the regularity of the minima of variational integrals*, Acta Math., to appear.

[25] Giaquinta, M., and Giusti, E.: *The singular set of the minima of certain quadratic functionals*, Analysis, to appear.

[26] Giaquinta, M., and Giusti, E.: *Quasi-minima*, to appear.

[27] Giaquinta, M., and Giusti, E.: *Differentiability of minima of non-differentiable functionals*, to appear.

[28] Giaquinta, M., and Hildebrandt, S.: *A priori estimates for harmonic mappings*, J. für Reine u. Angew. Math., to appear.

[29] Giaquinta, M., and Modica, G.: 1979, *Regularity results for some classes of higher order non linear elliptic systems*, J. für Reine u. Angew. Math. 311/312, pp.145-169.

[30] Giaquinta, M., and Modica, G.: 1979, *Almost-everywhere regularity results for solutions of non linear elliptic systems*, Manuscripta math. 28, pp.109-158.

[31] Giaquinta, M., and Modica, G.: 1982, *Non linear systems of the type of the stationary Navier-Stokes system*, J. für Reine u. Angew. Math. 330, pp.173-214.

[32] Giaquinta, M., and Nečas, J.: 1979, *On the regularity of weak solutions to non linear elliptic systems via Liouville's type property*, Comm. Math. Univ. Carolina 20, pp.111-121.

[33] Giaquinta, M., and Nečas, J.: 1980, *On the regularity of weak solutions to non linear elliptic systems of partial differential equations*, J. für Reine u. Angew. Math. 316, pp.140-159.

[34] Gilbarg, D., and Trudinger, N.S.: 1977, *'Elliptic partial differential equations of second order'*, Springer-Verlag, Heidelberg, New York.

[35] Giusti, E.: 1969, *Precisazione della funzioni* $H^{1,p}$ *e singolarità delle soluzioni deboli di sistemi ellittici non lineari*, Boll. UMI 2, pp.71-76.

[36] Giusti, E.: 1969, *Regolarità parziale delle soluzioni di sistemi ellittici quasi linear di ordine arbitrario*, Ann. S.N.S. Pisa 23, pp.115-141.

[37] Giusti, E.: These proceedings.

[38] Giusti, E, and Miranda, M.: 1968, *Un esempio di soluzioni discontinue per un problema di minimo relativo ad un integrale regolare del calcolo della variazioni*, Boll. UMI 2, pp.1-8.

[39] Giusti, E., and Miranda, M.: 1968, *Sulla regolarità delle soluzioni deboli di una classes di sistemi ellittici quasi lineari*, Arch. Rat. Mech. Anal. 31, pp.173-184.

[40] Hildebrandt, S.: These proceedings.

[41] Hildebrandt, S., Kaul, H., and Widman, K.O.: 1977, *An existence theorem for harmonic mappings of Riemannian manifolds*, Acta Math. 138, pp.1-16.

[42] Hildebrandt, S., and Widman, K.O.: 1975, *Some regularity results for quasilinear elliptic systems of second order*, Math. Z. 142, pp.67-86.

[43] Hildebrandt, S., and Widman, K.O.: 1977, *On the Hölder continuity of weak solutions of quasilinear elliptic system of second order*, Ann. S.N.S. Pisa 4, pp.145-178.

[44] Ivert, P.A.: 1979, *Regularitätsuntersuchungen von lösungen elliptischer Systeme von quasilinearen Differentialgleichungen zweiter Ordnung*, Manuscripta Math. 30, pp.53-88.

[45] Ivert, P.A.: to appear.

[46] Ladyzhenskaya, O.R., and Ural'tseva, N.N.: 1973, *'Linear and quasilinear elliptic equations'*, Moscow: Nauka (1964). Engl. Transl. Academic Press New York (1968). Second Russian edition: Nauka (1973).

[47] Luckhaus, S.: 1979, *Existence and regularity of weak solutions to the Dirichlet problem for semilinear elliptic systems of higher order*, J. für Reine u. Angew. Math. 306, pp. 192-207.

[48] Marcellini, P., and Sbordone, C.: *On the existence of minima of multiple integrals of the Calculus of Variations*, preprint.

[49] Maz'ya, V.G.: 1968, *Examples of non regular solutions of quasi-linear elliptic equations with analytic coefficients*, Funktsional' nyi Analiz. i Ego Prilosheniya 2,3, pp.53-57.

[50] Meyers, N.G.: 1963, *An L^p-estimate for the gradient of solutions of second order elliptic divergence equations*, Ann. S.N.S. Pisa (3) 17, pp.189-206.

[51] Meyers, N.G.: 1965, *Quasi-convexity and lower semicontinuity of multiple variational integrals of any order*, Trans. Amer. Math. Soc. 119, pp.125-149.

[52] Meyers, N.G., and Elcrat, A.: 1975, *Some results on regularity for solutions of non linear elliptic systems and quasi-regular functions*, Duke Math. I. 42, pp.121-136.

[53] Morrey, C.B.,jr.: 1938, *On the solutions of quasi-linear elliptic partial differential equations*, Trans. Amer. Math. Soc., 43 pp.126-166.

[54] Morrey, C.B.,jr.: 1940, *Existence and differentiability theorems for the solutions of variational problems for multiple integrals*, Bull. Amer. Math. Soc. 46, pp.439-458.

[55] Morrey, C.B.,jr.: 1943, *Multiple integral problems in the calculus of variations and related topics*, Univ. California Publ. Math. 1, pp.1-130.

[56] Morrey, C.B.,jr.: 1952, *Quasi-convexity and the lower semicontinuity of multiple integrals*, Pacific J. Math. 2, pp.25-53.

[57] Morrey, C.B.,jr.: 1954, *Second order elliptic systems of differential equations*, Ann. of Math. Studies No.33, Princeton Univ. Press, pp.101-159.

[58] Morrey, C.B.,jr.: 1958, *On the analyticity of the solutions of analytic non linear elliptic systems of artial differential equations I, II*, Amer. J. of Math. 80, pp.198-218, 219-234.

[59] Morrey, C.B.,jr.: 1959, *Second order elliptic equations in several variables and Hölder continuity*, Math. Z. 72, pp.146-164.

[60] Morrey, C.B.,jr.: 1960, *Multiple integral problems in the calculus of variations and related topics*, Ann.S.N.S. Pisa 14, pp.1-61.

[61] Morrey, C.B.,jr.: 1966, *'Multiple integrals in the calculus of variations'*, Springer-Verlag, Heidelberg, New York.

[62] Morrey, C.B.,jr.: 1968, *Partial regularity results for non linear elliptic system*, Journ. Math. and Mech. 17, pp. 649-670.

[63] Morrey, C.B.,jr.: 1970, *Differentiability theorems for non linear elliptic equations*, Actes Congrès intern Math.

[64] Moser, J.: 1960, *A new proof of De Giorgi's theorem concerning the regularity problem for elliptic differential equations*, Comm. Pure Appl. Math. 13, pp.457-468.

[65] Moser, J.: 1961, *On Harnack's theorem for elliptic differential equations*, Comm. Pure Appl. Math. 14, pp.577-591.

[66] Murat, F.: 1978, *Compacité par compensation II*, Proc. Int. Mat., 'Recent Methods in Non Linear Analysis', Rome, Pitagora Ed.

[67] Nash, J.: 1958, *Continuity of solutions of parabolic and elliptic equations*, Ann. J. Math. 80, pp.931-954.

[68] Nečas, J.: 1977, *Example of an irregular solution to a non linear elliptic system with analytic coefficients and conditions for regularity*. In: 'Theory of Non Linear Operators.' Abhandlungen Akac. Der Wissen. Der DDR. Proc. Summer School held at Berlin 1975.

[69] Nečas, J.: 1979, *On the regularity of weak solutions to variational equations and inequalities for non linear second order elliptic systems*, Equadiff. IV, Praha, Springer-Verlag Lecture Notes No. 703.

[70] Nirenberg, L.: 1959, *Remark on strongly elliptic partial differential equations*, Comm. Pure Appl. Math. 8, pp.649-675.

[71] Nirenberg, L.: 1959, *On elliptic partial differential equations*, Ann.S.N.S. Pisa 13, pp.115-162.

[72] Pepe, L.: 1971, *Risultati di regolarità parziale per le soluzioni* $H^{1,p}(\Omega)$ $1 < p < 2$ *di sistemi ellittici quasilineari*, Ann. Univ. Fenara 8.

[73] Serrin, J.: 1964, *Local behaviour of solutions of quasilinear elliptic equations*, Acta Math. 111, pp.247-302.

[74] Stampacchia, G.: 1966, *Equations elliptiques du second ordre à coefficients discontinues*, Les Presses de l'Univ. de Montréal.

[75] Stredulinsky, E.W.: 1980, *Higher integrability from reverse Hölder inequalities*, Indiana Univ. Math. I 29,3, pp.408-417.

[76] Uhlenbeck, K.: 1977, *Regularity for a class of nonlinear elliptic systems*, Acta Math., 138, pp.219-240.

[77] Widman, K.O.: 1971, *Hölder continuity of solutions of elliptic systems*, Manuscripto Math., 5, pp.299-308.

SOME ASPECTS OF THE REGULARITY THEORY FOR NONLINEAR ELLIPTIC SYSTEMS

Enrico Giusti

Universita Degli Studi,
Istituto Matematico "Ulisse Dini",
Viale Morgagni 67/A,
50134 Firenze, Italy.

1. INTRODUCTION

The problem of the regularity of solutions of nonlinear elliptic systems (and of nonlinear equations of order greater than two) in several variables has been successfully attacked only in recent years.

The celebrated theorem, due to De Giorgi [3] and Nash [25] who proved the Hölder-continuity of weak solutions of second order elliptic equations with discontinuous coefficients, has been the cornerstone of the regularity theory for nonlinear elliptic equations of second order. About ten years later, De Giorgi [4] constructed an example proving that his result could not be extended to systems of elliptic equations. Shortly thereafter M. Miranda and the writer [18] showed that De Giorgi's example could be modified to produce an elliptic system of nonlinear differential equations with analytic coefficients possessing a unique discontinuous solution. At about the same time similar examples were constructed by Maz'ya [22].

The example is the following: if we define

$$a_{ij}^{\alpha\beta}(x,u) = \delta_{ij}\,\delta^{\alpha\beta} + \left(\delta_j^\beta + \frac{4}{n-2}\,\frac{u_j\,u_\beta}{1+|u|^2}\right)\left(\delta_i^\alpha + \frac{4}{n-2}\,\frac{u_i\,u_\alpha}{1+|u|^2}\right)$$

the system

$$\int a_{ij}^{\alpha\beta}(x,u)\;D_\beta u^j D_\alpha \phi^i dx = 0 \qquad \forall\phi \in C_o(R^n;R^n) \tag{1.1}$$

J. M. Ball (ed.), Systems of Nonlinear Partial Differential Equations, 147–172.

admits among its solutions the (vector-valued) function $u = x|x|^{-1}$. This function, at least for sufficiently high dimensions, is the unique solution of the Dirichlet problem with $u = x$ on the boundary of the unit ball.

Several other examples of discontinuous (and even unbounded) solutions of elliptic systems have been constructed since. We shall only mention three of them:

1) Giaquinta [7] has shown that the Dirichlet problem for an elliptic system with zero boundary data may have discontinuities at the boundary.

2) Nečas [26], [27] (see also Giaquinta-Nečas [13] and Nečas-John-Stara [28]) has produced an elliptic system of the type

$$D^i a_i^\alpha (Du) = 0$$

possessing a Lipschitz-continuous non differentiable solution.

3) An unpublished result due to Souček shows that linear elliptic systems with discontinuous coefficients may have solutions with discontinuities in a dense set.

We remark that all these examples hold for systems in at least three independent variables. It is a well-known result that all the solutions of elliptic systems in the plane are regular.

The last example shows that the way followed in the case of a single second-order equation is barred when we are dealing with systems of more than one equation, or with equations of order greater than two. And in fact, the regularity results for general nonlinear systems are obtained using, and not in spite of, non-linearity.

Of course, it is not the question here to prove in general the regularity of the solutions, since the examples tell us that singularities may actually occur. One possibility is to look for special classes of elliptic systems having only regular solutions, or in other words to find reasonable conditions that imply regularity. This is the case of systems with diagonal principal part, for which we refer to Hildebrandt's lectures. Another interesting class has been studied by K. Uhlenbeck [30] (see also Ivert [20]). However, though of primary interest in their applications (surfaces of prescribed mean curvature, harmonic mappings) these systems have very special structure and are by no means representative of a general situation.

It is clear from the examples quoted that if one wants to give a general regularity theory for elliptic systems, one must take into account the existence of singular points. From this point of view the regularity of a solution is a property that holds only "generically"; more precisely, one tries to prove that any solution of an elliptic system is regular everywhere except in some small closed set Σ. This is what is called "partial regularity". The smallness of the singular set means in general that Σ has zero measure, but in many cases one can get estimates of its Hausdorff dimension.

Two methods have been developed so far in this direction. The first, introduced by Morrey [24] and later simplified and extended in [19] and [16], has its origin in an idea introduced by De Giorgi [2] in the theory of minimal surfaces (see also [1] and [17]), and is centred essentially on a "blow-up" procedure. The second, developed more recently [8], is more direct and relies heavily on a "reverse Hölder inequality" first proved by Gehring [6] in connection with quasiconformal mappings, and then extended to cover elliptic systems by Giaquinta and Modica [11].

Although overlapping considerably, the domains of the two methods do not coincide exactly, and therefore they are both of interest in partial regularity.

The main purpose of these lectures is to illustrate the results that can be obtained by means of the blow-up technique, as well as the limits of its applicability. To avoid being overwhelmed by technical details, we shall confine ourselves to the special (but in many respects typical) case of second order systems.

2. PARTIAL REGULARITY

We shall consider systems of the form

$$\sum_{i,j=1}^{n} \sum_{\alpha,\beta=1}^{n} \int_\Omega \left\{ A_{ij}^{\alpha\beta}(x,u)\, D_\beta u^j D_\alpha \phi^i + B_i(x,u,Du)\phi^i \right\} dx = 0 \qquad \forall\, \phi \in C_0^1(\Omega;\mathbb{R}^N) \tag{2.1}$$

with coefficients $A_{ij}^{\alpha\beta}$ continuous in $\Omega \times \mathbb{R}^N$, and satisfying for some $p > 2$:

$$v^{p-2}|\xi|^2 \leqslant A_{ij}^{\alpha\beta}(x,u)\, \xi_\alpha^i \xi_\beta^j \quad \forall\, \xi \in \mathbb{R}^{nN}, \quad \forall\, x,u \in \Omega \times \mathbb{R}^N \tag{2.2}$$

$$\left| A_{ij}^{\alpha\beta}(x,u) \right| \leqslant c_1\, v^{p-2} \tag{2.3}$$

where $V^2 = 1 + |u|^2$. Moreover, we shall suppose that

$$|B_i(x,u,Dx)| \leqslant V^{p-2} g(x) \tag{2.4}$$

In order to simplify the notation we shall write the system (2.1) in the form

$$\int_\Omega (A(x,u)DuD\phi + B(x,u,Du)\phi)\,dx = 0 \quad \forall\, \phi \in C_0^1(\Omega;\mathbb{R}^N) \tag{2.5}$$

Moreover, we shall stress the growth of the coefficients by setting

$$a_{ij}^{\alpha\beta} = V^{2-p} A_{ij}^{\alpha\beta} \quad ; \quad b_i = V^{2-p} B_i \tag{2.6}$$

so that the system (2.1) takes the final form

$$\int_\Omega V^{p-2}(a(x,u)DuD\phi + b(x,u,Du)\phi)\,dx = 0 \quad \forall\, \phi \in C_0^1(\Omega;\mathbb{R}^N) \tag{2.7}$$

The system is elliptic in the sense that

$$a(x,u)\ \xi\cdot\xi \geqslant |\xi|^2 \quad ; \tag{2.8}$$

in addition we suppose that

$$|a(x,u)| \leqslant c_1 \quad ; \quad |b(x,u,Du)| \leqslant g(x) \tag{2.9}$$

By a weak solution of (2.7) we mean a function $u \in L^p(\Omega;\mathbb{R}^N)$ such that $V^{p-2/2}Du \in L^2(\Omega)$, satisfying (2.7) for every test function $\phi \in C_0^1(\Omega;\mathbb{R}^N)$. It is easily seen that by closure (2.7) holds for every ϕ in $L^p(\Omega)$ with $V^{p-2/2}D\phi \in L^2(\Omega)$.

The following inequality is very useful in what follows. We remark that this is the only place where the strong ellipticity condition (2.8) is needed.

<u>2.1 Lemma</u> (Caccioppoli's inequality)

Let u *be a weak solution of* (2.7), *and suppose that conditions* (2.8) *and* (2.9) *hold. Then for every* $x_o \in \Omega$ *and for every* ρ, R $(0 < \rho < R < \mathrm{dist}(x_o, \partial\Omega))$ *we have*

$$\int_{B_\rho} V^{p-2}|Du|^2 dx \leqslant c_2 \left\{\left(1+\frac{1}{(R-\rho)^2}\right) \int_{B_R} V^{p-2}|u-u_R|^2 + \int_{B_R} V^{p-2}|b|^2 dx\right\} \tag{2.10}$$

where u_R is the average of u in $B_R = B(x_o,R)$:

$$u_R = \fint_{B_R} u dx = |B_R|^{-1} \int_{B_R} u dx \quad . \tag{2.11}$$

The proof of the lemma is obtained simply taking $\phi = \eta(u-u_R)$ in (2.7) (where as usual η is a smooth function with support in B_R, satisfying $0 \leqslant \eta \leqslant 1$ and $|D\eta| \leqslant 2(R-\rho)^{-1}$) and using repeatedly the inequality $2ab \leqslant \varepsilon a^2 + \varepsilon^{-1}b^2$.

The next lemma is a result concerning systems with constant coefficients.

2.2 Lemma

Let v *be a weak solution of the system*

$$\int (cDvD\phi + d\phi)dx = 0 \qquad \forall\ \phi \in C^1_0(B_1;\mathbb{R}^N) \tag{2.12}$$

with constant coefficients $c = c^{\alpha\beta}_{ij}$ *satisfying* (2.8) *and with* $d \in L^2(B_1)$. *Define:*

$$I(0,\rho) = \rho^{-n} \int_{B\rho} |u-u_{0,\rho}|^2 dx \tag{2.13}$$

$$D(0,\rho) = \rho^{-n} \int_{B\rho} |d|^2 dx \quad . \tag{2.14}$$

There exists a constant Λ, *depending only on* n, N *and* c_1, *such that for every* $\rho \leqslant 1/2$:

$$I(0,\rho) \leqslant \frac{\Lambda}{2}(\rho^2 I(0,1) + \rho^{2-n} D(0,1)). \tag{2.15}$$

Inequality (2.15) is well known for harmonic functions, in which case it holds with $\Lambda = 2$. In general it can be obtained as follows. We split $v = w + z$, where w is the unique solution of the Dirichlet problem for the system (2.12) with zero boundary data, and consequently z is a solution to the homogeneous system

$$\int cDzD\phi dx = 0 \qquad \forall\ \phi \in C^1_0(B_1;\mathbb{R}^N) \tag{2.16}$$

In particular, z is a smooth function in B_1, and every derivative of z satisfies the same system (2.16). Applying repeatedly the previous lemma to z and to its derivatives, and making use of the Sobolev imbedding theorem we get

$$\sup_{B_{1/2}} |Dz|^2 \leqslant c_3 \int_{B_1} |z-z_{0,1}|^2 dx$$

and therefore

$$\int_{B_\rho} |z-z_{0,\rho}|^2 dx \leqslant c_4 \rho^{n+2} \int_{B_1} |z-z_{0,1}|^2 dx$$

for every $\rho < 1/2$.
On the other hand we have for w the trivial estimate

$$\int_{B_\rho} |w-w_{0,\rho}|^2 dx \leqslant \int_{B_\rho} |w|^2 dx \leqslant \int_{B_1} |w|^2 dx \leqslant c_5 \int_{B_1} |Dw|^2 dx \leqslant \int_{B_1} |d|^2 dx$$

and (2.15) follows at once.

The next lemma is the essential step in the theory of partial regularity.

2.3 Lemma

For every $M > 0$ *and* τ, $0 < \tau < 1/2$, *there exist two constants* $\varepsilon_o = \varepsilon_o(M,\tau)$ *and* $R_o = R_o(M,\tau)$ *such that if* u *is a weak solution of the system* (2.7) *with conditions* (2.8), (2.9) *and if for some* $x_o \in \Omega$ *and some* $R < \min(R_o, \operatorname{dist}(x_o,\partial\Omega))$ *we have*

$$|u_{x_o,R}| \leqslant M \tag{2.17}$$

$$J(x_o,R) = R^{-n} \int_{B(x_o,R)} V^{p-2} |u-u_{x_o,R}|^2 dx < \varepsilon_o^2 \tag{2.18}$$

$$\mathcal{B}(x_o,R) = R^{-n} \int_{B(x_o,R)} V^{p-2} |b|^2 dx < \varepsilon_o^2 \tag{2.19}$$

then

$$J(x_o,R\tau) \leqslant \Lambda\, (\tau^2 J(x_o,R) + \tau^{2-n} \mathcal{B}(x_o,R)) \tag{2.20}$$

where Λ *is the constant appearing in lemma* 2.2.

The proof of the lemma proceeds by contradiction. Suppose the lemma is false. Then for some M and τ there would exist:

i) a sequence $\{x_k\}$ in Ω;

ii) two infinitesimal sequences $\{\varepsilon_k\}$ and $\{R_k\}$;

iii) a sequence $\{u^k\}$ of solutions of the system (2.7) such that

$$|u^k_{x_k,R_k}| \leqslant M \tag{2.21}$$

$$\max\,(J_k(x_k,R_k);\quad \mathcal{B}_k(x_k,R_k)) = \varepsilon_k^2 \tag{2.22}$$

and

$$J_k(x_k,\tau R_k) > \Lambda(\tau^2 J_k(x_k,R_k) + \tau^{2-n}\mathcal{B}_k(x_k,R_k)). \tag{2.23}$$

We now scale both the independent and the dependent variables by setting $x = x_k + R_k y$ and

$$v^k(y) = \varepsilon_k^{-1}(u^k(x_k+R_k y) - u^k_{x_k,R_k}) \tag{2.24}$$

$$d_k(y) = \varepsilon_k^{-1}R_k^2\, b(x_k+R_k y, u^k_{x_k,R_k} + \varepsilon_k v^k(y),\ \varepsilon_k R_k^{-1} Dv^k(y)) \tag{2.25}$$

The function v^k satisfies the system

$$\int W_k^{p-2}\,(a_k(y)Dv^k D\phi + d_k(y)\phi)\,dy = 0 \qquad \forall\, \rho \in C_0^1(B_1;\mathbb{R}^N) \tag{2.26}$$

where

$$W_k^2 = 1 + |\varepsilon_k v^k + u^k_{x_k,R_k}|^2 \tag{2.27}$$

and

$$a_k(y) = a(x_k+R_k y, u^k_{x_k,R_k} + \varepsilon_k v^k(y)). \tag{2.28}$$

Moreover we have, from (2.22) and (2.23):

$$\max (\mathcal{J}_k(0,1); \ \mathcal{D}_k(0,1)) = 1 \ , \tag{2.29}$$

$$\mathcal{J}_k(0,\tau) > \Lambda \ (\tau^2 \mathcal{J}_k(0,1) + \tau^{2-n} \mathcal{D}_k(0;1)) \ , \tag{2.30}$$

where

$$\mathcal{J}_k(0,\rho) = \rho^{-n} \int_{B\rho} W_k^{p-2} \ |v^k - v^k_{0,\rho}|^2 dy \tag{2.31}$$

and

$$\mathcal{D}_k(0,1) = \int_B W_k^{p-2} |d_k|^2 dy \ . \tag{2.32}$$

Let now $k \to \infty$. We may suppose, passing possibly to a subsequence, that

$$x_k \to x_o \quad ; \quad u^k_{x_k, R_k} \to \lambda \ , \tag{2.33}$$

and moreover:

$$v^k \to v \quad \text{weakly in } L^2(B_1) \ , \tag{2.34}$$

$$\varepsilon_k v^k \to 0 \quad \text{almost everywhere in } B_1 \ , \tag{2.35}$$

$$a_k(y) \to a(x_o, \lambda) \quad \text{a.e. in } B_1 \ , \tag{2.36}$$

$$W_k \to V_o = (1 + |\lambda|^2)^{1/2} \quad \text{strongly in } L^p(B_1) \ , \tag{2.37}$$

$$W_k^{(p-2)/2} \ v^k \to V_o^{(p-2)/2} \ v \quad \text{weakly in } L^2(B_1) \ , \tag{2.38}$$

$$W_k^{(p-2)/2} \ d_k \to V_o^{(p-2)/2} \ d \quad \text{weakly in } L^2(B_1) \ , \tag{2.39}$$

$$Dv^k \to Dv \quad \text{weakly in } L^2_{loc}(B_1) \ , \tag{2.40}$$

$$W_k^{(p-2)/2} \ v^k \to V_o^{(p-2)/2} \ v \quad \text{strongly in } L^2_{loc}(B_1) \ , \tag{2.41}$$

$$W_k^{(p-2)/2} \ Dv^k \to V_o^{(p-2)/2} \ Dv \quad \text{weakly in } L^2_{loc}(B_1) \ . \tag{2.42}$$

Finally, v is a weak solution of the system

$$\int (a(x_o,\lambda)DvD\phi \quad + \quad d\phi)\, dy = 0 \qquad \forall\ \phi \in C^1_0(B_1;\mathbb{R}^N) \tag{2.43}$$

and therefore by lemma 2.2:

$$I(0,\tau) \leqslant \frac{\Lambda}{2}\,(\tau^2\ I(0,1) + \tau^{2-n}D(0,1)). \tag{2.44}$$

On the other hand, passing to the limit in (2.30) and taking into account (2.38), (2.39) and (2.41) we get

$$I(0,\tau) \geqslant \Lambda(\tau^2\ I(0,1) + \tau^{2-n}D(0,1)) \tag{2.45}$$

and from (2.29), (2.30):

$$I(0,\tau) \geqslant \Lambda\ \tau^2\ V_o^{p-2} \tag{2.46}$$

The last two inequalities contradict (2.44), and this proves the lemma.

Suppose for a moment that $b = 0$, and that ε_o, R_o are independent of M. We can apply again and again the lemma 2.3 to get an estimate of the form

$$J(x_o,\tau^k R) \leqslant \Lambda^k\ \tau^{2k}\ J(x_o,R) \qquad .$$

This inequality, as we shall see, gives the required regularity result. In the general case we have to control the behaviour of the average of u in balls of decreasing radii, and to get rid of assumption (2.19) that forbids the iteration of the procedure. For that we need first two inequalities. The first is very simple:

$$|u_{x_o,\tau r}| \leqslant |u_{x_o,r}| + \tau^{-n/2}\ J^{\frac{1}{2}}(x_o,r) \quad . \tag{2.47}$$

The second is a little tricky, but not difficult:

$$\mathcal{B}(x_o,r) \leqslant K(r^{2p}\ \fint_{B_r} g^p dx)^{2/p}\ (J(x_o,r)+(1+|u_{x_o,r}|^2)^{p/2})^{1-2/p} \tag{2.48}$$

We shall assume now that the function g belongs to the space $L^{p,n-2p+p\sigma}$ for some $\sigma > 0$, or in other words that there exists a constant G such that

$$\int_{B_r} g^p dx \leqslant G^{p/2} r^{n-2p+p\sigma} \tag{2.49}$$

for every ball B_r contained in Ω. With this assumption we get from (2.48):

$$\mathcal{B}(x_o,r) \leqslant KG\, r^{2\sigma} (J(x_o,r)+(1+|u_{x_o,r}|^2)^{p/2})^{1-2/p} . \tag{2.50}$$

We choose now τ in such a way that $\tau^\sigma < 1/2$ and $2\Lambda\tau^{2-\sigma} \leqslant 1$.

2.4 Lemma

Let u *be a weak solution of the system* (2.7) *with conditions* (2.8), (2.9) *and* (2.49). *For* $M > 0$ *let* $\varepsilon_o(M)$ *and* $R_o(M)$ *be the constants appearing in lemma* 2.3 *(depending only on* M *since* τ *has been fixed) and let*

$$\eta_o(M) = \min (\varepsilon_o(M) \;;\; M\,\tau^{n/2}(1-\tau^{\sigma/2})/2\sqrt{2}) \quad , \tag{2.51}$$

$$\lambda_o(M) = KG\,(\eta_o^2 + (1+M^2)^{p/2})^{1-2/p} \quad , \tag{2.52}$$

$$r_o(M) = \min (R_o(M) \;;\; (\tau^n\, \eta_o^2\, (1-\tau^\sigma)/\lambda_o)^{1/2\sigma} . \tag{2.53}$$

If for some $x_o \in \Omega$ *and* $R < \min(r_o; \operatorname{dist}(x_o,\partial\Omega))$ *we have*

$$|u_{x_o,R}| \leqslant M/2 \quad , \tag{2.54}$$

$$J(x_o,R) < \eta_o^2 \quad , \tag{2.55}$$

then for every integer k:

$$J(x_o,\tau^k R) \leqslant \tau^{k\sigma} (\eta_o^2 + \tau^{-n} R^{2\sigma} \lambda_o \sum_{i=0}^{k-1} \tau^{i\sigma}) \tag{2.56}$$

For the proof, we set

$$u_k = |u_{x_o,\tau^k R}| \;;\; J_k = J(x_o,\tau^k R) \;;\; B_k = \tau^{-2k\sigma} R^{2\sigma} \mathcal{B}(x_o,\tau^k R);$$

$$V_k^2 = 1 + |u_k|^2 .$$

so that from (2.47) and (2.50) we get:

$$|u_{k+1}| \leqslant |u_k| + \tau^{-n/2} J_k^{1/2} \tag{2.57_k}$$

$$B_k \leqslant KG(J_k + V_k^p)^{1-p/2} \tag{2.58_k}$$

We shall prove by induction (2.56), which we rewrite as

$$(j_k) \qquad J_k \leqslant \tau^{k\sigma} (\eta_o^2 + \tau^{-n} R^{2\sigma} \lambda_o \sum_{i=0}^{k-1} i\sigma)$$

together with

$$(p_k) \qquad |u_k| \leqslant \frac{M}{2} + \tau^{-n/2} (\eta_o^2 + \tau^{-n} (1-\tau^{\sigma})^{-1} R^{2\sigma} \lambda_o \sum_{h=0}^{k-1} \tau^{h\sigma/2})$$

and

$$(q_k) \qquad B_k \leqslant \lambda_o \quad .$$

For that let us write down the auxiliary relations

$$(a_k) \qquad |u_k| \leqslant M$$

$$(b_k) \qquad \tau^{2k\sigma} R^{2\sigma} B_k \leqslant \eta_o^2$$

$$(c_k) \qquad J_k < \eta_o^2$$

$$(d_k) \qquad J_{j+1} \leqslant \tau^{\sigma} J_k + \tau^{(2k+1)\sigma - n} R^{2\sigma} B_k \quad .$$

It is a simple matter of calculations to verify that all these relations hold for $k = 0$. Actually, (j_o), (p_o), (a_o) and (c_o) follow at once from (2.54) and (2.55); (q_o) from (a_o), (c_o) and (2.52); (b_o) is a consequence of (q_o) and (2.53) and (d_o) follows from lemma 2.3.

Suppose now that they hold for some k. Then:

$(q_k),(d_k),(j_k) \Rightarrow (j_{k+1})$

$(p_k),(j_k),(2.57_k) \Rightarrow (p_{k+1})$

$(j_{k+1}),(2.53) \Rightarrow (c_{k+1})$

$(p_{k+1}),(2.51) \Rightarrow (a_{k+1})$

$(a_{k+1}),(c_{k+1}),(2.58_{k+1}) \Rightarrow (q_{k+1}) \Rightarrow (b_{k+1})$

$(a_{k+1}),(b_{k+1}),(c_{k+1})$ and lemma 2.3 $\Rightarrow (d_{k+1})$.

This concludes the proof of the lemma. The proof of the regularity theorem is now not difficult.

2.5 Theorem

Let u *be a weak solution of the system* (2.7) *with conditions* (2.8) *and* (2.9) *and with the function* g(x) *satisfying* (2.49). *Let* $x_o \in \Omega$ *be such that*

$$\sup_{0<R<\operatorname{dist}(x_o,\partial\Omega)} |u_{x_o,R}| < +\infty \tag{2.59}$$

$$\liminf_{R\to 0} J(x_o,R) = 0 \tag{2.60}$$

Then u *is Hölder-continuous in a neighbourhood of* x_o.

Proof

Let $M = 4 \sup_{0<R<\operatorname{dist}(x_o,\partial\Omega)} |u_{x_o,R}|$ and let $\eta_o(M)$, $r_o(M)$ be as in (2.51) and (2.53).

By (2.60) there exists $R < r_o$ such that $J(x_o,R) < \eta_o^2$. Since $u_{x,R}$ and $J(x,R)$ are continuous functions of x, there exists a neighbourhood U of x_o such that for every $x \in U$:

$$|u_{x,R}| \leqslant M/2$$

$$J(x,R) < \eta_o^2 \quad .$$

For these x we get from lemma 2.4:

$$J(x,t^kR) \leqslant \tau^{k\sigma} (\eta_o^2 + \tau^{-n}(1-\tau^\sigma)^{-1} \lambda_o R^{2\sigma}) \leqslant 2 \eta_o^2 \tau^{k\sigma}$$

and since $V \geqslant 1$:

$$I(x,\tau^kR) \leqslant 2 \eta_o^2 \tau^{k\sigma}$$

where

$$I(x,r) = \int_{B(x,r)} |u(y)-u_{x,r}|^2 dy \quad .$$

If $0 < r < R$, choosing k so that $\tau^{k+1} < r/R \leqslant \tau^k$ we get

$$I(x,r) \leqslant (\tau^kR/r)^n I(x,\tau^kR) \leqslant 2 \eta_o^2 \tau^{-n-\sigma} (r/R)^\sigma$$

where we have used the fact that the function of t:

$$\int_{B(x,r)} |u(y)-t|^2 dy$$

takes its minimum for $t = u_{x,r}$.
In conclusion, for every $x \in U$ and for every r, $0 < r < R$, we have:

$$I(x,r) \leqslant c_6 r^\sigma \quad ,$$

an estimate implying that $u \in C^{o,\sigma/2}(U)$.

It is clear from the above that the set Ω_o of the regular points of u is open. Let $\Sigma = \Omega - \Omega_o$ be the singular set. From the inequality

$$J(x_o,R) \leqslant (\fint_{B(x_o,R)} V^p dx)^{1-2/p} (\fint_{B(x_o,R)} |u-u_{x_o,R}|^p dx)^{2/p} \qquad (2.61)$$

and from Theorem 2.5 we conclude that Ω_o contains all the Lebesgue points of u, and therefore that Σ has zero measure. Actually, we may give a better estimate for the dimension of Σ.

We recall that the q-dimensional Hausdorff measure of a set E is

$$H_q(E) = \liminf_{\varepsilon\to 0} \left\{ \sum_{i=1}^{\infty} (\operatorname{diam} X_i)^q \; ; \; \bigcup_{i=1}^{\infty} X_i \supset E,\ \operatorname{diam} X_i < \varepsilon \right\}$$

and that the dimension of a set E is defined to be the infimum of those q for which $H_q(E) = 0$. We have the following results, which are proved in [15].

2.6 Lemma

Let $f \in L^1_{loc}(\Omega)$ *and for* $0 \leq q < n$ *let*

$$E_q = \left\{ x \in \Omega : \limsup_{r\to 0} r^{-q} \int_{B(x,r)} f(y)\,dy > 0 \right\}$$

Then

$$H_q(E_q) = 0 \qquad .$$

2.7 Lemma

Let $u \in H^{1,p}_{loc}(\Omega)$, *and let*

$$\Sigma_1 = \left\{ x \in \Omega : \sup_{0<r<\operatorname{dist}(x,\partial\Omega)} |u_{x,r}| = +\infty \right\}$$

Then $\dim(\Sigma_1) \leq n-p$.

With the help of these two lemmas we shall estimate the dimension of the singular set. We have:

$$\fint_{B(x_o,R)} V^p dx \leq c_7 \; (J(x_o,R) + V_R^p) \tag{2.62}$$

where $V_R^2 = 1 + |u_{x_o,R}|^2$. Moreover:

$$\int_{B(x_o,R)} |u-u_{x_o,R}|^p dx \leq c_8 \; R^2 \int_{B(x_o,R)} |u-u_{x_o,R}|^{p-2}|Du|^2 dx$$

$$\leq c_9 \; R^2 \; V_R^{\,p-2} \int_{B(x_o,R)} V^{p-2}|Du|^2 dx$$

and therefore from (2.61):

$$J(x_o,R) \leqslant c_{10} \, (J(x_o,R) + V_R^p)^{1-2/p} \times \\ \times \, (R^2 V_R^{p-2} \fint_{B(x_o,R)} V^{p-2} |Du|^2 dx)^{2/p} \tag{2.63}$$

Define now

$$\Sigma_2 = \left\{ x \in \Omega : \limsup_{R \to 0} R^{2-n} \int_{B(x,R)} V^{p-2} |Du(y)|^2 dy > 0 \right\} .$$

If a point x_o does not belong to $\Sigma_1 \cup \Sigma_2$ we have

$$\sup_{0<R<\mathrm{dist}(x_o,\partial\Omega)} |u_{x_o,R}| < +\infty$$

and

$$\lim_{R \to 0} R^{2-n} \int_{B(x_o,R)} V^{p-2} |Du|^2 dx = 0 .$$

From (2.63) we conclude that

$$\lim_{R \to 0} J(x_o,r) = 0$$

and hence from Theorem 2.5 u is regular near x_o. We have therefore

$$\Sigma \subset \Sigma_1 \cup \Sigma_2 .$$

On the other hand from lemma 2.7 we infer that

$$\dim (\Sigma_1) \leqslant n-2$$

whereas lemma 2.6 gives at once

$$H_{n-2}(\Sigma_2) = 0$$

so that in conclusion the (closed) singular set Σ satisfies

$\dim (\Sigma) \leqslant n-2.$

3. REGULARITY AND LIOUVILLE'S PROPERTY

In general it is not simple to give conditions, different from (2.59) and (2.60), assuring that a solution u of an elliptic system is regular near a point x_o. From this point of view, partial regularity is really a generic property, and it seems quite impossible to obtain explicit (e.g. algebraic) conditions that discriminate between regular and singular points.

In this section we shall take a different attitude, and instead of explicit conditions we shall discuss the relation between regularity and another property of the system in question, namely the Liouville property. This connection was pointed out by Frehse [5] and studied by Giaquinta and Nečas [12], [13] (see also [21] and [14]). We shall follow here the ideas of Meier [23].

We will consider systems of the type (2.7) with $p = 2$:

$$\int (a(x,u)DuD\phi + b(x,u,Du)\phi)\,dx = 0 \qquad \forall\ \phi \in C_0^1(\Omega,\mathbb{R}^N) \tag{3.1}$$

where a and b satisfy conditions (2.8) and (2.9). We say that the system (3.1) has the Liouville property at $x_o \in \Omega$ if the system

$$\int a(x_o,v)DvD\phi dx = 0 \qquad \forall\ \phi \in C_0^1(\Omega,\mathbb{R}^N) \tag{3.2}$$

has no bounded entire solutions in $\mathbb{R}^n$, other than the constant vectors.

3.1 Theorem

Let u *be a bounded solution of the system* (3.1), *with conditions* (2.8), (2.9) *and* (2.49). *Suppose that the system has the Liouville property at* x_o. *Then* x_o *is a regular point for* u.
The proof will be achieved by showing that

$$\liminf_{R\to 0} \fint_{B(x_o,R)} |u-u_{x_o,R}|^2 dx = 0 \tag{3.3}$$

and then applying theorem 2.5 (note that since u is bounded (2.59) is automatically satisfied).

We may suppose that $x_o = 0$ and that $B_1 = B(0,1) \subset \Omega$. For

$0 < R < 1$ let $x = Ry$ and $v_R(y) = u(Ry)$. The function v_R is a solution of the system:

$$\int (a(Ry,v_R)Dv_R D\phi + R^2 b(Ry,v_R,R^{-1}Dv_R)\phi)\,dy = 0 \qquad \forall\ \phi \in C^1_0(B_{1/R},\mathbb{R}^N) \tag{3.4}$$

in the ball $B_{1/R}$. From (2.9) and the Caccioppoli inequality (2.10) we get for $s < t < 1/R$:

$$\int_{B_s} |Dv_R|^2 dy \leqslant c(s,t) \int_{B_t} |v_R|^2 dy + R^4 \int_{B_t} g^2(Ry)\,dy \tag{3.5}$$

whereas from (2.49):

$$R^4 \int_{B_t} g^2(Ry)\,dy = R^{4-n} \int_{B_{tR}} g^2(x)\,dx \leqslant G\, t^{n-4+2\sigma} R^{2\sigma} . \tag{3.6}$$

From (3.5), (3.6), and the boundedness of v_R it follows that we can select a sequence $R_j \longrightarrow 0$ such that the sequence $v_j = v_{R_j}$ converges to some function v strongly in $L^2_{loc}(\mathbb{R}^n,\mathbb{R}^N)$ and almost everwhere, whereas $Dv_j \longrightarrow Dv$ weakly in L^2_{loc}. Moreover (3.6) implies that $b_j(y) = R^2_j\, b(R_jy,v_j,\ R^{-1}_j Dv_j) \longrightarrow 0$ in L^2_{loc}. Passing to the limit in (3.4) we conclude that v is a bounded entire solution of the system (3.2), and whence it is constant. From

$$\lim_{j\to\infty} \int_{B_1} |v_j - v|^2 dy = 0$$

we get

$$\lim_{k\to\infty} \fint_{B_{R_k}} |u(x)-v|^2 dx = 0$$

which implies (3.3) at once.

We remark that the converse is also true; namely that regularity implies the Liouville property. More precisely, if all the solutions of the system

$$\int a(x_o,u)DuD\phi dx = 0 \qquad \forall\ \phi \in C^1_0(\Omega,\mathbb{R}^N) \tag{3.7}$$

(with coefficients frozen at x_o) are regular at x_o, then the system has the Liouville property at x_o.

Suppose again that $x_o = 0$, and let u be a non-constant entire bounded solution of (3.7). Setting $u_k(x) = u(kx)$, and reasoning as above we can suppose (passing possibly to a subsequence) that $u_k \longrightarrow v$ strongly in L^2_{loc} and weakly in H^1_{loc}. The function v is a solution of the same system (3.7).

We shall prove that v is discontinuous at 0. For that we show that

$$\fint_{B_r} |v-v_r|^2 dx$$

does not tend to zero with r. We have

$$\fint_{B_r} |v-v_r|^2 dx = \lim_{k\to\infty} \fint_{B_{kr}} |u-u_{kr}|^2 dx$$

and hence from the Caccioppoli's inequality (2.10):

$$\fint_{B_r} |v-v_r|^2 dx \geqslant \liminf_{R\to\infty} R^{2-n} \int_{B_R} |Du|^2 \quad . \tag{3.8}$$

We follow now an idea of Meier [23]. Setting

$$\Phi(R) = R^{2-n} \int_{B_R} |Du|^2 dx$$

we have from Giaquinta and Giusti [8] (see also Giaquinta's notes) the inequality

$$\Phi(tR) \leqslant c_{11}\, t^2 \{1+t^{-n}\, \omega(c_{12}\, \Phi(R))\}\, \Phi(R) \quad , \quad 0 < t < 1 \tag{3.9}$$

where ω is the modulus of continuity of the coefficients.

Let now $2c_{11}t^2 = 1$, and let η_o be such that

$$t^{-n}\,\omega(c_{12}\eta) \leqslant 1 \qquad \forall\, \eta < \eta_o$$

Since $\Phi(R)$ is not identically zero, there exists $R_o > 0$ such that

$$\inf_{tR_o < R \leqslant R_o} \Phi(R) = \sigma_o > 0 \quad .$$

Let $R > R_o$. Then either $\Phi(R) \geqslant \eta_o$ or by (3.9) $\Phi(tR) \leqslant \Phi(R)$. By iteration we get in this case

$$\Phi(t^k R) \leqslant \Phi(R)$$

for every integer k. On the other hand for some k we have $tR_o < t^k R \leqslant R_o$ and whence in this case $\Phi(R) \geqslant \sigma_o$. In conclusion

$$\Phi(R) \geqslant \min(\eta_o, \sigma_o)$$

and from (3.8) we conclude that v is discontinuous at 0.

We remark that the above result is not exactly the converse of Theorem 3.1. Actually, from the existence of nonconstant bounded entire solutions for the system (3.7) we deduce the existence of discontinuous solutions for the same system (3.7), but not for the original system (3.1). This is in general a weaker statement, since the regularity of the solutions of the system (3.7) implies that of the solutions of (3.1), but not viceversa. We have therefore equivalence between regularity and Liouville property only for systems of the type

$$\int a(v)\,DvD\phi\,dx = 0 \qquad \forall\, \phi \in C^1_0(\Omega, \mathbb{R}^N)$$

with coefficients independent of x.

4. QUADRATIC FUNCTIONALS

The results of the two preceding sections hold for the minimum points of quadratic functionals:

$$F(u;\Omega) = \int_\Omega a^{\alpha\beta}_{ij}(x,u)\,D_\alpha u^i D_\beta u^j\,dx \tag{4.1}$$

where summation over repeated indices is understood. The

coefficients $a_{ij}^{\alpha\beta}$ are continuous in $\Omega \times \mathbb{R}^N$ and satisfy the ellipticity conditions

$$a(x,u)\xi\cdot\xi \geqslant |\xi|^2 \quad ; \quad |a(x,u)| \leqslant c_1 \quad . \tag{4.2}$$

More precisely we have:

4.1 Theorem

Let $u \in H^1_{loc}(\Omega,\mathbb{R}^N)$ *be a local minimum for the functional* F. *Then* u *is Hölder-continuous in an open set* $\Omega_o \subset \Omega$ *and we have*

$$\dim(\Omega-\Omega_o) < n-2 \quad . \tag{4.3}$$

We note that strict inequality holds in (4.3); this is also true for the solutions to elliptic systems.

4.2 Theorem

The functional (4.1) *has the Liouville property at* x_o *if and only if every local minimum of the frozen functional*

$$\int a(x_o,v)DvDvdx$$

is regular at x_o.

For the proofs of these theorems we refer respectively to [9] and [23]. Moreover we have:

4.3 Theorem

Let $u \in H^1_{loc}(\Omega,\mathbb{R}^N)$ *be a local minimum of* F. *Then* $u \in H^{1,q}_{loc}(\Omega,\mathbb{R}^N)$ *for some* $q > 2$, *and we have:*

$$\left(\fint_{B_{R/2}} |Du|^q dx\right)^{1/q} \leqslant c_{13} \left(\fint_{B_R} |Du|^2 dx\right)^{1/2} \tag{4.4}$$

for every ball $B_R \subset \Omega$.

The proof of Theorem 4.3 is given in [9].

The strict inequality in the estimate (4.3) poses immediately the question whether the dimension of the singular set Σ is less than or equal to $n - 3$.

The answer to this question is known up to now only in a special case; namely for the minima of functionals (and not for solutions to elliptic systems!) when the coefficients have the special form

$$a_{ij}^{\alpha\beta}(x,u) = g_{ij}(x,u)G^{\alpha\beta}(x) \quad . \tag{4.5}$$

Denoting by ω the modulus of continuity of the coefficients:

$$|a(x,u) - a(x',u')| \leq \omega(|x-x'|^2 + |u-u'|^2) \tag{4.6}$$

we shall suppose that

$$\int_0^1 \frac{\omega(t^2)}{t}\, dt < +\infty \quad .$$

In this case we have:

4.4 Theorem

Let u *be a bounded local minimum for the functional* (4.1) *with coefficients of the form* (4.5) *satisfying* (4.2). *Then the dimension of the singular set* Σ *cannot exceed* $n - 3$. *Moreover, if* $n = 3$, Σ *consists at most of isolated points.*

The reason of the special form (4.5) of the coefficients is that in this case we have the inequality (see [10])

$$\int_{\partial B} |u(Rx)-u(rx)|^2\, dH_{n-1} \leq c_{14} \log \frac{R}{r} (\Phi(R) - \Phi(r)) \tag{4.7}$$

where

$$\Phi(t) = t^{2-n} \exp\left\{c_{15} \int_o^t \frac{\omega(s^2)}{s}\, ds\right\} \int_{B_t} a(x,u) DuDudx \quad . \tag{4.8}$$

We shall limit ourselves to the case $n = 3$. We note first that by (4.7) the function $\Phi(t)$ is increasing, and therefore tends to a finite limit as $t \longrightarrow 0$.

Suppose now that u is a local minimum of F, possessing a sequence of singular points x_j convergent to 0. Let

$R_j = 2|x_j| < 1$. The function

$$u^j(x) = u(R_j x)$$

is a local minimum in the unit ball B of the functional

$$F^j(u,B) = \int_B a^j(x,u^j) Du^j Du^j dx$$

where

$$a^j(x,z) = a(R_j x, z) \quad .$$

Moreover each u^j has a singular point $y_j = R_j^{-1} x_j$ with $|y_j| = 1/2$. Since u, and therefore the u^j, are uniformly bounded, it is possible (using the analogue of the Caccioppoli inequality) to extract a subsequence (that we shall denote again with u^j) converging strongly in L^2_{loc} to a function v, minimising the limit functional

$$F_o(v,B) = \int_B a(0,v) DvDvdx \quad .$$

Since $|y_j| = 1/2$, we can also suppose that $y_j \longrightarrow y_o$. The point y_o must be a singular point for v.

Let now $0 < s < t < 1$, and let us write the inequality (4.7) with $r = sR_j$ and $R = tR_j$. We have

$$\int_{\partial B} |u^j(sx) - u^j(tx)|^2 \, dH_{n-1} \leqslant c_{14} \log \frac{t}{s} (\Phi(tR_j) - \Phi(sR_j)) \quad .$$

If we let $j \longrightarrow \infty$ the right-hand-side tends to zero and therefore for almost every value of s and t we get

$$\int_{\partial B} |v(sx) - v(tx)|^2 \, dH_{n-1} = 0$$

so that v is homogeneous of degree zero.

In conclusion, the whole segment joining 0 with y_o is made of singular points of v, contradicting the conclusion of

Theorem 4.3, and in particular the fact that the singular set has dimension strictly less than $3-2 = 1$. We shall not go into the details of the proof for dimension n greater than 3, and we refer directly to [10].

We observe that in spite of the special form of the coefficients, Theorem 4.4 applies to harmonic mappings of Riemannian manifolds, $u : M^n \longrightarrow M^N$, provided that every point of M^n has a neighbourhood which is mapped into a bounded coordinate chart of M^N (this holds in particular if M^N is the space $\mathbb{R}^N$ equipped with any Riemannian metric). In this case, if u minimises the energy:

$$E(u) = \int |du|^2$$

a representative of u in local coordinates is a local minimum of the functional

$$E(u;A) = \int_A g_{ij}(u)\, G^{\alpha\beta}(x)\, D_\alpha u^i D_\beta u^j \sqrt{G}dx$$

where $G_{\alpha\beta}$ and g^{ij} are the metric tensors of M^n and M^N respectively.

We mention that in the case of harmonic mappings, Theorem 4.4 has been proved by Schoen and Uhlenbeck [29] without the above restriction on the target manifold M^N.

REFERENCES

[1] Almgren, F.J. Jr.: 1968, *Existence and regularity almost everywhere of solutions to elliptic variational problems among surfaces of varying topological type and singularity structure*, Ann. of Math., 87, pp.321-391.

[2] De Giorgi, E.: 1961, *Frontiere orientate di misura minima*, Sem. Scuola Norm. Sup., Pisa.

[3] De Giorgi, E.: 1957, *Sulla differenziabilità e l'analiticità delle estremali degli integrali multipli regolari*, Mem. Acc. Sci. Torino (3), 3, pp.25-43.

[4] De Giorgi, E.: 1968, *Un esempio di estremali discontinue per un problema variazionale di tipo ellittico*, Boll. Un. Mat. Ital., (4), 1, pp.135-137.

[5] Frehse, J.: 1977, *Essential selfadjointness of singular elliptic operators*, Boll. Soc. Bras. Mat., 8, pp.87-107.

[6] Gehring, F.W.: 1973, *The L^p-integrability of the partial derivatives of a quasi conformal mapping*, Acta Math., 130, pp.265-277.

[7] Giaquinta, M.: 1978, *A counterexample to the boundary regularity of solutions to elliptic quasilinear systems*, Manus. Math., 14, pp.217-220.

[8] Giaquinta, M., and Giusti, E.: 1978, *Nonlinear elliptic systems with quadratic growth*, Manus. Math., 24, pp.323-349.

[9] Giaquinta, M., and Giusti, E.: 1982, *On the regularity of the minima of variational integrals*, to appear in Acta Math.

[10] Giaquinta, M., and Giusti, E.: 1981, *The singular set of the minima of certain quadratic functionals*, to appear.

[11] Giaquinta, M., and Modica, G.: 1979, *Regularity results for some classes of higher order nonlinear elliptic systems*, J. Reine u. Ang. Math., 311/312, pp.145-169.

[12] Giaquinta, M., and Nečas, J.: 1979, *On the regularity of weak solutions to nonlinear elliptic systems via Liouville's type property*, Comm. Math. Univ. Carol, 20, pp.111-121.

[13] Giaquinta, M., and Nečas, J.: 1979, *On the regularity of weak solutions to nonlinear elliptic systems of partial differential equations*, J. Reine u. Ang. Math., 316, 140-159.

[14] Giaquinta, M., Nečas, J., John, O., and Starà, J.: 1982, *On the regularity up to the boundary for second order nonlinear elliptic systems*, Pac. J. of Math., 99, pp.1-17.

[15] Giusti, E.: 1969, *Precisazione delle funzioni $H^{1,p}$ e singolarità delle soluzioni deboli di sistemi ellittici nonlineari*, Boll. Un. Mat. Ital., 2, pp.71-76.

[16] Giusti, E.: 1969, *Regolarità parziale delle soluzioni di sistemi ellittici quasilineari di ordine arbitrario*, Ann. Sc. Norm. Sup. Pisa, 23, pp.115-141; 1973, *Un'aggiunta alla mia nota*, Ann. Scuola Norm. Sup., Pisa, 27, pp.161-166.

[17] Giusti, E.: 1977, *Minimal surfaces and functions of bounded variation*, Notes on Pure Math., 10.

[18] Giusti, E., and Miranda, M.: 1968, *Un esempio di soluzioni discontinue per un problema di minimo relativo ad un integrale regolare del calcolo delle variazioni*, Boll. Un. Mat. Ital., 2, pp.1-8.

[19] Giusti, E., and Miranda, M.: 1968, *Sulla regolarità delle soluzioni deboli di una classe di sistemi ellittici quasi lineari*, Arch. Rat. Mech. & Anal., 31, pp.173-184.

[20] Ivert, P.A.: 1979, *Regularitätsuntersuchungen von Lösungen elliptischer systeme von quasilinearen differentialgleichungen zweiter ordnung*, Manus. Math., 30, pp.53-88.

[21] Kawohl, B.: 1979, *On Liouville theorem, continuity and Hölder continuity of weak solutions to some quasilinear elliptic systems*, preprint, Darmstad.

[22] Maz'ya, V.G.: 1968, *Examples of non regular solutions of quasilinear elliptic equations with analytic coefficients*, (Russian) Funkt. Anal. i Ego Pril., 2. pp.53-57.

[23] Meier, M.: 1982, *Liouville theorems, partial regularity, and Hölder continuity of weak solutions to quasilinear elliptic systems*, preprint, Bonn.

[24] Morrey, C.B. Jr.: 1968, *Partial regularity results for nonlinear elliptic systems*, J. Math. & Mech., 17, pp.649-670.

[25] Nash, J.: 1958, *Continuity of solutions of parabolic and elliptic equations*, Am. J. of Math., 80, pp.931-954.

[26] Nečas, J.: 1977, *'Example of an irregular solution to a nonlinear elliptic system with analytic coefficients and conditions for regularity'*, in Theory of Nonlinear Operators, Abhandlungen Akad. der Wiss. der DDR.

[27] Nečas, J.: 1977, *On the regularity of weak solutions to variational equations and inequalities for nonlinear second order elliptic systems*, Equadiff 4, Praha, Springer-Verlag Lecture Notes No. 703.

[28] Nečas, J., John, O., and Starà, J.: 1980, *Counterexample to the regularity of weak solutions of elliptic systems*, Comm. Math. Univ. Carol., 21, pp.145-154.

[29] Schoen, R., and Uhlenbeck, K.: 1982, *A regularity theory for harmonic maps*, J. Diff. Geo., 17, pp.307-335.

[30] Uhlenbeck, K.: 1977, *Regularity for a class of nonlinear elliptic systems*, Acta Math., 138, pp.219-240.

QUASILINEAR ELLIPTIC SYSTEMS IN DIAGONAL FORM

S. Hildebrandt

Mathematisches Institut der Universität Bonn
Wegelerstrasse 10
D-5300 Bonn
Federal Republic of Germany

Firstly examples of systems in diagonal form appearing in differential geometry are considered. Then a survey on regularity results is given. Finally two proofs of a regularity theorem are given in some detail.

1. INTRODUCTION

Giaquinta and Giusti have considered in their lectures elliptic systems of the form

$$\begin{bmatrix} L_{11} & \cdots & L_{1N} \\ \vdots & & \vdots \\ L_{N1} & \cdots & L_{NN} \end{bmatrix} \begin{bmatrix} u^1 \\ \vdots \\ u^N \end{bmatrix} = \begin{bmatrix} f^1 \\ \vdots \\ f^N \end{bmatrix}$$

for vector-valued functions

$$u(x) = (u^1(x),\dots,u^N(x))$$

of n variables $x = (x^1,\dots,x^n) \in \Omega$ (= bounded domain in $\mathbb{R}^n$) where

$$L_{ij} = -D_\beta\{a^{\alpha\beta}_{ij}(x,u,Du)D_\alpha\}$$

and

$$f^i = f^i(x,u,Du) \quad .$$

J. M. Ball (ed.), Systems of Nonlinear Partial Differential Equations, 173–217.

As we have seen in their lectures, weak solutions will, in general, be irregular, and one can only hope to obtain *partial regularity results*.

However, one might expect better results for systems having a particular structure. Special features of this kind often occur in equations and boundary conditions derived from physical or geometrical problems. Quite often, one encounters *systems in diagonal form*

$$\begin{bmatrix} L & & 0 \\ & \ddots & \\ 0 & & L \end{bmatrix} \begin{bmatrix} u^1 \\ \vdots \\ u^N \end{bmatrix} = \begin{bmatrix} f^1 \\ \vdots \\ f^N \end{bmatrix}$$

where

$$L = -D_\beta\{a^{\alpha\beta}(x,u,Du)\} \quad .$$

We shall simply write

$$Lu^i = f^i(x,u,Du) \quad , \qquad 1 \leq i \leq N \quad ,$$

or

$$Lu = f(x,u,Du) \quad , \qquad f = (f^1,\ldots,f^N) \tag{1.1}$$

Definition 1. We say that $u(x)$ is a weak solution of (1.1) in Ω if $u \in H^{1,2} \cap L^\infty(\Omega,\mathbb{R}^N)$, and if

$$\int_\Omega a^{\alpha\beta}(x,u,Du)D_\alpha u \cdot D_\beta\varphi\, dx = \int_\Omega f(x,u,Du) \cdot \varphi\, dx \tag{1.2}$$

holds for all $\varphi \in C_c^\infty(\Omega,\mathbb{R}^N)$.

In the following, we shall make the following

Assumptions. Let $u(x)$ be a weak solution of (1.1) such that $A^{\alpha\beta}(x) := a^{\alpha\beta}(x,u(x),Du(x))$ and $f^i(x,u(x),Du(x))$ are measurable functions on Ω satisfying

$$|u| \leq M \tag{A1}$$

$$|A^{\alpha\beta}| \leq \mu, \ \lambda|\xi|^2 \leq A^{\alpha\beta}\xi_\alpha\xi_\beta \quad \text{for all} \quad \xi \in \mathbb{R}^n \tag{A2}$$

where $M \geq 0$, $\mu \geq 0$, and $\lambda > 0$.

Moreover, setting

$$Q(x,p) := A^{\alpha\beta}(x)\, p^i_\alpha p^i_\beta \quad \text{for} \quad p = (p^i_\alpha) \in \mathbb{R}^{nN} , \tag{1.3}$$

we suppose the *two-sided condition*

$$|f(x,u,p)| \leq a\, Q(x,p) + b \tag{A3}$$

as well as a *one-sided condition*

$$u(x) \cdot f(x,u(x),\nabla u(x)) \leq a^* Q(x,p) + b^* \tag{A4}$$

for $x \in \Omega$ and $p \in \mathbb{R}^{nN}$, where a,a^*,b,b^* denote non-negative constants.

Note that (A1) and (A3) imply (A4) with $a^* = aM$ and $b^* = bM$. Hence we may always assume that

$$a^* \leq aM . \tag{A5}$$

However it is not useless to consider (A4) separately from (A4) because it will turn out that the existence of an $a^* < aM$ is yielding better results.

Before we are going to describe some of the results recently obtained for systems in diagonal form, we shall exhibit a few examples of such systems arising in differential geometry and in physics.

2. EXAMPLES OF SYSTEMS IN DIAGONAL FORM

2.1. Isothermic conjugate parameters on a surface

Let $X(u,v)$ be a regular surface S in $\mathbb{R}^3$ with positive Gauß curvature K. Introduce by

$$u = u(x,y) \quad , \quad v = v(x,y)$$

new parameters on S such that the coefficients of the second fundamental form with respect to x and y satisfy

$$L = N \quad , \quad M = O \quad .$$

Then $u(x,y)$ and $v(x,y)$ must satisfy the following system discovered by Darboux:

$$\begin{aligned} -\Delta u &= (\Gamma^1_{11}+\tfrac{1}{2}D_u \log K)|Du|^2+\Gamma^1_{22}|Dv|^2+(2\Gamma^1_{12}+\tfrac{1}{2}D_v \log K)Du\cdot Dv \\ -\Delta v &= \Gamma^2_{11}|Du|^2+(\Gamma^2_{22}+\tfrac{1}{2}D_v \log K)|D_v|^2+(2\Gamma^2_{12}+\tfrac{1}{2}D_u \log K)Du\cdot Dv \end{aligned} \tag{2.1}$$

where

$$D = (D_x, D_y) \ , \ D_x = \frac{\partial}{\partial x} \ , \ D_y = \frac{\partial}{\partial y} \ , \ D_u = \frac{\partial}{\partial u} \ , \ D_v = \frac{\partial}{\partial v} \ .$$

Obviously, one is only interested in solutions of these equations which are diffeomorphisms. That is, we are looking for solutions u,v of (2.1) whose Jacobian

$$Du \wedge Dv = \frac{\partial(u,v)}{\partial(x,y)}$$

does not vanish.

2.2. Elliptic Monge-Ampère equations

Let $z(u,v)$ be a solution of an elliptic Monge-Ampère equation

$$F(u,v,z,p,q,r,s,t) = 0$$

where

$$F := Ar + 2Bs + Ct + (rt-s^2) - E$$

and where A,B,C,E are solutions of $u,v,z,p=z_u,q=z_v$ such that

$$AC - B^2 + E > 0 \quad ,$$

and where we have set $r = z_{uu}$, $s = z_{uv}$, $t = z_{vv}$. We try to introduce new variables x,y such that the mapping $u = u(x,y)$, $v = v(x,y)$ is a diffeomorphism, and that

$$F_t \, du^2 - F_s \, du \, dv + F_r \, dv^2 = \lambda\{dx^2 + dy^2\}$$

holds. Then $u(x,y)$, $v(x,y)$ have to satisfy a system of the form

$$\begin{aligned} -\Delta u &= h_1|Du|^2 + h_2|Dv|^2 + h_3 Du\cdot Dv + h_4 Du\wedge Dv \\ -\Delta v &= h_1^*|Du|^2 + h_2^*|Dv|^2 + h_3^* Du\cdot Dv + h_4^* Du\wedge Dv \end{aligned} \tag{2.2}$$

where $h_1,\ldots,h_4$ are algebraic expressions of A,B,C,E, and their first order derivatives.

For the embedding problems of Weyl and Minkowski, the functions $h_1,\ldots,h_4$ depend only on u and v. In such cases, H. Lewy and E. Heinz were able to use coordinate transformations

$$(x,y) \to (u(x,y),v(x,y))$$

furnished by solutions of (2.2) to obtain important a priori estimates for the solutions $z(u,v)$ of the Monge-Ampère equation $F=0$.

2.3. Surfaces of prescribed mean curvature in $\mathbb{R}^3$

Let $u(x,y) = (u^1(x,y), u^2(x,y), u^3(x,y))$, $u \in C^2(\Omega, \mathbb{R}^3)$, be a representation of a regular C^2-surface S in $\mathbb{R}^3$ such that

$$|u_x|^2 = |u_y|^2 \quad , \quad u_x \cdot u_y = 0 \quad \text{in} \quad \Omega \; . \tag{2.3}$$

Then

$$\Delta u = 2H(x,y)\; u_x \wedge u_y \quad \text{in} \quad \Omega \tag{2.4}$$

and $H(x,y)$ is the mean curvature of S in $u(x,y)$. Conversely, let $H = H(u)$ be a continuous function on $\mathbb{R}^3$, and let Γ be a closed curve in $\mathbb{R}^3$. Then we consider the so-called

Problem of Plateau: *Determine a surface* S, *bounded by* Γ, *which has the mean curvature* $H(u)$ *in* $u \in \mathbb{R}^3$.

Analytically, this means the following:

Determine a C^2*-mapping* $u\colon \Omega \to \mathbb{R}^3$ *of a domain* Ω *in* $\mathbb{R}^2$, *of the conformal type of the unit disc, which maps* $\partial\Omega$ *topologically onto* Γ, *and satisfies* (2.3) *as well as*

$$\Delta u = 2H(u)\; u_x \wedge u_y \quad \textit{in} \quad \Omega. \tag{2.5}$$

Note that the only irregular points of the surface S, represented by $u(x,y)$, are branch points (i. e. points where $|u_x(x,y)| = 0$), and it turns out that these points are isolated.

It is not difficult to see that the equations (2.5) are the Euler equations of every functional

$$E(u) = \int_\Omega \{ |Du|^2 + Q(u) \cdot (u_x \wedge u_y) \}\, dx\, dy \tag{2.6}$$

where $Q = (Q^1, Q^2, Q^3)$ satisfies

$$\operatorname{div} Q = 4H \quad , \tag{2.7}$$

and the conformality relations (2.3) follow too from $\delta E = 0$ by carrying out suitable variations of the independent variables x,y.

If "*Plateau's boundary condition*" $u\colon \partial\Omega \to \Gamma$ is replaced by $u\colon \partial\Omega \to S$, thus S being a *supporting surface* of the "*free boundary*" $u|_{\partial\Omega}$ of u, then one obtains from $\delta E = 0$ in addition the free boundary condition

$$\cos \alpha = \frac{1}{2}\, Q \cdot \nu \quad . \tag{2.8}$$

Here ν denotes the unit normal on the supporting surface S, and α stands for the angle of intersection between the surface $u(x,y)$ and S.

For free boundary problems in capillarity, H, S, and α are prescribed, and one has to determine $u: \Omega \to \mathbb{R}^3$ satisfying (2.3), (2.5), and (2.6).

Of particular interest are

minimal surfaces (soap films): $H \equiv 0$,

surfaces of constant mean curvature (soap bubbles): $H \equiv \text{const}$,

capillarity surfaces: H is a linear function of u,

rotating fluids.

A natural approach to Plateau's and to free boundary problems of the kind discussed before is to minimize a functional (2.6) in a suitable class of mappings where Q is chosen in such a way that (2.7) and - if α is prescribed - that (2.8) holds.

2.4. Minimal surfaces in a Riemannian manifold

These are surfaces which locally can be written as

$$u(x,y) = (u^1(x,y),\ldots,u^N(x,y)) \quad , \quad N \geqq 2 \quad ,$$

where

$$g_{iK}(u)\, u_x^i\, u_x^K = g_{iK}(u)\, u_y^i\, u_y^K \, , \; g_{iK}(u)\, u_x^i\, u_y^K = 0 \tag{2.9}$$

and

$$-\Delta u^\ell = \Gamma^\ell_{iK}(u)[u_x^i\, u_x^K + u_y^i\, u_y^K] \quad , \quad 1 \leqq \ell \leqq N \quad , \tag{2.10}$$

holds. Here,

$$ds^2 = g_{iK}(u)\, du^i\, du^K$$

is the Riemannian metric of the manifold, in which the minimal surface sits, and Γ^ℓ_{iK} are the Christoffel symbols of second kind with respect to (g_{iK}).

One notes that (2.10) are the Euler equations of the generalized Dirichlet integral

$$E(u) = \int g_{ik}(u)\{u_x^i\, u_x^k + u_y^i\, u_y^K\}dx\, dy \tag{2.11}$$

and the conformality conditions (2.9) follow once more from $\delta E = 0$ and Plateau or free boundary conditions. One might look for stationary points of E with respect to various boundary conditions:

(i) u be bounded by one or several curves $\Gamma_1, \dots, \Gamma_k$;

(ii) u be bounded by surfaces, or by surfaces and arcs (movable or not);

(iii) compact or complete surfaces u.

In particular, the case $N = 2$ leads to generalizations of Riemann's mapping theorem (for instance, to global versions of Lichtenstein's mapping theorem).

2.5. Harmonic mappings

These are C^2-mappings $U: X \to M$ of a Riemannian manifold X into a Riemannian manifold M, $n = \dim X$, $N = \dim M$ which are stationary points of the Dirichlet integral

$$E(U) = \int_X \| dU \|^2 \, d\,\mathrm{vol} \quad . \tag{2.12}$$

If we locally write U as $u(x) = (u^1(x), \dots, u^N(x))$, $x = (x^1, \dots, x^n)$, by using local coordinates x and u on X and M with the line elements

$$d\sigma^2 = \gamma_{\alpha\beta}(x)\, dx^\alpha\, dx^\beta \quad \text{and} \quad ds^2 = g_{iK}(u)\, du^i\, du^k \quad ,$$

respectively, we may write

$$E(U) = \int_X g_{iK}(u)\, \gamma^{\alpha\beta}(x)\, D_\alpha u^i\, D_\beta u^k \sqrt{\gamma(x)}\, dx \quad , \tag{2.13}$$

where $(\gamma^{\alpha\beta}) = (\gamma_{\alpha\beta})^{-1}$ and $\gamma = \det(\gamma_{\alpha\beta})$, and the harmonicity of U means that $u(x)$ satisfies the Euler equations

$$\Delta_X u^\ell + \Gamma^\ell_{iK}(u)\, \gamma^{\alpha\beta}(x)\, D_\alpha u^i\, D_\beta u^k = 0 \tag{2.14}$$

where Δ_X denotes the Laplace-Beltrami operator

$$\Delta_X = \frac{1}{\sqrt{\gamma}} D_\beta [\sqrt{\gamma}\, \gamma^{\alpha\beta}\, D_\alpha] \tag{2.15}$$

on X expressed in local coordinates $x = (x^1, \dots, x^n)$.

By virtue of Nash's embedding theorem, we may assume that M sits as an N-dimensional submanifold in an (N+k)-dimensional Euclidean space $\mathbb{R}^{N+k}$. Then harmonic maps $X \to M$ can also be described as mappings $u(x) = (u^1(x), \dots, u^{N+k}(x))$ which make

$$\int_X \gamma^{\alpha\beta}(x)\, D_\alpha u^i\, D_\beta u^i \sqrt{\gamma(x)}\, dx$$

stationary under the side condition

$$u(\mathcal{X}) \subset \mathcal{M} \quad .$$

For instance, if $\mathcal{X} \subset \mathbb{R}^n$, say $\mathcal{X} = \Omega \cup \Sigma$, $\Sigma = \partial\Omega$, then harmonic maps are critical points of the ordinary Dirichlet integral

$$\int_\Omega |Du|^2 \, dx$$

subjected to the condition $u(\mathcal{X}) \subset \mathcal{M}$.

An interesting example is furnished by the n-sphere

$$\mathcal{M} = S^n = \{u \in \mathbb{R}^{n+1} : |u| = 1\} \subset \mathbb{R}^{N+1} \quad .$$

Then the corresponding Euler equations are

$$-\Delta u = u|Du|^2 \quad . \tag{2.16}$$

A simple computation shows that

$$u(x) = \left(\frac{x}{|x|}, 0\right)$$

is a classical solution of (2.16) in $B_1(0) - \{0\}$, where $B_1(0) = \{x: |x| < 1\}$. Furthermore we obtain that, for $n \geqq 3$, u is of class $H^{1,2} \cap L^\infty$ on $B_1(0)$, and that u is a weak solution of (2.16) (cf. De Giorgi [8]).

That is, harmonic maps $U: \mathcal{X} \to \mathcal{M}$ may have isolated singularities if $\dim \mathcal{X} \geqq 3$ although their energy $E(U)$ is finite, and there exist "weakly harmonic" but irregular harmonic maps. This contrasts with the two-dimensional situation where Sacks and Uhlenbeck [98] have proved that every isolated singularity of a harmonic map with finite energy can be removed.

On the other hand, Giaquinta-Giusti [18], [19] and Schoen--Uhlenbeck [100] have recently proved that the Hausdorff dimension of the set Σ of singularities of a mapping $U: \mathcal{X} \to \mathcal{M}$ minimizing (2.12) with respect to Dirichlet boundary conditions is at most $n - 3$, where $n = \dim \mathcal{X}$, provided that $U \in H^{1,2} \cap L^\infty$, and that Σ consists of isolated points if $n = 3$.

However, $u(x) = \left(\frac{x}{|x|}, 0\right)$ cannot yield an absolute minimum of the Dirichlet integral

$$E(v) = \int_{B_1(0)} |Dv|^2 \, dx \quad , \quad B_1(0) \subset \mathbb{R}^n$$

in the class

$$\mathcal{C} := \{v \in H^{1,2}(B_1(0),\mathbb{R}^{n+1}) : |v| = 1,\ v(x) = (x,0) \text{ for } |x| = 1\}$$

if $n = 3$. In fact, Jäger-Kaul and Giaquinta-Giusti have recently proved that u is not even a relative minimum of E in $\mathcal{C}$ if $3 \leq n \leq 6$. On the other hand, Jäger-Kaul [64] were able to show that u minimizes E in $\mathcal{C}$ provided that $n \geq 7$. Hence one might conjecture that the Hausdorff dimension the set of singularities of an energy minimizing bounded map $U: \mathcal{X} \to \mathcal{M}$ is at most $n - 7$.

Let us consider some *further examples of harmonic maps:*

$n = 1$: in this case, the harmonic maps $U: \mathcal{X} \to \mathcal{M}$ are geodesic in $\mathcal{M}$;

$n = 2$: two-dimensional minimal surfaces $U: \mathcal{X} \to \mathcal{M}$ are harmonic maps;

$n \geq 2$: consider the Gauß map $\mathcal{G}: \mathcal{X} \to G(n,p)$ of an n-dimensional oriented submanifold $\mathcal{X}$ of $\mathbb{R}^{n+p}$ into the Grassmannian $G(n,p)$ equipped with the canonical Cartan metric; then $\mathcal{G}$ is harmonic iff and only if the mean curvature vector field of $\mathcal{X}$ is parallel (cf. Ruh-Vilms [97]).

In particular, $\mathcal{G}$ is harmonic if $\mathcal{X}$ is a minimal submanifold of $\mathbb{R}^{n+p}$.

Moreover, if $n = 2$ and $p = 1$, then $\mathcal{G}: \mathcal{X} \to G(2,1)$ is harmonic if and only if $\mathcal{X}$ has constant mean curvature. We mention that the Gauß map $\mathcal{G}: \mathcal{X} \to G(n,p)$ is that correspondance associating to each point $P \in \mathcal{X}$ its (oriented) tangent plane $T_P\mathcal{X} \in G(n,p)$.

For this reason, Liouville theorems for harmonic maps will imply Bernstein theorems for minimal submanifolds (cf. Hildebrandt--Jost-Widman [58]), and removability of singular points of harmonic maps implies in certain cases the removability of singular points of solutions of the minimal surface system (cf. Meier [86]). We close our observations on harmonic maps with the following

Remark. If $n = N = 2$, $\mathcal{X} = \mathbb{R}^2$, and if K is constant, then the equations (2.1) agree with the equations for harmonic mappings.

2.6. Two important technical applications

If $z = z(x)$ is a solution of

$$A^{\alpha\beta}(x,z,Dz)\,z_{x^\alpha x^\beta} = B(x,z,Dz),$$

or of

$$F(x,z,Dz,D^2z) = 0 \quad ,$$

then $u := Dz$ or $u := D^2z$, respectively, satisfies a diagonal system of the form (1.1) where the right hand side $f(x,u,Du)$ is of quadratic growth in Du.

Suppose that we can derive C^α-estimates for u in terms of $|u|_{L^\infty}$. Then we obtain estimates of $|z|_{C^{1,\alpha}}$ or $|z|_{C^{2,\alpha}}$, respectively, in terms of $|Dz|_{L^\infty}$ or $|D^2z|_{L^\infty}$. Therefore, good C^α-estimates for solutions of (1.1) will immediately be useful for the study of general nonlinear equations (cf. also [48], pp. 521-523).

3. REGULARITY RESULTS AND COUNTEREXAMPLES

3.1. *Why is* $H^{1,2} \cap L^\infty$ *the home of the weak solutions of* (1.1)?

Why don't we take u in $H^{1,2}$? - Of the various possible answers we give just one. Let us consider the boundary value problem

$$\begin{aligned} -\Delta u &= |\nabla u|^2 \quad \text{in } B_R(0) \;, \; 0 < R < 1 \;, \\ u &= 0 \quad \text{on } \partial B_R(0) \;, \end{aligned} \tag{3.1}$$

for $n = 2$ which has the two weak solutions $u \equiv 0$ and $u(x) = \log\log\frac{1}{|x|} - \log\log\frac{1}{R}$. Thus there is no hope to prove a regularity result for weak solutions in $H^{1,2}$ which are unbounded, even in the case of one equation with two independent variables.

Moreover, Frehse [15] has pointed out that even the Euler equation of a variational integral may have an unbounded $H^{1,2}$-solution for $n = 2$, $N = 1$. For instance, $u(x) = 12 \log\log\frac{1}{|x|}$ is a weak $H^{1,2}$-solution of the Euler equation of

$$\int_{B_R(0)} [1+g(x,u)]\,|\nabla u|^2\, dx \quad , \quad 0 < R < 1 \quad , \tag{3.2a}$$

where

$$g(x,u) = \frac{1}{1+(\exp u)(\log |x|)^{-12}} \tag{3.2b}$$

Therefore, we shall understand weak solutions u of (1.1) as in Definition 1, that is, $u \in H^{1,2} \cap L^\infty$.

3.2. Theorem of Ladyshenskaya-Uralceva: N = 1

It turns out that every weak solution of a scalar equation (1.1) is regular without any quantitative restriction for the numbers a and M. This result is based on the following key theorem [75]:

Let u *be a solution of* (1.1) *in* Ω *satisfying* (A1), (A2), (A3), *and suppose that* $N = 1$. *Then* u *is Hölder continuous in* Ω.

In fact, this result even holds for equations of the form

$$-D_\alpha\{A^\alpha(x,u,Du)\} = f(x,u,Du) \quad , \tag{3.3}$$

if we assume "natural growth conditions" for f and A^α. Moreover, Ladyshenskaya and Uralceva have proved that absolute minima of

$$\int_\Omega F(x,u,Du)\, dx$$

which are in $H^{1,2}(\Omega,\mathbb{R})$, $N = 1$, have to be bounded on Ω if

$$m_1|p|^2 - k \leqq F(x,u,p) \leqq m_2|p|^2 + k \quad ,$$

$0 < m_1 \leqq m_2$, $k > 0$, and if u is bounded on $\partial\Omega$.

Thus the scalar case is satisfactory treated. For systems ($N > 1$) the situation is quite different.

3.3 The analogue of the L-U-Theorem does not hold for N > 1, even if n = N = 2

This can be seen from the following example due to Frehse [14]:

Let $n = N = 2$, $x = (x^1,x^2)$, $u = (u^1,u^2)$, and $0 < R < 1$. Then

$$u^1 = \sin(\sigma \log\log \frac{1}{|x|}) \quad , \quad u^2 = \cos(\sigma\log\log\frac{1}{|x|}), \sigma \neq 0, \tag{3.4}$$

is a weak solution of

$$\left.\begin{aligned} -\Delta u^1 &= (u^1 + \frac{1}{\sigma}u^2)|\nabla u|^2 \\ -\Delta u^2 &= (u^2 - \frac{1}{\sigma}u^1)|\nabla u|^2 \end{aligned}\right\} \text{ in } B_R(0) \quad . \tag{3.5}$$

which is discontinuous.

For $n = N \geqq 3$ we have seen that $u(x) = \frac{x}{|x|}$ is a weak solution of

$$-\Delta u = u|\nabla u|^2 . \tag{3.6}$$

3.4. Morrey's Theorem and Frehse's counterexample

Let $F(x,u,p)$ be a continuous function on $\Omega \times \mathbb{R}^N \times \mathbb{R}^{nN}$ satisfying

$$m_1|p|^2-k \leqq F(x,u,p) \leqq m_2|p|^2+k \; , \; 0 < m_1 \leqq m_2, \; k \geqq 0, \tag{3.7}$$

and set

$$I_\Omega(u) := \int_\Omega F(x,u,Du)\, dx \; . \tag{3.8}$$

Following Giaquinta and Giusti, we call u a quasiminimum (or Q-minimum) of the functional I_Ω if there is a number $Q \geqq 1$ such that

$$I_{\Omega'}(u) \leqq Q \cdot I_{\Omega'}(u+\varphi) \tag{3.9}$$

holds for all $\varphi \in H^{1,2}(\Omega,\mathbb{R}^N)$ with $\operatorname{supp}\varphi \subset \Omega' \subset\subset \Omega$. Let $n \geqq 2$, $N \geqq 1$. Then Morrey [90] has proved:

Suppose that $u \in H^{1,2}_{loc}(\Omega,\mathbb{R}^N)$ *is a Q-minimum of* I_Ω. *Then there exist constants* K *and* $\alpha \in (0,1)$ *depending only on* m_1, m_2, k, *and* Q, *such that*

$$\int_{B_R(x_o)} |Du|^2\, dx \leqq K \cdot (R/R_o)^2 \tag{3.10}$$

for $x_o \in \Omega$ *and for* $0 < R \leqq R_o \leqq \operatorname{dist}(x_o,\partial\Omega)$. *If* $n = 2$, *this implies* $u \in C^{0,\alpha}(\Omega,\mathbb{R}^N)$.

Proof by hole-filling [127]: For the sake of simplicity we assume $k = 0$, although this is not essential.
Let $\varphi \in H^{1,2}(\Omega',\mathbb{R}^N)$, and $\operatorname{supp}\varphi \subset \Omega' \subset\subset \Omega$. Then

$$m_1 \int_{\Omega'} |Du|^2 dx \leqq F_{\Omega'}(u) \leqq Q\, F_{\Omega'}(u+\;) \leqq m_2 Q \int_{\Omega'} |Du+D\varphi|^2\, dx$$

whence for $Q^* := 2(m_2/m_1)Q$

$$\int_{\Omega'} |Du|^2\, dx \leqq (Q^*/2) \int_{\Omega'} |Du + D\varphi|^2\, dx \tag{3.11}$$

holds.

Let $\Omega' = B_{2R} = B_{2R}(x_o) \subset\subset \Omega$. Then there is a function $\eta \in \operatorname{Lip}(B_{2R},\mathbb{R})$ with $0 \leqq \eta \leqq 1$, $\eta \equiv 1$ on B_R, $\eta \equiv 0$ on $\Omega - B_{2R}$, and $|D\eta| \leqq 1/R$. Moreover, there is a number $c > 0$, independent of R, such that *Poincaré's inequality*

$$\int_{T_{2R}} |u-\omega_R|^2 dx \leq cR^2 \int_{T_{2R}} |Du|^2 \, dx \tag{3.12}$$

holds, where $T_{2R} := B_{2R} - B_R$, and where ω denotes the mean value

$$\omega_R := \fint_{T_{2R}} u(x) \, dx \, . \tag{3.13}$$

Then we infer from (3.11) and (3.12) that

$$\int_{B_{2R}} |Du|^2 dx \leq (Q^*/2) \int_{B_{2R}} |(1-\eta)Du + D\eta(\omega_R-u)|^2 \, dx$$

$$\leq Q^* \int_{B_{2R}} (1-\eta)^2 |Du|^2 dx + Q^* \int_{B_{2R}} |D\eta|^2 |u-\omega_R|^2 \, dx$$

$$\leq Q^* \int_{T_{2R}} |Du|^2 \, dx + Q^* R^{-2} \int_{T_{2R}} |u-\omega_R|^2 \, dx$$

$$\leq Q^*(1+c) \int_{T_{2R}} |Du|^2 \, dx \quad .$$

Set $q := (1+c)\, Q^* - 1$, and

$$\Phi(R) := \int_{B_R} |Du|^2 \, dx \quad .$$

Then we arrive at

$$\Phi(R) \leq \frac{q}{1+q} \Phi(2R) \tag{3.14}$$

Iteration yields (3.10) with

$$\alpha := {}_2\log\,(1+\tfrac{1}{q}) > 0 \quad , \qquad \text{q.e.d.}$$

One might think that also stationary points u of the functional I_Ω are regular, at least, if one assumes besides (3.7) that the ellipticity condition

$$\lambda|\zeta|^2 \quad F_{p^i_\alpha p^K_\alpha}(x,u,p)\, \zeta^i_\alpha\, \zeta^K_\beta \leq \lambda'|\zeta|^2 \quad ,$$

$0 < \lambda \leq \lambda'$, holds, and that $u \in H^{1,2} \cap L^\infty(\Omega,\mathbb{R}^N)$. However, this is not the case. Frehse [17] has constructed an example of such an integral which possesses a discontinuous but bounded $u \in H^{1,2}$ as critical point where $n = N = 2$.

3.5. Conformally invariant integrals. Grüter's Theorem.

Despite of Frehse's counterexample, there are some results for the case $n = 2$ yielding the regularity of stationary points of I_Ω if one poses further restrictions on F besides (3.7).

For instance, by exploiting the isoperimetric inequality, Wente [120] has proved that, for $H \equiv \text{const}$, each $H^{1,2}(\Omega,\mathbb{R}^3)$-solution of (2.5) is regular. Tomi [116] proved that each $H^{1,2}\cap L^\infty$-solution of (2.5) is regular, if H is Lipschitz continuous.

Furthermore, Heinz [41] showed that each $H^{1,2}$-solution of (2.5) is in $C^2(\Omega-\Sigma,\mathbb{R}^3)$, where $\Omega-\Sigma$ is open and $\text{meas}\ \Sigma = 0$, provided that

$$\sup_{\mathbb{R}^3} [H(u)+(1+|u|)|DH(u)|] < \infty .$$

Grüter [22, 23, 24] pointed out that

(i) even the Hausdorff dimension of Σ is zero

and that

(ii) Σ has to be empty if it only consists of isolated points in Ω.

(The last remark is proved by a technique introduced by Sacks-Uhlenbeck [98] for the treatment of harmonic mappings).

This leads to the following <u>conjecture</u>:

All critical points $u \in H^{1,2}(\Omega,\mathbb{R}^N)$, $\Omega \subset \mathbb{R}^2$, *of*

$$I(u) = \int_\Omega F(u,Du)\,dx \tag{3.15}$$

are continuous maps if the integral I *is conformally invariant.*

Grüter has pointed out [25], that the integrands of conformally invariant integrals (3.15) have the form

$$F(u,p) = g_{ik}(u)\, p^i \cdot p^k + Q_{ik}(u)\, \det(p^i,p^k) \tag{3.16}$$

where (g_{ik}) is symmetric and positive definite, and (Q_{ik}) is skew-symmetric, if we assume that $F \in C^2$ and satisfies

$$m_1|p|^2 \leq F(u,p) \leq m_2|p|^2 \quad , \quad 0 < m_1 \leq m_2 \quad .$$

An important step towards the afore mentioned conjecture is the following result due to Grüter [23, 24]:

Suppose that $F(u,p)$ *is of the form* (3.16) *where* $Q_{ik} \in C^1$ *and*

$g_{ik} \in C^3$ *on* $\mathbb{R}^3$. *Moreover, let* $u \in H^{1,2} \cap L^\infty(\Omega, \mathbb{R}^N)$ *be a critical point of the integral* (3.15) *which, in addition, satisfies the conformality relations*

$$g_{ik}(u)\, u^i_{x^1} u^K_{x^1} = g_{iK}(u)\, u^i_{x^2} u^K_{x^2}\,,\quad g_{iK}(u)\, u^i_{x^1} u^K_{x^2} = 0 \text{ in } \Omega.$$

Then u *is continuous on* Ω *(which, in turn, implies higher regularity).*

The assumption $u \in H^{1,2} \cap L^\infty$ can be replaced by $u \in H^{1,2}$ if one adds some further conditions. For instance, if $N = 3$, $g_{iK} = \delta_{iK}$, and $H := \frac{1}{8}\left[\frac{\partial Q_{23}}{\partial u^1} + \frac{\partial Q_{31}}{\partial u^2} + \frac{\partial Q_{12}}{\partial u^3}\right]$, it suffices to assume $u \in H^{1,2}(\Omega, \mathbb{R}^3)$ and $H(u) \in L^\infty(\Omega, \mathbb{R})$. Further details can be found in [23] and [24].

Grüter's technique has recently been used in [9] and [26] to prove boundary regularity of minimal surfaces with free boundaries. The proof is indirect and does not lead to a priori estimates. In fact, one can in certain situations see that a priori estimates do not exist (cf. [49]).

3.6. Some regularity results for $n \geq 2$

Suppose now that u is a weak solution of (1.1) in $\Omega \subset \mathbb{R}^n$ satisfying (A1), (A2), (A3), (A4). As we have seen, it is no loss of generality to assume

$$a^* \leq aM \quad . \tag{A5}$$

Let us now consider the three conditions

$$(1)\quad aM < 1\ ,\quad (2)\quad aM + a^* < 2\ ,\quad (3)\quad a^* < 1 \quad .$$

Clearly, (1) ⇒ (2) ⇒ (3), because of (A5). Hence, a regularity theorem with the assumption (3) is the strongest, and it is the weakest with (1), while (2) yields a result somewhere between (1) and (3).

One verifies without difficulty that $v := |u|^2$ is a subsolution for the operator $L = -D_\beta\{A^{\alpha\beta}D_\alpha\}$ on Ω, if (3) holds. *Hence one might conjecture that* (3) *is sufficient to prove continuity of* u, whence also (1) and (2) would imply regularity. Note that the assumptions (1), (2), (3) are optimal in the sense that one can, in general, not prove that u is regular if one of the three conditions (1), (2), (3) is violated. For instance, if $n = N \geq 3$, the solution $u(x) = |x|^{-1}x$ of

$$-\Delta u = u|Du|^2$$

yields an example where $a = a^* = M = 1$, i.e., (1), (2), and (3) just fail to hold.

Further weight to the afore stated conjecture is added by the following results proved in [54]:

Suppose that u *satisfies* $a^* < 1$. *Then the following statements hold:*

(i) $\bar{u}(x) := \lim_{R\to 0} ⨍_{B_R(x)} u(y)\,dy$ *exists for all* $x \in \Omega$.

Hence we have a representation of u *by an* ACL-*function* $\bar{u}$ *satisfying*

$$\bar{u}(x) = \lim_{R\to 0} ⨍_{B_R(x)} \bar{u}(y)\,dy \quad \textit{for all} \quad x \in \Omega.$$

For this reason, we shall not anymore distinguish between u *and its particular representation* $\bar{u}$.

(ii) *Each point* x *of* Ω *is a Lebesgue point of* u, *i.e.,*

$$\lim_{R\to 0} ⨍_{B_R(x)} |u(x) - u(y)|^2\,dy = 0 \quad \textit{for all} \quad x \in \Omega.$$

(iii) *If* v *is a solution of* $Lv = 0$ *in* Ω *such that* $u - v \in \mathring{H}^{1,2}(\Omega, \mathbb{R}^N)$, *then* v *is Hölder continuous in* Ω, *and*

$$u(x) = v(x) + \int_\Omega G(y,x)\, f(y,u(y),Du(y))\,dy \quad \textit{for all} \quad x \in \Omega,$$

where $G(x,y)$ *is Green's function for* L *in* Ω.

(iv) u *is continuous on a dense open subset of* Ω.

(v) *For all* $y \in \Omega$, *the following estimate holds:*

$$\int_\Omega G(x,y)\,|Du(x)|^2\,dx \le \frac{\mathrm{osc}_\Omega^2 |u|^2}{2\lambda(1-a^*)} + K \cdot [\mathrm{meas}\ \Omega]^{2/n}$$

where $K = K(\lambda,\mu,M,b^*)$ *is zero if* $b^* = 0$.

<u>Despite these most promising results, the conjecture is false if $n \ge 3$!</u>

This follows from examples by Meier [83] and Struwe [114]. Meier showed that $a^* < 1$, $b = b^* = 0$ does not imply the Liouville property for (1.1) (i. e. there are equations of this form which possess weak solutions $u \in H^{1,2}_{loc} \cap L^\infty(\mathbb{R}^n, \mathbb{R}^N)$ on $\mathbb{R}^n$ which are non-constant and satisfy $a^* < 1$ and $b = b^* = 0$). However, Ivert [61] noted that one can derive from a priori estimates (which depend only on $a, a^*, \lambda, \mu, M, n, N$) by a simple scaling argument the Liouville property. Thus there exist no a priori estimates of the modulus of continuity for weak solutions if $n \geq 3$. Recently, Struwe has even produced an irregular solution for $n \geq 3$ with $a^* < 1$.

Let us sketch how Meier's example is constructed: Firstly, he construes a function $v \in C^2(\mathbb{R}^n)$, $n \geq 3$, with $v \not\equiv \text{const}$, $1 \leq v \leq c$, $\int_{\mathbb{R}^n} |Dv|^2 dx < \infty$, $\Delta v \leq 0$, and $\Delta v \equiv 0$ for $|x| \geq 1$, as well as $|Dv| > 0$ for $|x| \geq 1$. Secondly, he sets $u(x) = (u^1(x), u^2(x))$, where

$$u^1 := v(x)$$
$$u^2 := v(x - x_o)$$

and where x_o is a fixed point in $\mathbb{R}^n$ with $|x_o| = 2$. Then, for each $R > 0$, there is a number $\alpha(R) > 0$ such that

$$|Du(x)|^2 \geq \alpha(R) \quad \text{if} \quad |x| \leq R \quad .$$

Thus the functions h^1, h^2, defined by

$$h^1(x) := \frac{-\Delta v(x)}{|Du(x)|^2} \quad , \quad h^2(x) := \frac{-\Delta v(x - x_o)}{|Du(x)|^2} \quad ,$$

are continuous on $\mathbb{R}^n$ and satisfy $h^i \leq 0$ as well as $h^i(x) \equiv 0$ for $|x| \geq 3$, $i = 1,2$.

Thence,

$$1 \leq u^i(x) \leq c \quad \text{and} \quad -\beta \leq h^i(x) \leq 0$$

for some number $\beta > 0$, where $x \in \mathbb{R}^n$ and $i = 1,2$, and we obtain

$$-\Delta u^1 = h^1(x)\ |Du|^2 =: f^1(x, Du)$$
$$-\Delta u^2 = h^2(x)\ |Du|^2 =: f^2(x, Du)$$

and

$$u^1(x)f^1(x,Du(x)) + u^2(x)f^2(x,Du(x)) \leqq 0$$

that is, $a^* = 0$.

What about the case $n = 2$?
Meier [83] proved that in this case one can obtain the Liouville property by assuming $a^* < 1$, $b = b^* = 0$, and Wiegner [132] was able to show that $a^* < 1$ even implies interior regularity of weak solutions. Thus it is quite astounding that $a^* < 1$ does not yield boundary regularity on account of a counterexample which has recently been found by Meier [88].

What is true for $n \geqq 3$?
The following result grew out of a succession of papers by Hildebrandt-Widman and by Wiegner ([53], [129], [130], [54]; cf. also [48], section 6, for a detailed historical account):

Theorem. *The condition* $aM + a^* < 2$ *(and hence also* $aM < 1$*) implies Hölder continuity of* u *in the interior, and also at the boundary of* Ω *if we assume that* $\partial\Omega$ *and* $u|_{\partial\Omega}$ *are sufficiently regular. In fact, one can derive a priori estimates for the Hölder norms on interior subdomains and in the neighborhood of boundary points.*

In section IV we shall present a proof of this theorem based upon the approach of [54, 57]. We follow the exposition in [21] very closely except that we prove the smallness of
$R^{2-n} \int_{B_R(x_o)} Q(x,Du)\,dx$ for some small R by a reasoning which employs Green's function for L. This argument replaces the somewhat circumstantial approach in the proof of Lemma 4.4 in [54] by a straight forward attack.

Moreover, we shall describe in section V the new proof of the afore stated theorem which recently was obtained by Caffarelli [6]. We shall only consider interior regularity.

4. INTERIOR REGULARITY BY MEANS OF THE GREEN FUNCTION.

We assume once more that u is a weak solution of (1.1) in $\Omega \subset \mathbb{R}^n$ satisfying (A1-4), and that w.l.o.g. $a^* \leqq aM$. Then we state first some properties of Green's function $G(x,y)$ for the operator $L = -D_\beta[A^{\alpha\beta}(x)D_\alpha]$, where $A^{\alpha\beta}(x) := a^{\alpha\beta}((x),u(x),Du(x))$, the proof of which can be found in [27].

There is a function $G(\cdot,y) \in \mathring{H}^{1,1}(B_{2R}(x_o),\mathbb{R})$ on each ball

$B_{2R}(x_o)$ in Ω and for each $y \in B_{2R}(x_o)$ such that

$$\int A^{\alpha\beta} D_\alpha \varphi \, D_\beta G(\cdot,y) \, dx = \varphi(y) \quad \text{for all} \quad \varphi \in C_c^\infty(B_{2R}(x_o),y).$$

Moreover, there exist numbers K_1, K_2, $K_3 > 0$ depending only on $n \geq 3$, λ, and μ, but not on x_o and R, such that

$$G(x,y) \geq K_1 |x-y|^{2-n} \quad \text{for} \quad x,y \in B_R(x_o) \quad , \tag{G1}$$

$$0 \leq G(x,y) \leq K_2 |x-y|^{2-n} \quad \text{for} \quad x,y \in B_{2R}(x_o) \quad , \tag{G2}$$

$$0 \leq G^\rho(x,y) \leq K_2 |x-y|^{2-n} \quad \text{for} \quad x,y \in B_{2R}(x_o) \quad , \tag{G3}$$

where

$$G^\rho(\cdot,y) := \fint_{B_\rho(y)} G(\cdot,z) \, dz \quad ,$$

and where $\rho > 0$ is sufficiently small, say $\rho < 2R - |y-x_o|$. Furthermore, $G^\rho(\cdot,y) \in \mathring{H}^{1,2} \cap L^\infty(B_{2R}(x_o),\mathbb{R})$,

$$\lim_{\rho \to +0} G^\rho(x,y) = G(x,y) \quad , \quad x \neq y \quad , \tag{G4}$$

and

$$\int_{T_{2R}(x_o)} |D\, G^\rho(x,y)|^2 dx \leq K_3 \, R^{2-n} \quad \text{for} \quad |y-x_o| < R/2. \tag{G5}$$

In addition,

$$\int_{B_{2R}} A^{\alpha\beta}(x) \, D_\alpha \varphi(x) \, D_\beta G^\rho(x,y) \, dx = \fint_{B_\rho(y)} \varphi(x) \, dx. \tag{G6}$$

Finally we have

$$\sup_{B_{2R}(x_o)} v \leq K_4 \cdot \inf_{B_{2R}(x_o)} v \tag{H}$$

for each weak solution $v \geq 0$ of $Lv = 0$ in $B_{2R}(x_o) \subset \Omega$.

Lemma 1. *There is a number* $c > 0$ *depending only on* a^*, b^*, λ, μ, n, *and* M *such that the following holds:*
For each $\varepsilon > 0$ *and for each ball* $B_R(x_o) \subset \Omega$ *with* $R_o^2 < \frac{c}{2\varepsilon}$,

there exists an index $i_o \in \{1,2,\dots,p\}$ *with* $p := \left[\frac{2cM^2}{\varepsilon}\right] + 1$ *such that*

$$R^{2-n} \int_{B_R(x_o)} Q(x,Du)\,dx < \varepsilon$$

where $R := 2^{-i_o} R_o$, *provided that* $a^* < 1$.

<u>Proof</u>. We insert $\varphi = u\, G^\rho(\cdot,y)$, $y \in B_{2R}(x_o)$, in (1.2) and subtract from the resulting equation the relation

$$\int_{B_{2R}(x_o)} A^{\alpha\beta} D_\alpha w\, D_\beta G^\rho(\cdot,y)\,dx = 0 \quad ,$$

where w is the solution of

$$Lw = 0 \quad \text{in} \quad B_{2R}(x_o) \quad , \quad w - u \in \mathring{H}^{1,2}(B_{2R}(x_o),\mathbb{R}) \quad .$$

Then

$$2 \int_{B_{2R}(x_o)} [A^{\alpha\beta} D_\alpha u\, D_\beta u - u\cdot f(x,u,Du)] G^\rho(x,y)\,dx$$

$$\leqq \fint_{B_\rho(y)} (w - |u|^2)\,dx \quad ,$$

and (A3) yields

$$2(1-a^*) \int_{B_{2R}(x_o)} Q(x,Du) G^\rho(x,y)\,dx$$

$$\leqq \fint_{B_{2R}(x_o)} (w-|u|^2)\,dx + 2b^* \int_{B_{2R}(x_o)} G^\rho\,dx \quad .$$

By letting ρ tend to zero, Fatou's lemma implies that

$$\int_{B_{2R}(x_o)} Q(x,Du) G(x,y)\,dx \leqq \frac{w(y) - |u(y)|^2}{2(1-a^*)} + KR^2 \tag{4.1}$$

where K always denotes a number independent of R. By virtue of (G1) and of the maximum principle, we infer that

$$R^{2-n} \int_{B_R(x_o)} Q(x,Du)\,dx \leqq \frac{M(2R) - M(R)}{2(1-a^*)K_1} + KR^2 \tag{4.2}$$

where we have set

$$M(R) := \sup_{B_R(x_o)} |u|^2 .$$

Set $c := \max \left\{ \frac{1}{2(1-a^*)K_1}, K \right\}$. Then we arrive at

$$R^{2-n} \int_{B_R(x_o)} Q(x,Du)dx \leq c[M(2R)-M(R)] + cR^2 . \tag{4.3}$$

For $M_i := M(2^{-i}R_o)$, $i = 1,2,\ldots,$ we find that

$$(M_o-M_1) + (M_1-M_2) + \ldots + (M_{p-1}-M) \leq M^2 .$$

Thus we may conclude:
For each $p = 1,2,3,\ldots,$ there is an index $i_o \in \{1,2,\ldots,p\}$ such that $M_{i_o-1} - M_i \leq M^2/p$.

Then we obtain the statement of the lemma since

$$c[M_{i_o-1}-M_i] + cR^2 \leq \frac{cM^2}{p} + cR_o^2 < \frac{\varepsilon}{2} + \frac{\varepsilon}{2} = \varepsilon .$$

From (4.1), we obtain the following

Corollary 1. *If* $a^* < 1$, *then for* $y \in B_{2R}(x_o)$:

$$\int_{B_{2R}(x_o)} Q(x,Du)G(x,y)dx \leq \frac{\operatorname{osc}_{B_{2R}(x_o)} |u|^2}{2(1-a^*)} + KR^2 , \tag{4.4}$$

and for $y \in B_R(x_o)$:

$$\int_{B_R(x_o)} |x-y|^{2-n}|Du(x)|^2dx \leq \frac{\operatorname{osc}_{B_{2R}(x_o)} |u|^2}{2\lambda(1-a^*)K_1} + KR^2 , \tag{4.5}$$

if $n \geq 3$.

Lemma 2. *There exist numbers* $\sigma \in (0,1)$ *and* $K > 0$ *depending only on* n, N, λ, μ, M, a, b, $\operatorname{diam} \Omega$, *and on* $\operatorname{osc}_\Omega u$, *such that the Hölder seminorm* $[u]_{\sigma,\overline{\Omega'}}$ *is estimated by*

$$[u]_{\sigma,\overline{\Omega'}} \leq Kd^{-\sigma}$$

for each $\Omega' \subset\subset \Omega$ *with* $\text{dist}(\Omega', \partial\Omega) \geq d$, *provided that*

$$a \cdot \text{osc}\, u < 1 \quad .$$

The number K *does not depend on* diam Ω *if* $b = 0$.

Proof (for $n \geq 3$): Suppose that $B_{2R} := B_{2R}(x_o) \subset \Omega$, and set $T_{2R} := B_{2R} - B_R$, $\omega_R := \fint_{T_{2R}} u(x)dx$. Let η be a Lipschitz function with $0 \leq \eta \leq 1$, $|D\eta| \leq 2/r$, $\eta(x) \equiv 1$ for $|x-x_o| \leq 5R/4$, and $\eta(x) \equiv 0$ for $|x-x_o| \geq 7R/4$. Finally let $G^\rho(\cdot, y)$ be the regularized Green function for L on B_{2R}, $y \in B_{2R}$, where $0 < \rho < R$.

Inserting $\varphi = [u-\omega_R]G^\rho(\cdot, x_o)\eta^2$ into (1.2), we obtain

$$(1-a\cdot\text{osc}_\Omega u) \int_{B_{2R}} Q(x,Du)G^\rho(x,x_o)\eta^2\, dx + \frac{1}{2} \fint_{B\ (x_o)} |u-\omega_R|^2\, dx$$

$$\leq I + II + KR^2$$

where

$$I = -\int_{T_{2R}} A^{\alpha\beta}\, D_\alpha u\cdot (u-\omega_R)2\eta\, D_\beta\eta\, G^\rho(\cdot, x_o)\, dx$$

and

$$II = \int_{T_{2R}} A^{\alpha\beta}\, D_\alpha\eta\, D_\beta G^\rho(\cdot, x_o)\, \eta\, |u-\omega_R|^2\, dx$$

since

$$\int_{B_{2R}} A^{\alpha\beta} D_\alpha[\eta^2|u-\omega_R|^2]\, D_\beta G^\rho(\cdot, x_o)dx = \fint_{B\ (x_o)} |u-\omega_R|^2\, dx \quad .$$

By applying Schwarz' inequality and Poincaré's inequality (3.12), we get

$$|I| \leq K \int_{T_{2R}} |Du|^2\eta^2 G(\cdot, x_o)dx + \int_{T_{2R}} |u-\omega_R|^2 G^\rho(\cdot, x_o)|D\eta|^2\, dx$$

$$\leq K\, R^{2-n}\Big\{\int_{T_{2R}} |Du|^2\, dx + R^{-2}\int_{T_{2R}} |u-\omega_R|^2\, dx\Big\}$$

$$\leq K\, R^{2-n} \int_{T_{2R}} Q(x,Du)\, dx$$

and

$$|II| \leq R^{n-2} \int_{T^*_{2R}} |DG^\rho(\cdot,x_o)|^2 |u-\omega_R|^2 dx + R^{2-n} KR^2 \int_{T_{2R}} |u-\omega_R|^2 dx ,$$

since $D\eta \equiv 0$ on $T^*_{2R} := B_{\frac{7R}{4}} - B_{\frac{5R}{4}}$. We note that $T^*_{2R} \subset\subset T_{2R}$, and that $\mathrm{dist}(T^*_{2R}, \partial T_{2R}) = R/4$. Moreover, $\Psi(x) := G^\rho(x,x_o)$ is a solution of $L\Psi = 0$ in T_{2R}. Hence taking (3.12) and the Caccioppoli-Moser inequality into account, we get

$$\int_{T_{2R}} |u-\omega_R|^2 |DG^\rho(\cdot,x_o)|^2 \, dx$$

$$\leq KR^{-2} \int_{T_{2R}} |u-\omega_R|^2 |G^\rho(\cdot,x_o)|^2 dx + K \int_{T_{2R}} |G^\rho(\cdot,x_o)|^2 |Du|^2 \, dx$$

$$\leq KR^{2(2-n)} \int_{T_{2R}} |Du|^2 \, dx ,$$

whence also

$$|II| \leq KR^{2-n} \int_{T_{2R}} Q(x,Du) \, dx .$$

Thus we conclude that

$$\int_{B_R} Q(x,Du) G^\rho(x,x_o) dx \leq KR^{2-n} \int_{T_{2R}} Q(x,Du) \, dx + KR^2 ,$$

and therefore

$$\int_{B_R} Q(x,Du) G(x,x_o) \, dx \leq KR^{2-n} \int_{T_{2R}} Q(x,Du) \, dx + KR^2 ,$$

if we let ρ tend to zero and take Fatou's lemma into account. But this inequality implies that

$$\int_{B_R} |x-x_o|^{2-n} |Du(x)|^2 dx \leq K \int_{B_{2R}-B_R} |x-x_o|^2 |Du(x)|^2 dx + KR^2 .$$

By "filling the hole" on the right hand side, we arrive at

$$\Phi(R) \leq \theta \cdot \Phi(2R) + \theta \cdot R^2$$

where

$$\theta := \frac{K}{K+1} \in [1/2,1) \quad ,$$

and

$$\Phi(R) := \int_{B_R} |x-x_o|^{2-n} |Du(x)|^2 \, dx \quad .$$

Then, for

$$\Phi^*(R) := \Phi(R) + R^2 \quad ,$$

we arrive at

$$\Phi^*(R) \leqq \theta \cdot \Phi^*(2R) \quad .$$

By iteration of this inequality, we may conclude that

$$\Phi^*(R) \leqq 2^{\gamma} \Phi^*(R_o)(R/R_o)^{\gamma} \quad \text{for} \quad R \in (0,R_o) \quad ,$$

where

$$R_o := \frac{1}{2} \operatorname{dist}(x_o,\partial\Omega) \quad , \quad \gamma := {}_2\log(1/\theta) \quad .$$

Then,

$$\begin{aligned} &\int_{B_R(x_o)} |x-x_o|^{2-n} |Du(x)|^2 \, dx \\ &\leqq 2^{\gamma} \left\{ \int_{B_{R_o}(x_o)} |x-x_o|^{2-n} |Du(x)|^2 dx + R_o^2 \right\} \left(\frac{R}{R_o} \right)^{\gamma} \quad \text{for} \quad R \in (0,R_o], \end{aligned} \tag{4.6}$$

and we have proved the Hölder continuity of u in Ω with respect to the exponent $\sigma := \gamma/2$, by virtue of a well known theorem due to Morrey.

Finally, let $\bar{u} := (u-\omega_R)$. Then

$$L\bar{u} = \bar{f}(x,\bar{u},D\bar{u}) \quad \text{in} \quad \Omega$$

where $\bar{f}(x,\bar{u},p) := f(x,\bar{u}+\omega_R,p)$ satisfies

$$|\bar{f}(x,\bar{u}(x),p)| \leqq a \cdot Q(x,p) + b \quad .$$

The assumption $a \cdot \operatorname{osc}_{\Omega} u < 1$ implies that

$$a^*(\bar{u}) := a \cdot |\bar{u}|_{L^\infty(\Omega)} < 1$$

whence, by Corollary 1,

$$\int_{B_{R_o}(x_o)} |x-x_o|^{2-n}|Du(x)|^2 dx \leq \frac{2M}{2\lambda(1-a\cdot \operatorname{osc}_\Omega|u|)K_1} + KR_o^2 , \quad (4.7)$$

where K depends only on n, M, b, λ, and μ.

The desired a priori estimate now follows from (4.6) and (4.7).

We conclude this section with the <u>proof of the Theorem stated in 3.6</u> (for $n \geq 3$):

We insert in (1.2) the test function

$$\varphi = [u-t\omega_R]G^\rho(\cdot,y)\eta^2$$

where B_{2R}, ω_R, $G^\rho(\cdot,y)$, η, and T_{2R} are defined in the same way as in the proof of Lemma 2, and where $0 < \rho < 1$, $0 \leq t \leq 1$, $B_{2R} \subset \Omega$.

Then we obtain (with $\int = \int_{B_{2R}}$)

$$\int A^{\alpha\beta} D_\alpha u\cdot D_\beta u \; G^\rho(\cdot,y)\eta^2 dx + \frac{1}{2}\int a^{\alpha\beta} D_\alpha[|u-t\omega_R|^2]D_\beta G^\rho(\cdot,y)\eta^2 dx$$
$$= \int(u-t\omega_R)\cdot f(x,u,Du)G^\rho(\cdot,y)\eta^2 dx$$
$$- \int A^{\alpha\beta} D_\alpha u\cdot(u-t\omega_R)2\eta \, D_\beta\eta \; G^\rho(\cdot,y)\, dx \quad .$$

By virtue of the identity

$$\frac{1}{2}\eta^2 D_\alpha|u-t\omega_R|^2 = \frac{1}{2} D_\alpha[\eta^2|u-t\omega_R|^2] - \frac{1}{2}(D_\alpha\eta^2)|u-\omega_R|^2$$
$$- D_\alpha\eta^2 \; \omega_R\cdot(u-\omega_R)(1-t) - \frac{1}{2} D_\alpha\eta^2|\omega_R|^2(1-t)^2 \quad ,$$

we get

$$\frac{1}{2}\int A^{\alpha\beta} D_\alpha[|u-t\omega_R|^2]D_\beta \; G^\rho(\cdot,y)\eta^2 \, dx$$
$$= \frac{1}{2} ⨍_{B_\rho(y)} |u-t\omega_R|^2 \, dx - \frac{1}{2}|\omega_R|^2(1-t)^2 + \mathfrak{J}_1$$

with

$$\mathcal{J}_1 := -\int A^{\alpha\beta}\{\omega_R\cdot(u-\omega_R)(1-t)+\tfrac{1}{2}|u-\omega_R|^2\}D_\alpha\eta^2\; D_\beta G^\rho(\cdot,y)\;dx\;.$$

Employing (3.12) and (G5), we see that

$$\begin{aligned}|\mathcal{J}_1| &\leq KR^{-1}\int_{T_{2R}} |DG^\rho(\cdot,y)|\,|u-\omega_R|^2\;dx\\ &\leq K\{R^{n-2}\gamma(R)\int_{T_{2R}}|DG^\rho(\cdot,y)|^2dx+R^{-n}\;\gamma^{-1}(R)\int_{T_{2R}}|u-\omega_R|^2dx\}\\ &\leq K\;\gamma(R)\end{aligned}$$

with

$$\gamma(R) := \{R^{2-n}\int_{T_{2R}} Q(x,Du)dx\}^{1/2}\quad .$$

Moreover,

$$\mathcal{J}_2 := \int_{T_{2R}} A^{\alpha\beta}\;D_\alpha u\cdot(u-t\omega_R)2\eta\;D_\beta\eta\;G^\rho(\cdot,y)\;dx$$

can be estimated by

$$|\mathcal{J}_2| \leq KR^{1-n}\int_{T_{2R}} |Du|\;dx \leq K\;\gamma(R)\quad .$$

Thus we obtain

$$\begin{aligned}&\int_{B_{2R}}[Q(x,Du)-(u-t\omega_R)\cdot f(x,u,Du)]G^\rho(x,y)\eta^2\;dx\;+\\ &+\frac{1}{2}\fint_{B_\rho(y)}|u(x)-t\omega_R|^2dx\leq\frac{1}{2}(1-t)^2|\omega_R|^2+K\;\gamma(R)\quad .\end{aligned}\tag{4.8}$$

In addition, $|\omega_R| \leq M$, and

$$\int_{B_{2R}} G(x,y)\;dx \leq KR^2\quad .$$

Setting $t = 0$, we have

$$\int_{B_{2R}}(1-a^*)Q(x,Du)G^\rho(x,y)\eta^2dx\leq\frac{1}{2}M^2+K\{\gamma(R)+R^2b^*\}.$$

By means of Fatou's lemma, we arrive at

$$2(1-a^*) \int_{B_{2R}} Q(x,Du)G(x,y)\eta^2 dx \leq M^2 + K\{\gamma(R)+R^2\, b^*\} \tag{4.9}$$

letting $\rho \to 0$. By Lebesgue's theorem on dominated convergence and by (4.9), the inequality (4.8) yields

$$\begin{gathered} 2 \int_{B_{2R}} [1-a^*-atM]Q(x,Du)G(x,y)\eta^2\, dx + |u(y)-t\omega_R|^2 \\ \leq (1-t)^2 M^2 + K_o\, \gamma(R) + K_o'\, R^2 \end{gathered} \tag{4.10}$$

and also

$$\begin{gathered} 2 \int_{B_{2R}} [1-a|u-t\omega_R|]Q(x,Du)G(x,y)\eta^2 dx + |u(y)-t\omega_R|^2 \\ \leq (1-t)^2 M^2 + K_o\, \gamma(R) + K_i'\, R^2 \end{gathered} \tag{4.11}$$

for almost all $y \in B_{R/2}(x_o)$, and with numbers K_o, K_o' independent of R. Furthermore, $K_o' = 0$ if $b = b^* = 0$.

Since $a^* \leq aM$ and $a^* + aM < 2$, we have $a^* < 1$. Set

$$\nu := 2 - a^* - aM \quad \text{and} \quad t_o := \frac{1-a^*}{2-a^*} \quad .$$

Then we have

$$\nu > 0 \quad \text{and} \quad 0 < t_o < 1.$$

Consequently,

$$1 - a^* - t_o aM = t_o \nu > 0 \quad .$$

On account of (4.10), we obtain

$$\sup_{y \in B_{R/2}} |u(y)-t_o\omega_R|^2 \leq (1-t_o)^2 M^2 + K_o\, \gamma(R) + K_o' R^2 \quad . \tag{4.12}$$

Then we define $\tau > 0$ by

$$\tau := \min\left\{\frac{1}{4aM}, \frac{\nu}{2(2-a^*)aM}\right\} \quad .$$

Now let κ be the smallest integer such that $\kappa\tau \geq 1 - t_o$. We set

$$t_i = t_o + i\tau \quad \text{for} \quad 0 \leq i \leq \kappa-1 \quad \text{and} \quad t_\kappa = 1.$$

By Harnack's principle [91], we see that

$$G(x,y) \leq \sup_{z \in B_{j+1}} G(x,z) \leq K_4 \inf_{z \in B_{j+1}} G(x,z) \leq K_4\, G(x,x_o)$$

$$\text{for} \quad x \in B - B_j \; , \; y \in B_{j+1} \; , \; j = 0,1,2,\ldots,\kappa, \tag{4.13}$$

$$\text{with} \quad B := B_{2R}(x_o) \; , \; B_j := B_{2^{-j-1}R}(x_o) \quad .$$

By means of (4.11) and (4.13), we conclude that

$$2 \int_{B_j} [1-\alpha|u-t_{j+1}\omega_R|]Q(x,Du)G(x,y)\,dx + |u(y)-t_{j+1}\omega_R|^2$$

$$\leq (1-t_{j+1})^2 M^2 + K_o\, \gamma(R) + K_o' R^2 + 4aMK_4 \int_{B-B_j} Q(x,Du)G(x,x_o)\,dx$$

for a.a. $y \in B_{j+1}$. For $x \in B - B_j$ we have

$$0 \leq G(x,x_o) \leq K_2|x-x_o|^{2-n} \leq K_2(2^{-j-1}R)^{2-n} \quad ,$$

whence

$$0 \leq G(x,x_o) \leq 2^{(n-2)(\kappa+1)} K_2 R^{2-n} \quad \text{for} \quad x \in B-B_j, \; 0 \leq j \leq \kappa.$$

Setting

$$K_o'' := K_o + 2^{(n-2)(\kappa+1)+2}\, a\, M\, K_2\, K_4 \quad ,$$

we arrive at

$$\begin{aligned} &2 \int_{B_j} [1-a|u-t_{j+1}\omega_R|]Q(x,Du)G(x,y)\,dx + |u(y)-t_{j+1}\omega_R|^2 \\ &\leq (1-t_{j+1})^2 M^2 + K_o' R^2 + K_o''\, \gamma(R) \end{aligned} \tag{4.14}$$

for a.a. $y \in B_{j+1}$ if $\gamma(R) \leq 1$. By choosing ρ_o such that

$$\rho_o^2 \leq \frac{\tau^2 M^2}{2K_o'} \quad , \tag{4.15'}$$

and by setting

$$\delta := \min\{1, \frac{\tau^2 M^2}{2K_o''}\} \tag{4.15''}$$

we get

$$K'_o R^2 + K''_o \delta \leqq \tau^2 M \quad \text{for all} \quad R \in (0,\rho_o]. \tag{4.16}$$

Let $\varepsilon := \delta^2$, and apply Lemma 1. Also, set

$$R_o := \min\{\rho_o, \frac{1}{2}\operatorname{dist}(x_o,\partial\Omega), \frac{c}{4\varepsilon}\} \quad . \tag{4.17}$$

Then, for $R = 2^{-i_o} R_o$, $i_o \in \{1,2,\ldots,p\}$, $p := \left[\frac{2cM^2}{\varepsilon}\right] + 1$, we obtain that

$$\gamma(R) < \delta \leqq 1 \quad . \tag{4.18}$$

Now we shall demonstrate that

$$\sup_{y \in B_i} |u(y)-t_i\omega_R|^2 \leqq (1-t_i)^2 M^2 + \tau^2 M^2 \tag{4.19}$$

for $i = 0,1,\ldots,\kappa$.

Firstly, we note that (4.19) holds for $i = 0$ because of (4.12) and (4.17).

Suppose now that (4.19) is true for $i = j$ where $j \leqq \kappa-1$. Then we shall verify (4.19) for $i = j+1$. By induction, we have on B_j the estimate

$$1-a|u-t_{j+1}\omega_R| \geqq 1-a|u-t_j\omega_R|-aM\tau \geqq 1-a(1-t_o)M-2aM\tau$$

since

$$(1-t_j)^2 M^2 + \tau^2 M^2 < (1-t_o)^2 M^2 + \tau^2 M^2 < (1-t_o+\tau)^2 M^2 \quad .$$

Hence

$$1 - a|u - t_{j+1}\omega_R| \geqq 1 - \frac{aM}{2-a^*} - 2aM\tau$$

$$\geqq aM\{\frac{\nu}{(2-a^*)aM} - 2\tau\} \geqq 0 \quad \text{a.a. on } B_j \quad ,$$

since

$$\tau \leqq \frac{\nu}{2(2-a^*)aM} \quad .$$

Thus we find

$$\sup_{y \in B_{j+1}} |u(y)-t_{j+1}\omega_R|^2 \leqq (1-t_{j+1})^2 M^2 + \tau^2 M^2$$

taking (4.14), (4.17), and $R = 2^{-i_o} R_o$ into account. Hence,

$$\sup_{y\in B_\kappa} |u(y)-\omega_R| \leq \tau M \quad .$$

Because of $\tau \leq \frac{1}{4aM}$, we achieve that

$$\sup_{y\in B_\kappa} |u(y)-\omega_R| \leq \frac{1}{4a} \quad ,$$

and consequently

$$\sup_{x,y\in B_\kappa} |u(x)-u(y)| \leq \sup_{x\in B_\kappa} |u(x)-\omega_R| + \sup_{y\in B_\kappa} |u(y)-\omega_R| \leq \frac{1}{2a}$$

that is,

$$a \cdot \operatorname{osc}_{B_\kappa} u \leq 1/2 \quad .$$

By employing Lemma 2, we obtain the interior estimate for u as stated in the Theorem. Similarly, one may deduce the boundary estimate.

5. CAFFARELLI's APPROACH

We assume again that u is a weak solution of (1.1) in $\Omega \subset \mathbb{R}^n$ satisfying (A1-4) and $a^* \leq aM$.

The main idea of Caffarelli's approach is contained in the following

Proposition. *Suppose that* $aM + a^* < 2$, $b = b^* = 0$. *Then there exists a number* $\delta \in (0,1)$, *depending only on* n, μ/λ, *and on* a^* *such that*

$$u(B_R(x_o)) \subset B_{(1-\delta)M}(\delta\bar{u}) \quad , \quad \bar{u} := \fint_{B_{2R}(x_o)} u(x)\,dx \quad , \qquad (5.1)$$

holds for each ball $B_{2R}(x_o) \subset \Omega$.

Proof. By a simple scaling argument, we may w.l.o.g. assume that

$$a = 1$$

(replace u by $v = au$; then $M(v) = aM$, $a(v) = 1$, $a^*(v) = a^*$). Hence, for

$$\ell := \frac{aM+a^*}{2} = \frac{M+a^*}{2} \quad ,$$

we get $2a^* \leq M + a^* = 2\ell < 2$ or

$$a^* \leq \ell < 1.$$

We shall also (w.l.o.g.) assume that $x_o = 0$ and $R = 1$. Then we insert in (1.2) the test function

$$\varphi = (u + \xi)\eta$$

where $\eta \in C_c^\infty(B_2, \mathbb{R})$, $B_2 = B_2(x_o)$, and $\xi \in \mathbb{R}^N$ is a constant vector.

Then we obtain

$$\int A^{\alpha\beta} D_\alpha u \cdot D_\beta u \, \eta \, dx + \int A^{\alpha\beta} D_\alpha\left\{\frac{1}{2}|u|^2 + \xi \cdot u\right\} D_\beta \eta \, dx$$

$$= \int (u+\xi) \cdot f(x,u,Du)\eta \, dx$$

$$\leq \int [a^* + |\xi|] Q(x,Du)\eta \, dx$$

or

$$\int [1-a^*-|\xi|] Q(x,Du) \, \eta \, dx + \int A^{\alpha\beta} D_\alpha\left[\frac{1}{2}|u|^2 + \xi \cdot u\right] D_\beta \eta \, dx \leq 0.$$

We choose ξ such that

$$|\xi| = 1 - a^*.$$

Then we conclude that

$$\int_{B_2} A^{\alpha\beta} D_\alpha h \, D_\beta \eta \, dx \leq 0 \quad \text{for all} \quad \eta \in C_c^\infty(B_2, \mathbb{R})$$

where

$$h := \frac{1}{2} M^2 + (1-a^*)M - \left[\frac{1}{2}|u|^2 + \xi \cdot u\right] \geq 0.$$

By Harnack's inequality [91] for supersolutions of L in B_2 we infer that there is a number $\delta_1 \in (0,1)$ depending on n and μ/λ such that

$$h(x) \geq \inf_{B_1} h \geq \delta_1 \fint_{B_2} h \, dx \geq \delta_1[(1-a^*)M - \xi \cdot \bar{u}]$$

holds for a.a. $x \in B_1$, or

$$\begin{aligned} &\frac{1}{2}(M-|u|)(M+|u|) + (1-a^*)M - \xi \cdot u \\ &\geq \delta_1[(1-a^*)M - \xi \cdot \bar{u}] \quad \text{a.e. on } B_1. \end{aligned} \tag{5.2}$$

Then, for a.a. $x \in B_1$, we can argue as follows:
Fix $x \in B_1$. If $u(x) \neq 0$, we choose

$$\xi := \frac{u(x)}{|u(x)|}(1-a^*) \quad .$$

Then we define

$$r := \frac{|u(x)|}{M}$$

and $\theta \in [0,2\pi)$ is the angle between $u(x)$ and $\bar{u}$. Hence

$$\xi \cdot u(x) = Mr(1-a^*) \text{ , and } \xi \cdot \bar{u} = (1-a^*)|\bar{u}| \cos\theta .$$

Therefore we conclude from (5.2) that

$$M(1-r)\left[\frac{1}{2}(M+|u(x)|)+(1-a^*)\right] \geq M\,\delta_1(1-a^*)\left(1-\frac{|\bar{u}|}{M}\cos\theta\right) \quad .$$

Note that

$$\frac{1}{2}(M+|u(x)|) + (1-a^*) \leq M + (1-a^*) < (2-a^*) + (1-a^*) = 3-2a^* .$$

Therefore

$$1-r \geq \delta_2\left[1 - \frac{|\bar{u}|}{M}\cos\theta\right] \quad ,$$

for every δ_2 with $0 < \delta_2 \leq \frac{1-a^*}{3-2a^*}\delta_1 < \frac{1}{2}$, whence

$$r - r^2 \geq \delta_2 r - \delta_2 r\frac{|\bar{u}|}{M}\cos\theta .$$

Let us subtract from this inequality the trivial relation

$$1 - r \geq \delta_2 - \delta_2 r \quad .$$

Then

$$1 - r^2 \geq \delta_2\left[1 - r\frac{|\bar{u}|}{M}\cos\theta\right] \quad ,$$

or

$$1 - \delta_2 + (\delta_2/2)^2 \geq r^2 - \delta_2 r\frac{|\bar{u}|}{M}\cos\theta + \left(\frac{\delta_2|\bar{u}|}{2M}\right)^2 \quad ,$$

since $|\bar{u}| \leq M$. Multiplication of the last inequality by M^2 then yields

$$|u(x) - \delta_2\bar{u}|^2 \leq M^2(1-\delta_2)^2 \quad .$$

Let us set

$$\delta := \frac{1-a^*}{3-2a^*}\,\delta_1 \quad .$$

Then we obtain $\delta \in (0,1)$, and

$$|u(x) - \delta\bar{u}|^2 \leq M^2(1-\delta)^2 \tag{5.3}$$

if $u(x) \neq 0$, and, for $u(x) = 0$, the equation (5.3) follows from

$$|\delta\bar{u}| \leq \delta M \leq (1-\delta)M$$

q.e.d.

Conclusions and Remarks.

1. By finitely many iterations, one can make the diameter of the image set of $B_R(x_o) \subset \Omega$ for fixed $x_o \in \Omega$ and some $R > 0$ as small as one pleases. That is, for each $\varepsilon > 0$ and for each $x_o \in \Omega$ there is an R such that

$$\operatorname{osc}_{B_R(x_o)} u < \varepsilon .$$

Hence Lemma 2 of 4.6 yields an Hölder estimate for u.

2. By carrying out an arbitrary number of iterations, one can also directly derive an Hölder estimate as well as a *Liouville Theorem* if $aM + a^* < 2$.

3. *Proof of the fact, that one can carry out the iteration procedure.* Form $v := u - \delta\bar{u}$. Then

$$Lv = g(x,v,Du) \quad \text{in} \quad \Omega,$$

where $g(x,v,p) := f(x,v+\delta\bar{u},p)$. Let us compute the numbers $M(v)$, $a(v)$, and $a^*(v)$ for v:

$$|v| \leq (1-\delta)M \text{ , hence } M(v) \leq (1-\delta)M; \; a(v) = a \quad ,$$

and $a^*(v) = a^* + \delta Ma$, since

$$v\cdot g(x,v,p) \leq u\cdot f(x,u,p) + \delta M|f(x,u,p)| \leq [a^*+\delta Ma]\underset{\sim}{Q}(x,p) \quad .$$

Thus we obtain $b(v) = b^*(v) = 0$ and

$$a^*(v) + a(v)M(v) \leqq a^* + \delta Ma + a(1-\delta)M = a^* + aM < 2 \quad ,$$

and the Proposition can be applied to v too.

4. One can also find modifications of the reasoning used in the proof of the Proposition which work if $b, b^* \neq 0$, or if we consider u in the neighborhood of points $x_o \in \partial\Omega$. However, then the arguments become somewhat cumbersome and loose their geometrical simplicity. Nevertheless, Caffarelli's approach seems to be much simpler than the reasoning described in section 4. On the other hand, there appear to exist difficulties if one wants to carry over Caffarelli's ideas to treat harmonic mappings. The reason for this failure is, that one has to find a priori estimates in terms of upper bounds κ on the sectional curvature K of the target manifold. Unfortunately, the condition $K \leqq \kappa$ only leads to one-sided conditions for solutions $u(x)$ of (2.14) but not to two-sided conditions. On the other hand, we have seen in 3.6 that one-sided conditions will, in general, not imply regularity of weak solutions. Nevertheless, the special structure of the system (2.14) enables one to prove estimates for its solution by employing the technique of 4 while - so far - the reasoning of this section has failed to give the same strong results. Thus both methods are presently of value, and the reader might use section 4 as introduction to the treatment of harmonic maps. For a detailed exposition, we refer to the paper [20] by Giaquinta and Hildebrandt, as well as to a different approach by Sperner [104], and to interesting gradient estimates by Jost and Karcher [70].

6. FURTHER RESULTS

The lack of space forbids us to dwell more extensively on the geometrical examples described in section I. Instead we give a few hints which may guide the reader to the literature:

6.1 H-surfaces

"Small" solutions of the Plateau problem have been found by Heinz [34], Werner [125], Hildebrandt [43], [44], [45], Wente [120], [121], [122], [124], Gulliver-Spruck [32], [33], Hildebrandt -Kaul [52], Gulliver [28], Steffen [105], [106], [107], [108]. "Large" solutions of the Plateau problem have been found by Struwe [114], Steffen [106], [109], and by Brezis-Coron [5].

Other unstationary H-surfaces have been found by Heinz [39], [40] and Ströhmer [111], [112], [113]. A free boundary value problem for H-surfaces has been treated by Hildebrandt [45 II] and by Hildebrandt-Jäger [51].

Results on minimal surfaces cannot be discussed here because there are too many of them. We refer to the four survey articles by Hoffman-Osserman [59], Böhme [4], and Hildebrandt [49], [50], concerning two dimensional minimal surfaces, and, besides, to the well known books by Courant [7], Osserman [93], and Nitsche [92].

6.2. Harmonic maps

An excellent survey is given by Eells-Lemaire [11]. Since the appearance of this article, various new results have been found. Some of them have been described in the reports by Eells-Lemaire [12], Eells [13], Hildebrandt [47], [48], and Jost [69].

Of particular interest are results about the existence of two-dimensional harmonic diffeomorphisms which are mostly based upon the fundamental work of Heinz concerning solutions of systems (2.2) which are diffeomorphisms, cf. [35], [36].

We particularly mention the papers by Schoen-Yau [99], Jost [66], [68], and Jost-Schoen [72].

Free boundary problems for harmonic maps have been treated by Baldes [2], and by Gulliver-Jost [30]. The last authors also found an example showing that, in comparison with the Dirichlet problem, further conditions will be needed to prove existence of solutions.

BIBLIOGRAPHY

[1] Alt, H.W., *Verzweigungspunkte von H-Flächen I, II,* Math. Z. 127, 333-362 (1972), Math. Ann. 201, 33-56 (1973).

[2] Baldes, A., *Harmonic mappings with partially free boundary,* manuscripta math., to appear.

[3] Böhme, R., S. Hildebrandt, and E. Tausch, *The two-dimensional analogue of the catenary,* Pacific J. Math. 88, 247-278 (1980).

[4] Böhme, R., *New results on the classical problem of Plateau on the existence of many solutions,* Séminaire Bourbaki 1981/82, no. 579.

[5] Brezis, H., and J.-M. Coron, *Sur la conjecture de Rellich pour les surfaces à courbure moyenne prescrite,* preprint Sept. 1982.

[6] Caffarelli, L.A., *Regularity theorems for weak solutions of some nonlinear systems,* Comm. Pure Appl. Math., to appear.

[7] Courant, R., *Dirichlet's principle, conformal mapping, and minimal surfaces*, New York, Interscience (1950).

[8] De Giorgi, E., *Un esempio di estremali discontinue per un problema variazionale di tipo ellitico*, Boll. Un. mat. Ital. IV. Ser. 1, 135-138 (1968).

[9] Dziuk, G., *Über die Stetigkeit teilweise freier Minimalflächen*, manuscripta math. 36, 241-251 (1981).

[10] Eells, J., and J.H. Sampson, *Harmonic mappings of Riemannian manifolds*, American J. Math. 86, 109-160 (1964).

[11] Eells, J., and L. Lemaire, *A report on harmonic maps*, Bull. London Math. Soc. 10, 1-68 (1978).

[12] Eells, J., and L. Lemaire, *Selected topics in harmonic maps*, N.S.F. Conference Board Math. Sci. (1982).

[13] Eells, J., *Regularity of certain harmonic maps*, Proceedings of the Durham Symposium on Global Riemannian Geometry (July 1982); Preprint Aug. 1982.

[14] Frehse, J., *A discontinuous solution of a mildly nonlinear elliptic system*, Math. Z. 134, 229-230 (1973).

[15] Frehse, J., *A note on the Hölder continuity of solutions of variational problems*, Abhandlung. Math. Sem. Hamburg 43, 59-63 (1975).

[16] Frehse, J., *On two-dimensional quasi-linear elliptic systems*, manuscripta math. 28, 21-50 (1979).

[17] Frehse, J., *Un problème variationnel bidimensionel possédant des extremales bornées et discontinues*, C.R. Acad. Sc. Paris.

[18] Giaquinta, M., and E. Giusti, *On the regularity of the minima of variational integrals*, Acta Math., to appear.

[19] Giaquinta, M., and E. Giusti, *The singular set of the minima of quadratic functionals*, preprint no. 453, SFB 72, Bonn 1981, to appear in Analysis.

[20] Giaquinta, M., and S. Hildebrandt, *A priori estimates for harmonic mappings*, Journal für die Reine u. Angew. Math., to appear.

[21] Giaquinta, M., and S. Hildebrandt, *Estimation a priori des solutions faibles de certains systemes non lineaires elliptiques*, Sêminaire Goulaouic-Meyer-Schwartz, 17 March 1981, Exposê no. XVII.

[22] Grüter, M., *Ein Hebbarkeitssatz für isolierte Singularitäten*, Bonn, SFB 72, Preprint Nr. 177 (1978).

[23] Grüter, M., *Über die Regularität schwacher Lösungen des Systems* $\Delta x = 2H(x)x_u \wedge x_v$. Dissertation, Düsseldorf 1979.

[24] Grüter, M., *Regularity of weak* H*-surfaces*, Journal für die Reine Angew. Math. 329, 1-15 (1981).

[25] Grüter, M., *A note on variational integrals which are conformally invariant*, preprint 1982.

[26] Grüter, M., S. Hildebrandt, and J.C.C. Nitsche, *On the boundary behavior of minimal surfaces with a free boundary which are not minima of the area*, manuscripta math. 35, 387-410 (1981).

[27] Grüter, M., and K.-O. Widman, *The Green function for uniformly elliptic equations*, manuscripta math. 37, 303-342 (1982).

[28] Gulliver, R., *The Plateau problem for surfaces of prescribed mean curvature in a Riemannian manifold*, Journal Diff. Geometry 8, 317-330 (1972).

[29] Gulliver, R., *Regularity of minimizing surfaces of prescribed mean curvature*, Ann. of Math. 97, 275-305 (1973).

[30] Gulliver, R., and J. Jost, *Oral cummunication*, July 1982.

[31] Gulliver, R., R. Osserman, and H. Royden, *A theory of branched immersion of surfaces*, American Journ. Math. 95, 750-812 (1973).

[32] Gulliver, R., and J. Spruck, *The Plateau problem for surfaces of prescribed mean curvature in a cylinder*, Inventiones math. 13, 169-178 (1971).

[33] Gulliver, R., and J. Spruck, *Existence theorems for parametric surfaces of prescribed mean curvature*, Indiana Univ. Math. J. 22, 445-472 (1972).

[34] Heinz, E., *Über die Existenz einer Fläche konstanter mittlerer Krümmung bei vorgegebener Berandung*, Math. Ann. 127 258-287 (1954).

[35] Heinz, E., *On certain nonlinear elliptic differential equations and univalent mappings*, J. Analyse math. 5, 197-272 (1956/57).

[36] Heinz, E., *Existence theorems for one-to-one mappings associated with elliptic systems of second order, I, II*, J. Analyse math. 15, 325-352 (1965); 17, 145-184 (1966).

[37] Heinz, E., *A-priori-Abschätzungen für isometrische Einbettungen zweidimensionaler Riemannscher Mannigfaltigkeiten in dreidimensionale Riemannsche Räume*, Math. Z. 100, 1-16 (1967).

[38] Heinz, E., *On the nonexistence of a surface of constant mean curvature with finite area and prescribed rectifiable boundary*, Arch. Rat. Mech. Anal. 35, 249-252 (1969).

[39] Heinz, E., *On surfaces of constant mean curvature with polygonal boundaries*, Arch. Rat. Mech. Anal. 36, 335-347 (1970).

[40] Heinz, E., *Unstable surfaces of constant mean curvature*, Arch. Rat. Mech. Anal. 38, 257-267 (1970).

[41] Heinz, E., *Ein Regularitätssatz für schwache Lösungen nichtlinearer elliptischer Systeme*, Göttinger Nachrichten, Nr. 1 (1975).

[42] Heinz, E., and S. Hildebrandt, *The number of branch points of bounded mean curvature*, J. Diff. Geometry 4, 227-235 (1970).

[43] Hildebrandt, S., *Über Flächen konstanter mittlerer Krümmung*, Math. Z. 112, 107-144 (1969).

[44] Hildebrandt, S., *On the Plateau problem for surfaces of constant mean curvature*, Comm. Pure Appl. Math. 23, 97-114 (1970).

[45] Hildebrandt, S., *Randwertprobleme für Flächen mit vorgeschriebener mittlerer Krümmung und Anwendungen auf die Kapillaritätstheorie I, II*, Math. Z. 112, 205-213 (1969), Arch. Rat. Mech. Anal. 39, 275-293 (1970).

[46] Hildebrandt, S., *Maximum principles for minimal surfaces and for surfaces of continuous mean curvature*, Math. Z. 128, 253-269 (1972).

[47] Hildebrandt, S., *Liouville theorems for harmonic mappings and an approach to Bernstein theorems*, Annals of Math. Studies 107-131, Princeton Univ. Press, Seminar on Diff. Geometry, 1982.

[48] Hildebrandt, S., *Nonlinear elliptic systems and harmonic mappings*, Proc. of the 1980 Beijing Symposium on Diff. Geom. and Diff. Equ., Vol. 1, pp. 481-615, Science Press, Beijing, China 1982.

[49] Hildebrandt, S., *Some results on minimal surfaces with free boundaries*, Lecture in honour of L. Cesari, Bologna 1982, to appear.

[50] Hildebrandt, S., *Partielle Differentialgleichungen und Differentialgeometrie*, Hauptvortrag auf der DMV-Tagung, Bayreuth 1982, to appear in Jahresber. der DMV.

[51] Hildebrandt, S., and W. Jäger, *On the regularity of surfaces with prescribed mean curvature at a free boundary*, Math. Z. 118, 289-3o8 (1970).

[52] Hildebrandt, S., and H. Kaul, *Two-dimensional variational problems with obstructions, and Plateau's problem for H-surfaces in a Riemannian manifold*, Comm. Pure Appl. Math. 25, 187-223 (1972).

[53] Hildebrandt, S., and K.-O. Widman, *Some regularity results for quasilinear elliptic systems of second order*, Math. Z. 142, 67-86 (1975).

[54] Hildebrandt, S., and K.-O. Widman, *On the Hölder continuity of weak solutions of quasilinear elliptic systems of second order*, Ann. Scuola Norm. Sup. Pisa (IV), 4, 145-178 (1977).

[55] Hildebrandt, S., H. Kaul, and K.-O. Widman, *An existence theory for harmonic mappings of Riemannian manifolds*, Acta Math. 138, 1-16 (1977).

[56] Hildebrandt, S., and K.-O. Widman, *Variational inequalities for vector-valued functions*, J. für Reine u. Angew. Math. 309, 191-220 (1979).

[57] Hildebrandt, S., and K.-O. Widman, *Sätze vom Liouvilleschen Typ for quasilineare elliptische Gleichungen und Systeme*, Nachr. Akad. Wiss. Göttingen, II. Math.-Phys. Klasse, Nr. 4, 41-59, Jahrgang 1979.

[58] Hildebrandt, S., J. Jost, and K.-O. Widman, *Harmonic mappings and minimal surfaces*, Inventiones math. 62,269-298 (1980).

[59] Hoffman, D., and R. Osserman, *The geometry of the generalized Gauss map*, Memoirs of the Amer. Math. Soc. Nov. 1980, Vol. 28, No. 236.

[60] Hoffman, D., and R. Osserman, *The area of the generalized Gaussian image and the stability of minimal surfaces in* S^n *and* $\mathbb{R}^n$, preprint.

[61] Ivert, P.A., *On quasilinear elliptic systems in diagonal form*, Math. Z. 170, 283-286 (1980).

[62] Jäger, W., *Ein Maximumprinzip für ein System nichtlinearer Differentialgleichungen*, Göttinger Nachrichten Nr. 11, Jahrgang 1976.

[63] Jäger, W., and H. Kaul, *Uniqueness and stability of harmonic maps and their Jacobi fields*, manuscripta math. 28, 269-291 (1979).

[64] Jäger, W., and H. Kaul, *Lecture in Oberwolfach July 1982*, Meeting on the Calculus of Variations.

[65] Jost, J., *Eine geometrische Bemerkung zu Sätzen über harmonische Abbildungen, die ein Dirichletproblem lösen*, manuscripta math. 32, 51-57 (1980).

[66] Jost, J., *Eineindeutigkeit harmonischer Abbildungen*, Diplomarbeit Bonn (1979). A shortened version has appeared in: Journ. für die Reine u. Angew. Math. 324, 141-153 (1981).

[67] Jost, J., *Ein Existenzbeweis für harmonische Abbildungen, die ein Dirichletproblem lösen, mittels der Methode des Wärmeflusses*, manuscripta math. 34, 17-25 (1981).

[68] Jost, J., *Existence proofs for harmonic mappings with the help of a maximum principle*, preprint 1982.

[69] Jost, J., *Harmonic mappings between surfaces*, to appear.

[70] Jost, J., and H. Karcher, *Geometrische Methoden zur Gewinnung von a-priori-Schranken für harmonische Abbildungen*, to appear in: manuscripta math.

[71] Jost, J., and M. Meier, *Boundary regularity for minima of certain quadratic functionals*, Manuskript.

[72] Jost, J., and R. Schoen, *On the existence of harmonic diffeomorphisms between surfaces*, Inventiones math. 66, 353-359 (1982).

[73] Jost, J., and S.T. Yau, *Harmonic mappings and Kähler manifolds*, Math. Annalen, to appear.

[74] Kaul, H., *Das Dirichletproblem für elliptische Differentialoperatoren mit Variationsstruktur auf Riemannschen Mannigfaltigkeiten*, Bonner Math. Schrift. 94 (1977).

[75] Ladyzenskaya, O.A., and N.N. Ural'ceva, *Linear and quasilinear elliptic equations*, New York and London, Academic Press (1968) (English translation of the first Russian edition 1964).

[76] Ladyzenskaya, O.A., and N.N. Ural'ceva, *Linear and quasilinear elliptic equations*, Second Russian Edition, Moscow: Nauka (1973).

[77] Lawson, H.B., and R. Osserman, *Non-existence, non-uniqueness and irregularity of solutions to the minimal surface system*, Acta Math. 139, 1-17 (1977).

[78] Lemaire, L., *Applications harmoniques de surfaces Riemanniennes*, Journ. Diff. Geom. 13, 51-78 (1978).

[79] Lemaire, L., *Minima and critical points of the energy in dimension two*, Springer Lecture Notes 838, 187-193 (1981).

[80] Lemaire, L., *Existence des applications harmoniques et courbure des variétés*, Séminaire Bourbaki, no. 553 (1979/80).

[81] Littman, W., G. Stampacchia, and H.F. Weinberger, *Regular points for elliptic equations with discontinuous coefficients*, Ann. Scuola Norm. Sup. Pisa, Sci. fis. mat. III. Ser. 17, 43-77 (1963).

[82] Meier, M., *Reguläre und singuläre Lösungen quasilinearer elliptischer Gleichungen und Systeme*, Dissertation, Bonn (1978).

[83] Meier, M., *Liouville theorems for nonlinear elliptic equations and systems*, manuscripta math. 29, 207-228 (1979).

[84] Meier, M., *Boundedness and integrability properties of weak solutions of quasilinear elliptic systems*, Journal für

die Reine u. Angew. Math. 333, 191-220 (1982).

[85] Meier, M., *Removable singularities for weak solutions of quasilinear elliptic systems*, Preprint 1982.

[86] Meier, M., *Removable singularities of bounded harmonic mappings and minimal submanifolds*, Preprint 1982.

[87] Meier, M., *Liouville theorems, partial regularity, and Hölder continuity of weak solutions to quasilinear elliptic systems*, Preprint 1982.

[88] Meier, M., *Lecture in Oberwolfach, July 1982*, Meeting on the Calculus of Variations.

[89] Morrey, C.B., *The problem of Plateau on a Riemannian manifold*, Ann. of Math. 49, 807-851 (1948).

[90] Morrey, C.B., *Multiple integrals in the calculus of variations*, Springer, Berlin-Heidelberg-New York (1968).

[91] Moser, J., *On Harnack's theorem for elliptic differential equations*, Comm. Pure Appl. Math. 14, 577-591 (1961).

[92] Nitsche, J.C.C., *Vorlesungen über Minimalflächen*, Springer, Berlin-Heidelberg-New York 1975.

[93] Osserman, R., *A survey of minimal surfaces*, Von Nostrand, London, Toronto, Melbourne 1969.

[94] Osserman, R., *Minimal varieties*, Bull. Amer. Math. Soc. 75, 1092-1120 (1969).

[95] Osserman, R., *Minimal surfaces, Gauss maps, total curvature, eigenvalue estimates and stability*, The Chern Symposium 1979, Springer 1980.

[96] Ruh, E.A., *Asymptotic behaviour of non-parametric minimal hypersurfaces*, J. Diff. Geometry 4, 509-513 (1970).

[97] Ruh, E.A., and J. Vilms, *The tension field of the Gauss map*, Trans. Amer. Math. Soc. 149, 569-573 (1970).

[98] Sacks, J., and K. Uhlenbeck, *The existence of minimal immersions of 2-spheres*, Annals of Math. 113, 1-24 (1981).

[99] Schoen, R., and S.T. Yau, *Harmonic maps and the topology of stable hypersurfaces and manifolds with nonnegative Ricci curvature*, Comm. Math. Helv. 39, 333-341 (1976).

[100] Schoen, R., and K. Uhlenbeck, *A regularity theory for harmonic maps*, Journal of Diff. Geometry, to appear.

[101] Schoen, R., and K. Uhlenbeck, *Boundary regularity and miscellaneous results on harmonic maps*, Preprint 1982.

[102] Sperner, E. jr., *Zur Regularitätstheorie gewisser Systeme elliptischer Differentialgleichungen*, Dissertation, Bonn (1975).

[103] Sperner, E. jr., *Ein Regularitätssatz für Systeme nichtlinearer elliptischer Differentialgleichungen*, Math. Z. 146, 57-68 (1976).

[104] Sperner, E. jr. ,*A priori gradient estimates for harmonic mappings*, Preprint 1982.

[105] Steffen, K., *Ein verbesserter Existenzsatz für Flächen konstanter mittlerer Krümmung*, Manuscripta math. 6, 105-139 (1972).

[106] Steffen, K., *Flächen konstanter mittlerer Krümmung mit vorgebenem Volumen oder Flächeninhalt*, Arch. Rat. Mech. Anal. 49, 99-128 (1972).

[107] Steffen, K., *Isoperimetric inequalities and the problem of Plateau*, Math. Ann. 222, 97-144 (1976).

[108] Steffen, K., *On the existence of surfaces with prescribed mean curvature and boundary*, Math. Z. 146, 113-135 (1976).

[109] Steffen, K., *On the uniqueness of surfaces of prescribed constant mean curvature spanning a given contour*, Preprint, Sept. 1982.

[110] Steffen, K., and H.C. Wente, *The non-existence of branch points in solutions to certain classes of Plateau type variational problems*, Math. Z. 163, 211-238 (1978).

[111] Ströhmer, G., *Instabile Minimalflächen in Riemannschen Mannigfaltigkeiten nichtpositiver Schnittkrümmung*, Journal für die Reine u. Angew. Math. 315, 16-39 (1980).

[112] Ströhmer, G., *Instabile Flächen vorgeschriebener mittlerer Krümmung*, Math. Z. 174, 119-133 (1980).

[113] Ströhmer, G., *Instabile Lösungen der Eulerschen Gleichungen gewisser Variationsprobleme*, Archive Rat. Mech. Anal. 79, 219-239 (1982).

[114] Struwe, M., *A counterexample in elliptic regularity theory*, manuscripta math. 34, 85-92 (1981).

[115] Tomi, F., *Ein einfacher Beweis eines Regularitätssatzes für schwache Lösungen gewisser elliptischer Systeme*, Math. Z. 112, 214-218 (1969).

[116] Tomi, F., *Bemerkungen zum Regularitätsproblem der Gleichungen vorgeschriebener mittlerer Krümmung*, Math. Z. 132, 323-326 (1973).

[117] Uhlenbeck, K., *Harmonic maps; a direct method in the calculus of variations*, Bull. Amer. Math. Soc. 76, 1082-1087 (1970).

[118] Uhlenbeck, K., *Regularity for a class of non-linear elliptic systems*, Acta Math. 138, 219-240 (1977).

[119] von Wahl, W., and M. Wiegner, *Über die Hölderstetigkeit schwacher Lösungen semilinearer elliptischer Systeme mit einseitiger Bedingung*, manuscripta math. 19, 385-399 (1976).

[120] Wente, H.C., *An existence theorem for surfaces of constant mean curvature*, J. math. Analysis Appl. 26, 318-344 (1969).

[121] Wente, H.C., *A general existence theorem for surfaces of constant mean curvature*, Math. Z. 120, 277-288 (1981).

[122] Wente, H.C., *The Dirichlet problem with a volume constraint*, manuscripta math. 11, 141-157 (1974).

[123] Wente, H.C., *The differential equation* $\Delta x = 2H(x_u \wedge x_v)$ *with vanishing boundary values*, Proc. A.M.S. 50, 131-137 (1975).

[124] Wente, H.C., *Large solutions to the volume constrained Plateau problem*, Archive Rat. Mech. Anal. 75, 59-77 (1980).

[125] Werner, H., *Das Problem von Douglas für Flächen konstanter mittlerer Krümmung*, Math. Ann. 133, 303-319 (1957).

[126] Widman, K.-O., *The singularity of the Green function for non-uniformly elliptic partial differential equations with discontinuous coefficients*, Uppsala University (1970).

[127] Widman, K.-O., *Hölder continuity of solutions of elliptic systems*, manuscripta math. 5, 299-308 (1971).

[128] Wiegner, M., *Über die Regularität schwacher Lösungen gewisser elliptischer Systeme*, manuscripta math. 15, 365-384 (1975).

[129] Wiegner, M., *Ein optimaler Regularitätssatz für schwache Lösungen gewisser elliptischer Systeme*, Math. Z. 147, 21-28 (1976).

[130] Wiegner, M., *A-priori Schranken für Lösungen gewisser elliptischer Systeme*, manuscripta math. 18, 279-297 (1976).

[131] Wiegner, M., *Das Existenz- und Regularitätsproblem bei Systemen nichtlinearer elliptischer Differentialgleichungen*, Habilitationsschrift, Bochum (1977).

[132] Wiegner, M., *On two-dimensional elliptic systems with a one-sided condition*, Math. Z. 178, 493-500 (1981).

TOPICS IN BIFURCATION THEORY

David G. Schaeffer

Duke University, Durham, NC 27706

LECTURE 1: A BRIEF INTRODUCTION TO BIFURCATION THEORY

An alternative title for this lecture might be "What is bifurcation?" We shall answer that question by example rather than definition. Specifically we consider two examples, the celebrated elastica, discussed by Euler in 1744, and the Bénard problem, to which we return in the third lecture. The style of these lectures is rather informal--we attempt to convey the ideas of the subject and omit many proofs. We do not try, even remotely, for completeness in our references.

The elastica is a model for the buckling of a beam under compression. (See Figure 1.1.) Personally I usually think of squeezing a plastic ruler in my hands, but the apparatus in the experiment of Stein [1959] was capable of generating a quarter of a million pounds of thrust. In the elastica it is assumed that the beam is incompressible but capable of bending, the potential energy being proportional to

$$\int_0^L \psi(s)^2 \, ds$$

where s is arc length and $\psi(s)$ the curvature. Let us describe the configuration of the beam by $\psi(s)$, the angle the center line of the beam makes with the horizontal (See Figure 1.2.) Also we normalize the length of the rod to π. It is shown in Reiss' article in Keller and Antman [1969] that equilibria of the rod are described by the two point boundary problems

$$\frac{d^2\psi}{ds^2} + \lambda \sin \psi = 0, \quad 0 < s < \pi \tag{1.1a}$$

$$\psi'(0) = \psi'(\pi) = 0$$

J. M. Ball (ed.), Systems of Nonlinear Partial Differential Equations, 219–262.

Figure 1.1: The elastica

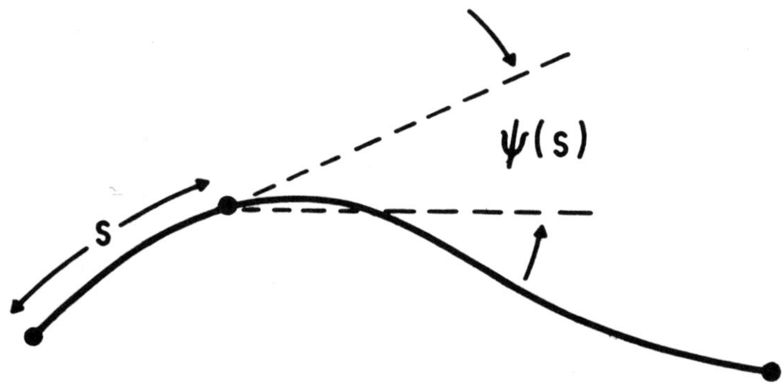

Figure 1.2: Definition of $\psi(s)$

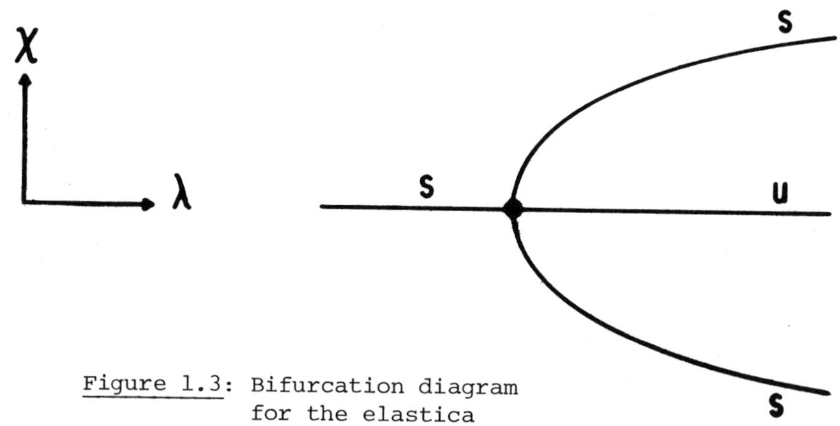

Figure 1.3: Bifurcation diagram for the elastica

where λ is the compressive load. Here we are assuming simply supported boundary conditions. This means that the ends are hinged, i.e., free to rotate.

The ideas of bifurcation theory come across more clearly if we reformulate (1.1) in an abstract way. Let

$$\mathfrak{X} = \{u \in C^2(0,\pi) : u'(0) = u'(\pi) = 0\}$$

where $C^2(0,\pi)$ denotes the set of twice continuously differentiable functions on $[0,\pi]$. Define a map

$$\Phi : \mathfrak{X} \times \mathbb{R} \to C^0(0,\pi)$$

by

$$\Phi(u,\lambda) = u'' + \lambda \sin u. \tag{1.2}$$

Observe that $\Phi(0,\lambda) = 0$ for all $\lambda \in \mathbb{R}$; this is the undeflected configuration of the beam. Bifurcation theory is concerned with the existence of other solutions. Because of the difficulty of such questions, often one can only answer a more restricted question: are there other solutions <u>close to the undeformed solution</u>? The restricted question may be approached through the implicit function theorem. Specifically, let $(d\Phi)_{u,\lambda} : \mathfrak{X} \to C^0(0,\pi)$ be the linearization of Φ with respect to its first argument at the point (u,λ); in symbols

$$(d\Phi)_{u,\lambda} \cdot v = \lim_{h \to 0} h^{-1}\{\Phi(u + hv,\lambda) - \Phi(u,\lambda)\}.$$

The operator (1.2), like the other operators we consider in these lectures, is Fréchet differentiable, so that $(d\Phi)_{u,\lambda}$ is a bounded linear operator. By the implicit function theorem one may solve the equation

$$\Phi(u,\lambda) = 0 \tag{1.3}$$

for u as a smooth function of λ locally near any value of λ for which $(d\Phi)_{u,\lambda}$ is invertible. However $u \equiv 0$, not depending on λ, is one solution of (1.3). Thus when $(d\Phi)_{u,\lambda}$ is invertible, this must be the only solution.

Let us investigate (1.2) with these techniques. The linearization of (1.2) is readily computed to be

$$(d\Phi)_{0,\lambda} \cdot v = v'' + \lambda v. \tag{1.4}$$

This operator is invertible unless the eigenvalue problem

$$v'' + \lambda v = 0, \qquad 0 < s < \pi$$

$$v'(0) = v'(\pi) = 0$$

has a solution, i.e., unless $\lambda = 0, 1^2, 2^2, 3^2, \ldots$. In particular for $0 < \lambda < 1$ the zero solution of (1.2) is locally unique. (Remark: It can be shown in this case that even globally the zero solution of (1.2) is unique.)

The analysis via the implicit function theorem suggests that something special happens at $\lambda = 1$. Here $(d\Phi)_{0,1}$ is not invertible. However $\ker(d\Phi)_{0,1}$ is only one dimensional, spanned by $\cos s$; likewise $\mathrm{range}(d\Phi)_{0,1}$ has codimension 1. We investigate this special point further by looking for solutions of (1.2) in the form

$$u = x \cos + \tilde{u}. \tag{1.5}$$

Here x is an unknown coefficient multiplying the cosine, not a spatial coordinate. The idea of the ansatz (1.5) is that the interesting phenomena are associated with fact that $(d\Phi)_{0,1}$ has a kernal; thus we single out this direction for special treatment in (1.5). All remaining components of u are to be included in the remainder $\tilde{u}$. To make the representation (1.5) unique we require that

$$\langle \tilde{u}, \cos \rangle = 0 \tag{1.6}$$

where $\langle \cdot, \cdot \rangle$ denotes the L^2 inner product,

$$\langle u, v \rangle = \int_0^\pi u(s)v(s)ds.$$

We claim that $\tilde{u}$ is uniquely determined by the equation

$$\Phi(x \cos + \tilde{u}, \lambda) = 0 \tag{1.7}$$

Indeed $\tilde{u}$ is overdetermined by (1.7)--the condition (1.6) reduces by one the dimension of the space in which $\tilde{u}$ lies and we must similarly restrict the equation (1.7) in order to achieve an invertible linearization. Technically this is formulated as follows. Let E be the self-adjoint projection in $L^2(0,\pi)$ onto the orthogonal complement of cos. (Remark: The one dimensional complementary projection $I - E$ has the explicit representation

$$(I - E)v(s) = \frac{2}{\pi} \{\int_0^\pi \cos t \; v(t)dt\} \cos(s).)$$

We split (1.7) into two equivalent equations

$$E\, \Phi(x \cos + \tilde{u}, \lambda) = 0 \tag{1.8a}$$

$$(I - E)\Phi(x \cos + \tilde{u},\lambda) = 0. \tag{1.8b}$$

Now we have achieved our goal. The linearization of (1.8a) at $u = 0$, $\lambda = 1$ equals

$$(d\Phi)_{0,1}: [\ker(d\Phi)_{0,1}] \to \text{range}(d\Phi)_{0,1}. \tag{1.9}$$

By general Banach space theory (1.9) is invertible, so it follows from the implicit function theorem that (1.8a) may be uniquely solved locally for $\tilde{u}$ as a function of x and λ, say $\tilde{u} = W(x,\lambda)$.

To complete our reduction of the problem let us define

$$g(x,\lambda) = (I - E)\Phi(x \cos + W(x,\lambda),\lambda). \tag{1.10}$$

Since the range of $I - E$ is one dimensional we may regard g as a map $g: \mathbb{R} \times \mathbb{R} \to \mathbb{R}$. (More accurately, g is only defined on a neighborhood of the origin in $\mathbb{R} \times \mathbb{R}$.) Thus the equation

$$g(x,\lambda) = 0 \tag{1.11}$$

is simply a single scalar equation for one scalar unknown x, depending on a parameter. The following theorem summarizes the information obtainable with these techniques.

<u>Theorem 1.1</u>: There is a neighborhood of $(0,0)$ in $C^2(0,\pi) \times \mathbb{R}$ in which the solutions of (1.2) are in one-to-one correspondence with the solutions of (1.11), the correspondence being provided by

$$(x,\lambda) \leftrightarrow (x \cos + W(x,\lambda),\lambda).$$

<u>Remark</u>: This reduction of the infinite dimensional problem to finite dimensions (here, one dimension) is an important general technique of bifurcation theory which is usually called the <u>Liapunov-Schmidt reduction</u>.

In the typical application of the Liapunov-Schmidt reduction there is no hope of obtaining an explicit formula for the function $g(x,\lambda)$ defined by (1.10). The problem is that $W(x,\lambda)$ is defined implicitly by an equation as complex as the original problem. However it is usually possible to compute some of the low order derivatives of g, and often this suffices. For example, it may be computed for the elastica that at $x = 0$, $\lambda = 1$ we have

$$g = g_x = g_{xx} = g_\lambda = 0 \tag{1.12a}$$

$$g_{xxx} = -\frac{3\pi}{8}, \quad g_{\lambda x} = \frac{\pi}{2}. \tag{1.12b}$$

(Remark: The equalities in (1.12a) admit simple intuitive explanations. We have $g(0,\lambda) \equiv 0$, coming from the undeformed solution of (1.3), so of course g and g_λ vanish. The equality $g_x = 0$ is related to the fact that x is a coordinate along $\ker(d\Phi)_{0,1}$. For the final equality, note that (1.2) is odd in u; moreover this parity is inherited by $g(x,\lambda)$. Thus g is odd in x and all even order derivatives vanish at $x = 0$. By contrast, (1.12b) can only be obtained by explicit computation.)

We have from (1.12) that

$$g(x,\lambda) = -ax^3 + b(\lambda - 1)x + \text{hot},$$

where $a,b > 0$ and hot indicates higher order terms. If we neglect those higher order terms, we may solve (1.11) explicitly: namely either $x = 0$ or $\lambda = 1 + a/b\, x^2$. These solutions are graphed in Figure 1.3. By Theorem 1.1, Figure 1.3 also provides an enumeration of solutions of the full equation $\Phi(u,\lambda) = 0$, provided the higher order terms in g do not change the qualitative structure of the zero set of g. The following theorem makes precise the sense in which the higher order terms have no effect.

Theorem 1.2: Let $g\colon \mathbb{R}^1 \times \mathbb{R}^1 \to \mathbb{R}^1$ satisfy

$$g = g_x = g_{xx} = g_\lambda = 0; \quad g_{xxx} < 0, \quad g_{\lambda x} > 0 \tag{1.13}$$

at $x = 0$, $\lambda = 0$. Then there exists a nonsingular change of coordinates on $\mathbb{R} \times \mathbb{R}$ near zero of the form $(X(x,\lambda),\Lambda(\lambda))$ and a positive function $s(x,\lambda)$ such that

$$g(x,\lambda) = s(x,\lambda)h(X(x,\lambda),\Lambda(\lambda)) \tag{1.14}$$

near zero, where $h(x,\lambda) = -x^3 + \lambda x$.

The following remarks may help the reader understand this result. The function h differs from g in that higher order terms have been neglected and the coefficients of the remaining terms have been scaled to unit magnitude. (Note also we have translated the bifurcation point from $\lambda = 1$ to $\lambda = 0$.) Since we are only interested in solutions of the equation $g(x,\lambda) = 0$, the nonzero function $s(x,\lambda)$ in (1.14) will have no effect. Thus (1.14) says that the diffeomorphism (X,Λ) maps the solution set of g into that of h, and the latter may of course be determined explicitly. Moreover the form of the change of coordinates preserves the slices of constant λ; this is appropriate since λ is set externally and not affected by x while the internal variable x is crucially affected by λ.

This theorem anticipates Lecture 2, where we will discuss it further. In particular we will take (1.14) as a definition: two functions g ahd h are <u>contact equivalent</u> if they may be related as in (1.14)--in other words, if one may pass from one to the other by a change of coordinates.

We conclude our discussion of the elastica with two remarks. First, at each of the values $\lambda = 1^2, 2^2, 3^2, \ldots$ new equilibrium solutions bifurcate from the trivial solution. (Keller and Antman [1969]). The second remark concerns stability, which can only be defined if an equation governing the evolution of the system is given. The lectures were too brief to permit a serious discussion of this topic. Loosely speaking an equilibrium solution u_0 is stable if the solution of the evolution equation with arbitrary initial data close to u_0 tends to u_0 as $t \to \infty$. For the elastica stability or instability of a solution branch is indicated on Figure 1.3 by "s" or "u", respectively. The proof of this involves some fairly subtle and not widely appreciated issues--see J. Maddocks [1982], for a discussion.

The second example we consider is the classical Bénard problem; i.e., convection in a planar layer of fluid heated from below. In the original experiments, Bérnard observed hexagonal convection cells. However in his apparatus the top of the fluid was a free surface. This configuration is harder to control experimentally and harder to model theoretically. More recently experimenters have worked with a fluid in contact with a good conductor of heat from both above and below. In this configuration it appears that rolls are the preferred solution. See Koschmieder [1974] for a good review of the experimental situation and Schluter, Lortz, and Busse [1965] for a theoretical analysis.

We consider the traditional model of this problem via the Navier-Stokes equations with the Boussinesq approximation. This means the fluid is treated as incompressible except for buoyancy; i.e., warmer fluid expands, becoming lighter, and therefore experiences an upward force. The state of the fluid is characterized by a velocity field $\vec{v}(x,y,z,t)$ and a temperature $T(x,y,z,t)$. The fluid is heated from below, so even in the absence of motion there is a temperature gradient. Thus we write

$$T(x,y,z,t) = Rz + \theta(x,y,z,t)$$

where R is the Raleigh number and θ measures the deviation of the temperature from the pure conduction state. The Boussinesq equations are

$$\frac{1}{P}\left\{\frac{\partial \vec{v}}{\partial t} + (v \cdot \nabla)\vec{v}\right\} = -\nabla p + \theta \vec{g} + \Delta \vec{v} \qquad (1.15a)$$

$$\operatorname{div} \vec{v} = 0 \qquad (1.15b)$$

$$\frac{\partial \theta}{\partial t} + (v \cdot \nabla)\theta = Rv_3 + \Delta\theta \qquad (1.15c)$$

Here P is the Prandtl number (the dimensionless ratio of viscosity to thermal conductivity), p is the pressure, $\vec{g}$ a unit vector in the z-direction representing gravity, v_3 the third component of $\vec{v}$, and Δ denotes the Laplace operator. Equations (1.15) are imposed on the domain

$$\Omega = \{(x,y,z) \in \mathbb{R}^3 : 0 < z < \pi\}.$$

Boundary conditions for these equations must be imposed on the faces $z = 0,\pi$. The most realistic boundary conditions for modelling the experiments with rigid, conducting boundaries are

$$\vec{v} = 0, \quad \theta = 0 \quad \text{on} \quad z = 0,\pi. \tag{1.16}$$

However numerical computation is required to solve the equations with these boundary conditions. Therefore in these lectures we will consider the simpler conditions

$$\frac{\partial}{\partial z} v_1 = \frac{\partial}{\partial z} v_2 = v_3 = \theta = 0 \quad \text{on} \quad z = 0,\pi \tag{1.17}$$

which describe a sort of stress free surface. The qualitative results are not affected by the change in boundary conditions. See Schluter, Lortz, and Busse [1965] or Chandrasekhar [1961] for a treatment of (1.16); these references also contain a careful derivation of the governing equations (1.15).

Note that $\vec{v} = 0$, $\theta = 0$ is an equilibrium solution of (1.15), (1.17), the pure conduction state. Bifurcation theory is concerned with the question of whether there are other equilibrium solutions. Equation (1.15), with terms containing a time derivative deleted, characterizes equilibrium solutions. We write this abstractly as

$$\Phi(u,R) = 0 \tag{1.18}$$

where $u = (\vec{v},\theta)$ is a shorthand for the state vector of the system. We discuss the uniqueness question for (1.18) through its linearization. Here we are proceeding formally since we have not chosen specific spaces for the domain and range of Φ.

On deleting time derivatives and linearizing (1.15) around the zero solution we obtain the system

$$\Delta\vec{v} - \nabla p + \theta\vec{g} = 0 \tag{1.19a}$$

$$\text{div } v = 0 \tag{1.19b}$$

$$\Delta\theta + Rv_3 = 0. \tag{1.19c}$$

This is a linear system of pde with constant coefficients on a domain which is invariant under translations in x and y. Thus we look for solutions with the form

$$e^{i(kx+\ell y)} f(z). \tag{1.20}$$

Moreover (1.19) is invariant under rotations in the x,y plane. Thus there are solutions of the form (1.20) if and only if there are solutions of the form

$$e^{i k' x} f(z) \tag{1.21}$$

where $k' = \sqrt{k^2 + \ell^2}$. In our analysis below we omit the prime in (1.21).

Note that (1.21) does not depend on y. Let us write out (1.19) componentwise, omitting all y-derivatives.

$$\Delta v_1 - \frac{\partial p}{\partial x} = 0 \tag{1.22a}$$

$$\Delta v_2 = 0 \tag{1.22b}$$

$$\Delta v_3 - \frac{\partial p}{\partial z} + \theta = 0 \tag{1.22c}$$

$$\Delta \theta + R\, v_3 = 0 \tag{1.22d}$$

$$\frac{\partial v_1}{\partial x} + \frac{\partial v_3}{\partial z} = 0 \tag{1.22e}$$

Equation (1.22b), together with the boundary condition (1.17), implies that v_2 is a constant. By passing to a moving frame, if necessary, we may assume that $v_2 = 0$. To satisfy (1.22e) we express the velocity in terms of the stream function ψ:

$$v_1 = \frac{\partial \psi}{\partial z}, \quad v_3 = -\frac{\partial \psi}{\partial x}.$$

Finally we eliminate the pressure by taking $\frac{\partial}{\partial z}$ of (1.22a) minus $\partial/\partial x$ of (1.22c). This yields the system

$$\Delta^2 \psi - \frac{\partial \theta}{\partial x} = 0 \tag{1.23a}$$

$$\Delta \theta - R \frac{\partial \psi}{\partial x} = 0. \tag{1.23b}$$

We now substitute the form (1.21) of the solution into (1.23). More accurately, to avoid working with complex numbers we insert the correct phase relationship explicitly and look for a solution of (1.23) with the form

$$\psi = \sin kx\, \tilde{\psi}(z), \quad \theta = \cos kx\, \tilde{\theta}(z). \tag{1.24}$$

We find that (1.24) satisfies (1.23) if $\tilde{\psi}, \tilde{\theta}$ satisfy the ode

$$\begin{pmatrix} (D^2 - k^2)^2 & k \\ -Rk & D^2 - k^2 \end{pmatrix} \begin{pmatrix} \tilde{\psi} \\ \tilde{\theta} \end{pmatrix} = 0 \tag{1.25}$$

where $D = d/dz$.

To satisfy (1.17) we must require that

$$\psi = d^2\psi/dz^2 = \theta = 0 \quad \text{for} \quad z = 0,\pi. \tag{1.26}$$

The beauty of these boundary conditions is that the natural ansatz to substitute into (1.25), namely e^{inz}, automatically leads to functions which satisfy (1.26), provided n is an integer. By comparison, with (1.16) one must take linear combinations of several complex exponentials to satisfy the boundary conditions, and an explicit solution is no longer possible.

We now look for solutions of (1.25) depending on z as sin z, n an integer. We find that there is a nonzero solution of this form if and only if

$$\det \begin{pmatrix} (-n^2 - k^2)^2 & k \\ -Rk & -n^2 - k^2 \end{pmatrix} = 0$$

which implies that

$$R = \frac{(n^2 + k^2)^3}{k^2} . \tag{1.27}$$

Define $R_c(k)$ as the RHS of (1.27) when $n = 1$. (Sub-"c" is a nmemonic for critical.) The graph of this function is sketched in Figure 1.4.

It is time to interpret our calculations. We are considering the possibility of bifurcation from the zero solution of an equation $\Phi(u,R) = 0$. We have shown that the linearization of the equation admits bounded exponential solutions if and only if

$$R \geq \min_k R_c(k) \tag{1.28}$$

In other words, if $R < \min R_c(k)$ the linearization $d\Phi$ is invertible, at least on any space amenable to Fourier analysis. This indicates that the zero solution is isolated.

It is instructive to compare the present situation with that of the beam, considered above. There we found that for $0 < \lambda < 1$ $d\Phi$ was invertible, for $\lambda = 1$ dim ker $d\Phi = 1$, and for $1 < \lambda < 4$

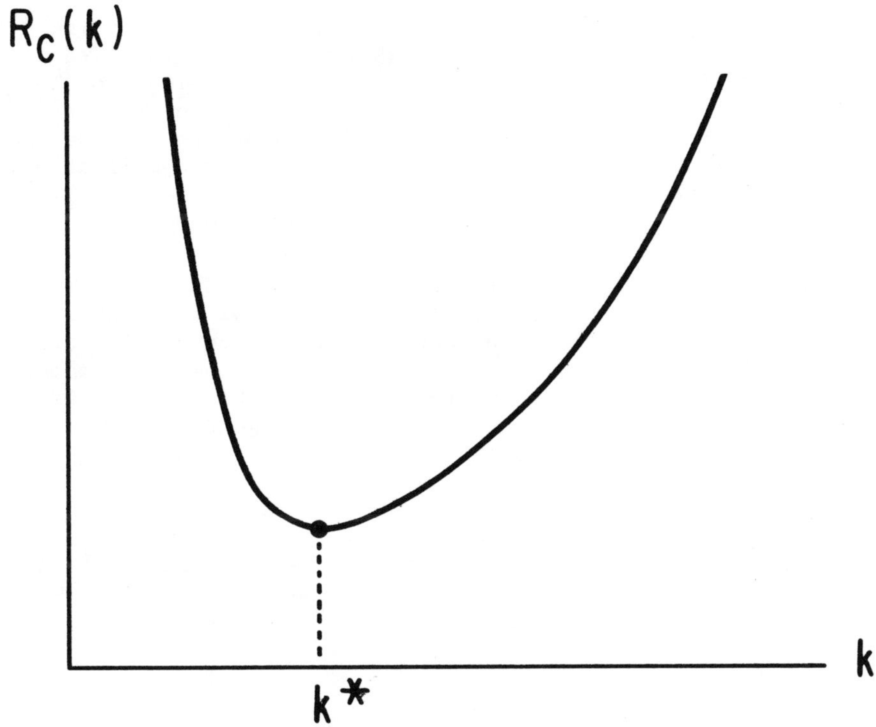

Figure 1.4: The critical Raleigh number as a function of wave number

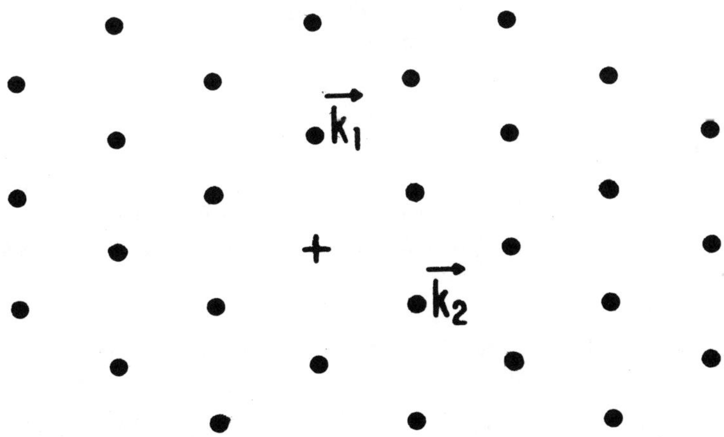

Figure 1.5: The lattice of wave numbers

$d\Phi$ was again invertible. Here we find that for $0 < R < \min R_c(k)$ $d\Phi$ is invertible, but that for any $R \geq \min R_c(k)$ the kernal of $d\Phi$ is nontrivial. Indeed the kernal is infinite dimensional--we found a solution with the form e^{ikx}, but rotation of the axes leads to new solutions. (This is the reverse of the process from which (1.21) was derived from (1.20).) Yet another complication of the present situation is that range $d\Phi$ is not closed. All these difficulties stem from working on a noncompact domain.

The standard mathematical way of dealing with these difficulties calls upon the experimental fact that the observed flows seem to be periodic. Thus one restricts attention by fiat to periodic solutions. In this way one achieves a compact domain. This approach ignores the possible effects of the other modes and leaves open a number of important questions. We follow this approach somewhat reluctantly, but viable alternatives are lacking.

The most heavily investigated question in this whole subject concerns the competition between rolls and hexagons. Therefore in choosing a space of periodic functions it is desirable that the space contain solutions of both types. We do that as follows. Let

$$\vec{e}_1 = \frac{4\pi}{\sqrt{3}\,k^*}(1,0), \qquad \vec{e}_2 = \frac{2\pi}{\sqrt{3}\,k^*}(1,\sqrt{3})$$

where k^* is the value of k at which $R_c(k)$ achieves its minimum. We consider functions periodic with respect to the grid generated by $\vec{e}_1$ and $\vec{e}_2$; i.e., functions such that

$$T_{e_1} u = T_{e_2} u = u$$

where T denotes translation. Such periodic functions may be expanded in a double Fourier series

$$\sum_{j_1, j_2 \in \mathbb{Z}} C_{j_1, j_2}\, e^{i(j_1\vec{k}+j_2\vec{k}_2)\cdot(x,y)} \tag{1.29}$$

where the fundamental wave numbers for this grid are

$$\vec{k}_1 = k^*(0,1), \qquad \vec{k}_2 = \frac{k^*}{2}(\sqrt{3},-1).$$

Note that $\vec{k}_1$ and $\vec{k}_2$ both have length k^* and make an angle of 120° with one another. The lattice of wave numbers which occurs in (1.29) is sketched in Figure 1.5.

Now that we have restricted to periodic functions there are only discrete values of R for which $d\Phi$ fails to be invertible, namely

$$\bigcup_{j_1, j_2} R_c(j_1\vec{k}_1 + j_2\vec{k}_2) \; .$$

However $R_c(\vec{k})$ only depends on $|\vec{k}|^2$ and min $R_c(\vec{k})$ occurs for $|\vec{k}| = k^*$. Thus $d\Phi$ on the restricted space is invertible for $R < R(k^*)$, is also invertible for $R(k^*) < R < R(k^*) + \varepsilon$ for some $\varepsilon > 0$, while for $R = R(k^*)$, $d\Phi$ has a 6 dimensional kernal spanned by the six terms in (1.29) whose wave numbers have length k^*.

The pde $\Phi(u,R) = 0$ describing the existence of spatially periodic, steady flows may be reduced to a finite system of equations by the Liapunov-Schmidt technique. Here the reduction leads to a 6×6 system of equations $g(x,R) = 0$ where

$$g\colon \mathbb{R}^6 \times \mathbb{R} \to \mathbb{R}^6 \, .$$

The reduced problem has been discussed by Schluter, Lortz, Busse [1965] among others. We will return to this problem in Lecture 3. Following Sattinger [1979] we will study the problem with group theory methods.

LECTURE 2: UNFOLDINGS

David G. Schaeffer

Duke University, Durham, NC 27706

In this lecture we develop the theory of unfoldings or perturbations of a bifurcation problem. The underlying mathematics here was first conjectured by R. Thom and made rigorous by J. Mather. (See Guillemin and Golubitsky [1973] for references.) The application to bifurcation theory has been pursued by the author and M. Golubitsky in a series of papers. We emphasize that the concepts in this lecture are local ones. Thus when we speak of a smooth function we understand it to be restricted to a suitably small neighborhood. In technical language we are dealing with germs. To smooth the exposition we frequently will not be explicit about this assumption.

We begin this lecture by describing another configuration of the elastica which also leads to a pitchfork bifurcation, the shallow arch illustrated in Figure 2.1. This system is governed by essentially the same equations as those in Lecture 1, except that here the following two constraints play an essential role: the two ends must have the same y-coordinates and x-coordinates differing by L, the distance between the two pins. (See Thompson and Hunt [1973] for greater detail.) We take λ, the suspended load, as the bifurcation parameter. This problem leads to a subcritical pitchfork bifurcation as illustrated in Figure 2.2--the term subcritical indicates that the bifurcating solution branches exist for $\lambda < \lambda_0$ rather than $\lambda > \lambda_0$ as in Lecture 1. In this situation the bifurcating solution branches are unstable. Thus if λ is increased beyond λ_0 the system will jump to a new equilibrium far from the bifurcation point. The following remarks may be helpful for the reader wondering where this new equilibrium is located in Figure 2.2.

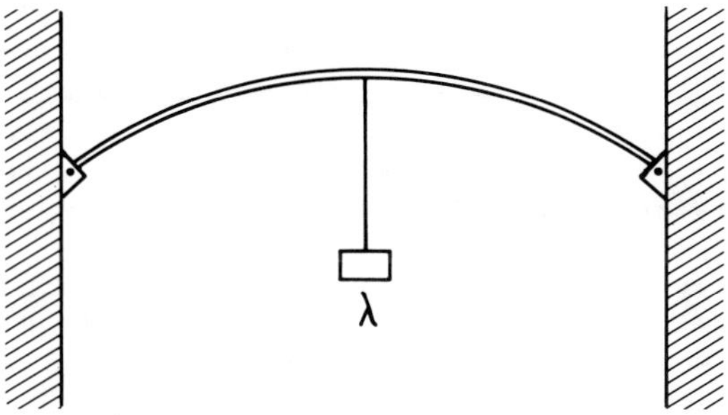

Figure 2.1: The shallow arch

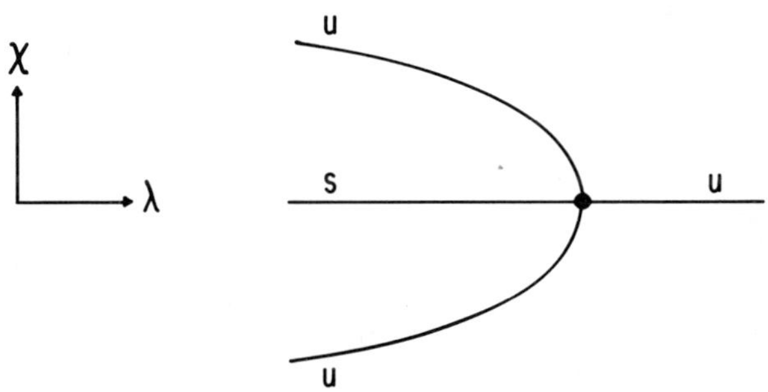

Figure 2.2: Bifurcation diagram for the shallow arch

Figure 2.3: Asymmetric buckling

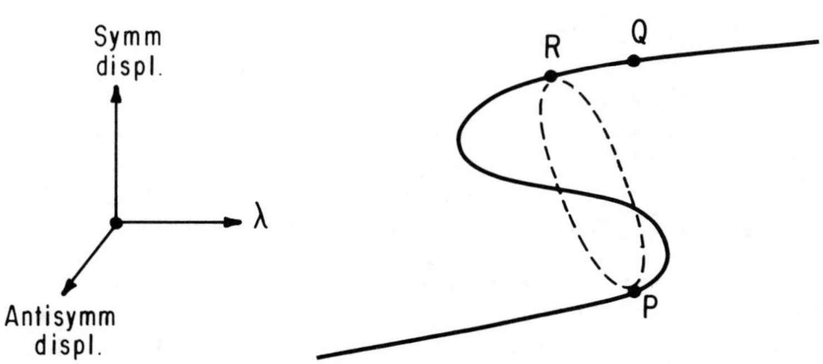

Figure 2.4: Expanded bifurcation diagram

Unlike the situation which occurred in Lecture 1, the line of "trivial" solutions in Figure 2.2 does not represent a pure, undeformed state. Rather this line represents solutions of the equations which are symmetric about the midpoint of the arch but which vary with λ. When instability first occurs, the bifurcation is into a mode which is antisymmetric with respect to reflection about the midpoint of the beam. The (unstable) equilibrium configuration corresponding to a typical point off the axis in the bifurcation diagram of Figure 2.2 is illustrated in Figure 2.3. By contrast the new equilibrium solution after snap through is again symmetric. Thus three dimensions are required for even a symbolic representation of this situation: one axis for λ, one for symmetric displacements, one for antisymmetric displacements. This is given in Figure 2.4. Here the solid line represents symmetric configurations, the dotted line, asymmetric. The point P is the bifurcation point of Figure 2.2; Q is the new equilibrium after snap through. The asymmetric states which bifurcate at P rejoin the symmetric branch at R; note that $\lambda_R < \lambda_Q = \lambda_P$. Again we refer to Thompson and Hunt [1973] for details.

This example has a number of implications for engineering practice. The first is that instability occurs first in a asymmetric mode. If one considered only symmetric configurations of this structure, the computed carrying load of this structure would be too high. The second is the issue of imperfection sensitivity, to which we now turn.

Figure 2.2 is the graph of the equation $g(x,\lambda) = 0$ where g is the function which results from the Liapunov-Schmidt reduction of this problem. We may perform an appropriate change of coordinates so that $g(x,\lambda) = x^3 + (\lambda - \lambda_0)x$. (Cf. Theorem 1.2.) As an extremely crude way of assessing the effect of imperfections on this problem, let us merely add a constant term, say

$$x^3 + (\lambda - \lambda_0)x + \alpha = 0 \; . \qquad (2.1)$$

For $\alpha > 0$ this deforms Figure 2.2 as indicated in Figure 2.5; for $\alpha < 0$ the mirror image, reflected about the λ axis, obtains. This behavior may be understood as follows. Near the origin the equation is dominated by the terms $\lambda x + \alpha = 0$ which determine in which way the four half lines of $\lambda x = 0$ split apart on making $\alpha \neq 0$. Far from the origin the terms $x^3 + \lambda x$ dominate, so here Figures 2.2 and 2.5 resemble one another.

Let us attempt to determine the λ-coordinate of the point P in Figure 2.5, as this is the quantity of principal significance; namely the load at which snap through occurs. This point is characterized the equations

$$g = g_x = 0. \qquad (2.2)$$

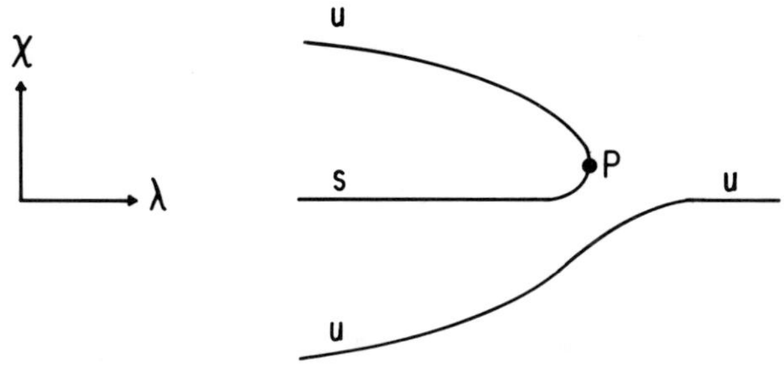

Figure 2.5: A perturbed bifurcation diagram

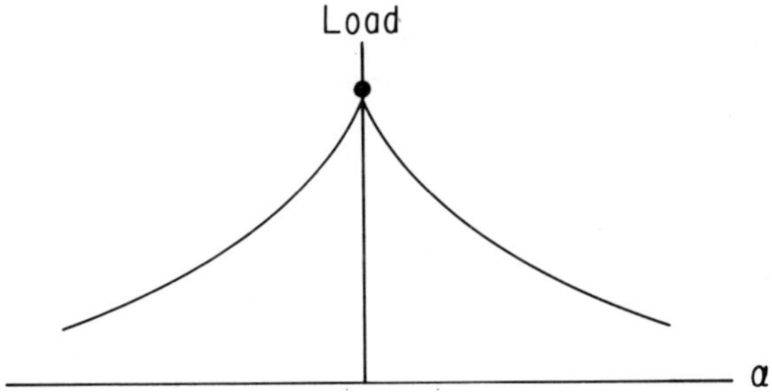

Figure 2.6: Carrying load as a function of α. (Theoretical)

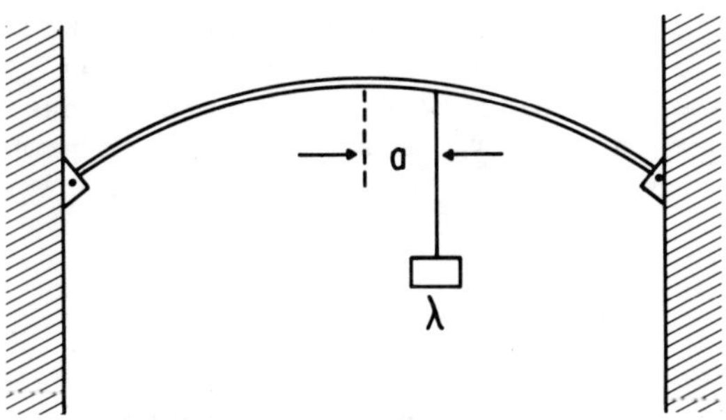

Figure 2.7: Asymmetric loading

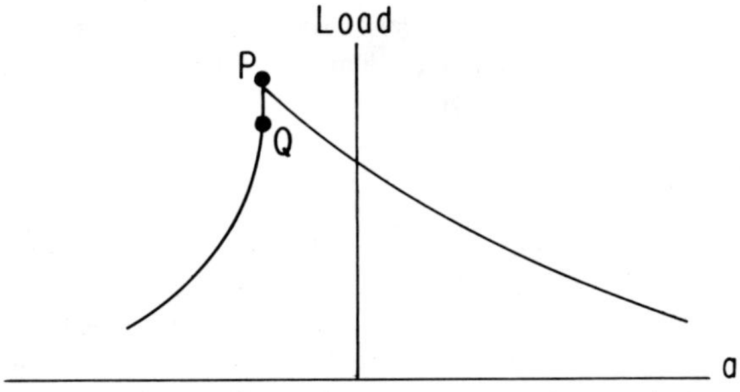

Figure 2.8: Carrying load as a function of a. (Experimental)

(The first equality says that P is in the zero set of g, the second that the tangent is vertical at P.) On substituting (2.1) into (2.2) we deduce from the second equation that $\lambda - \lambda_0$ is proportional to x^2, and from the first that α is proportional to x^3. It follows that

$$\lambda_0 - \lambda = C\,\alpha^{2/3} \tag{2.3}$$

for some constant C which turns out to be positive. Equation (2.3) is graphed in Figure 2.6. Note that the carrying load is very sensitive to small changes in α near $\alpha = 0$; the derivative in fact tends to ∞ as $\alpha \to 0$. In effect the shallow arch is only as strong as it is because it is perfectly symmetric; speaking anthropomorphically, the system doesn't buckle sooner because it can't decide which way to break symmetry.

Roorda [1965] has performed some most interesting experiments on the shallow arch in connection with imperfection sensitivity. As indicated in Figure 2.7 he applied the load a distance a off center for various values of a and measured the carrying load. His results are sketched in Figure 2.8. Note that the maximum load occurs slightly off center and that the cusp is not quite symmetric. It turns out that understanding these small discrepancies opens up fundamental issues for us.

The mathematical formulation of the unperturbed problem via the Liapunov-Schmidt reduction led to an equation $g(x,\lambda) = 0$. This analysis can easily be extended to Roorda's case, carrying along the parameter a. This leads to a reduced problem of the form

$$g(x,\lambda,a) = 0 \tag{2.4a}$$

where $g(x,\lambda,0)$ is the original reduced equation. (In the jargon of singularity theory, (2.4a) is an unfolding of the original equation.) However (2.4a) is still not an adequate model for the true experiment in that it is assumed that the arch is perfectly uniform along its length, that the two pins at ends are frictionless, etc. In the true experiment these idealized assumptions will be invalidated in innumerable small ways. Thus we propose that the most realistic mathematical model should be

$$g(x,\lambda,a) + \varepsilon p(x,\lambda,a) = 0 \tag{2.4b}$$

where the εp term is a small <u>but</u> <u>unknown</u> perturbation. Indeed a different choice of p would be required if one attempted to repeat the experiment, no matter how hard one tried to duplicate it exactly. Thus for making connections with experiment it is important to understand the effect of a small but arbitrary perturbation on a bifurcation problem.

The general theory of Golubitsky and Schaeffer [1979a,b] is well adapted to handle such perturbations of a bifurcation problem. This theory studies qualitative properties, by which we mean those properties which are unchanged by coordinate transformations. More precisely, two (reduced) bifurcation problems $g,h: \mathbb{R} \times \mathbb{R} \to \mathbb{R}$ are <u>contact equivalent</u> if they may be related as in (1.14). We shall now present two central results of the general theory, with sketches of their proofs, that are relevant for interpreting the experiments with the shallow arch. Here in quoting these results we restrict ourselves to the special case of the pitchfork bifurcation. Afterwards we indicate briefly how the general theorems differ from the special case; we refer to the original papers for greater detail.

<u>Theorem 2.1</u>: $G(x,\lambda;\alpha,\beta) = x^3 + \lambda x + \alpha + \beta x^2$ is a universal unfolding of $g(x,\lambda) = x^3 + \lambda x$.

The essential content of this result is the following. Given any small perturbation p, there exist values $\alpha(\varepsilon)$ and $\beta(\varepsilon)$ of the parameters such that for small ε, $g + \varepsilon p$ is contact equivalent to $G(\cdot,\cdot;\alpha(\varepsilon),\beta(\varepsilon))$. Thus, up to contact equivalence, the effect of an arbitrary small perturbation is included in a distinguished, two parameter family of perturbations. The next result discusses how the distinguished family of bifurcation problems $G(\cdot,\cdot;\alpha,\beta)$ depends on the two parameters α and β.

<u>Theorem 2.2</u>: The solution set

$$\{(x,\lambda): G(x,\lambda;\alpha,\beta) = 0\}$$

depends on α,β as indicated in Figure 2.9.

In Figure 2.9 the β axis and the curve $\alpha = \beta^3/27$ divide the α,β plane into four regions. Any choice of α,β within a given region leads to a bifurcation problem contact equivalent to the one sketched inside the region.

Let us discuss how these concepts apply to the experiment of Roorda. After the Liapunov-Schmidt reduction this problem is described mathematically by (2.4b). When $a = \varepsilon = 0$, this bifurcation problem is equivalent to $x^3 + \lambda x = 0$. By Theorem 2.1 there exist parameter values $\alpha(a,\varepsilon)$, $\beta(a,\varepsilon)$ for small a and ε such that $g + \varepsilon p$, the perturbed problem in (2.4b), is equivalent to $G(\cdot,\cdot;\alpha(a,\varepsilon),\beta(a,\varepsilon))$, where G is defined in the theorem. We are interested in (2.4b) for fixed ε as a varies. Theorem 2.2 suggests that we look at the curve in the α,β plane that $\alpha(a,\varepsilon)$, $\beta(a,\varepsilon)$ describes as a varies. Generically this curve will not pass through the origin; rather it will have the appearance sketched in Figure 2.10. Thus it is immediately clear that the response of the perturbed system should not be symmetric in a. This line of

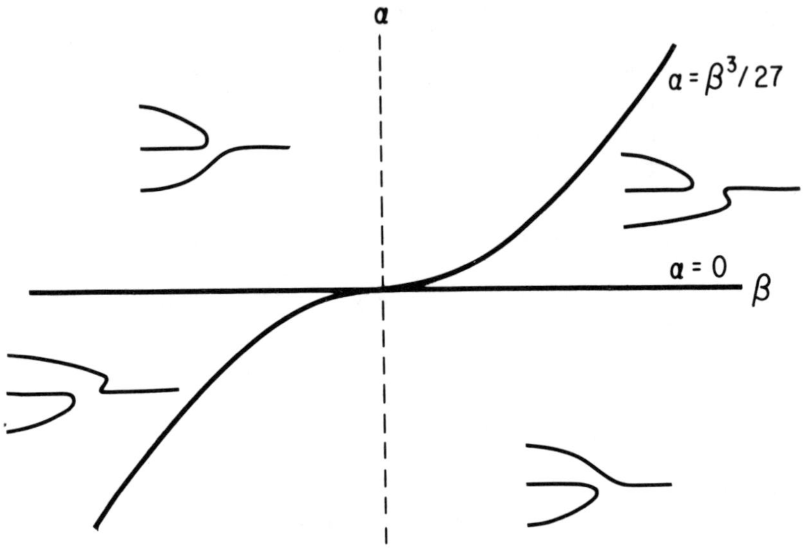

Figure 2.9: Graph of the parameter space (Cf. Theorem 2.2)

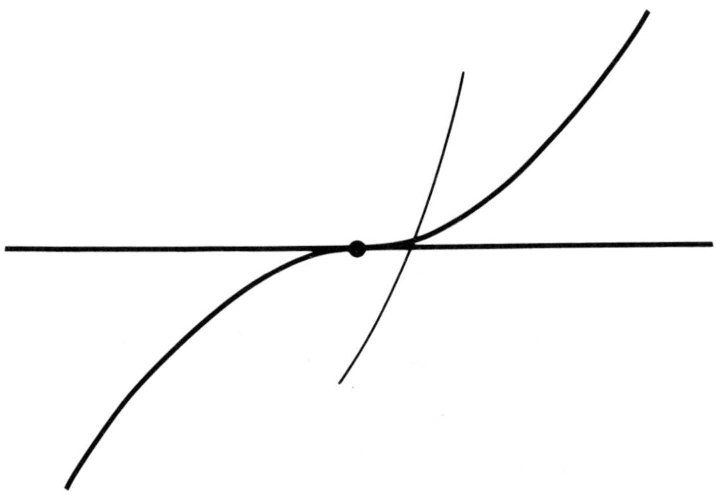

Figure 2.10: A generic curve in parameter space

analysis can be pursued considerably further. For example, Figure 2.10 suggests that there should be three distinct ranges of a in which different qualitative behavior obtains; those three ranges correspond to

$$a < a_P; \quad a_P < a < a_Q, \quad a_Q < a$$

in Figure 2.8. However we leave it for the reader to continue this analysis; we now turn to the proofs of Theorems 2.1 and 2.2.

The answer to the following question forms the basis for the proof of Theorem 2.1. For what perturbations p is $g + \varepsilon p$ contact equivalent to g itself <u>to first order in ε</u>? Here by equivalent to first order in ε we mean that

$$g(x,\lambda) + \varepsilon p(x,\lambda) = s(x,\lambda,\varepsilon) g(X(x,\lambda,\varepsilon), \Lambda(\lambda,\varepsilon)) + \mathcal{O}(\varepsilon^2) \ . \tag{2.5}$$

This equation may be analyzed by a simple Taylor series argument. The general construction is to mod out by all the perturbations p which do not change g to first order; the perturbations which remain are precisely those needed in the universal unfolding. The proof that this construction suffices depends on the Malgrange Preparation Theorem; see Golubitsky and Schaeffer [1979a] for details.

Let us consider this construction in somewhat greater detail. We expand everything in powers of ε:

$$s(x,\lambda,\varepsilon) = 1 + \varepsilon s_1(x,\lambda) + \mathcal{O}(\varepsilon^2)$$

$$X(x,\lambda,\varepsilon) = x + \varepsilon X_1(x,\lambda) + \mathcal{O}(\varepsilon^2)$$

$$\Lambda(\lambda,\varepsilon) = \lambda + \varepsilon \Lambda_1(\lambda) + \mathcal{O}(\varepsilon^2).$$

Then to satisfy (2.5) we need only match the zeroth and first powers of ε; moreover the zeroth power is automatically matched. Hence (2.5) is satisfied if and only if

$$p = s_1 g + g_x X_1 + g_\lambda \Lambda_1. \tag{2.6}$$

Thus the fundamental question has been rephrased: For what perturbations p are there functions $s_1(x,\lambda)$, $X_1(x,\lambda)$, and $\Lambda_1(\lambda)$ such that the representation (2.6) is possible? In our subsequent discussion of (2.6) we omit the subscript "1". For the time being we ignore the third term involving $\Lambda_1(\lambda)$, which contributes much less as a consequence of being independent of x. Thus we ask what p's can be represented in the form

$$p = s g + g_x X \tag{2.7}$$

for some s and X?

The following observation is fundamental, both for the proofs and for carrying out computations. If $p(x,\lambda)$ admits the representation (2.7), so does $f(x,\lambda)p(x,\lambda)$ for any smooth function f; indeed one may take fs and fX as the appropriate functions in (2.7). Similarly, if p_1 and p_2 admit the representation (2.7), so does $p_1 + p_2$. In other words, the functions admitting the representation (2.7) form an <u>ideal</u> in the algebra of smooth functions.

In our specific example $g(x,\lambda) = x^3 + \lambda x$, (2.7) becomes

$$p = s(x,\lambda)(x^3 + \lambda x) + (3x^2 + \lambda)X(x,\lambda) . \tag{2.8}$$

<u>Lemma 2.3</u>: $x^3, \lambda x, \lambda^2$ may all be written in the form (2.8).

<u>Proof</u>: First we consider x^3. Indeed we have

$$x^3 = -\frac{1}{2}(x^3 + \lambda x) + \frac{x}{2}(3x^2 + \lambda);$$

i.e., x^3 admits the representation (2.8) with $s = -1/2$, $X = x/2$. The proof for λx and λ^2 is similar and is left to the reader.

Now, given any smooth function $p(x,\lambda)$ there exist constants a,b,c,d such that

$$p = a + bx + cx^2 + d\lambda + p_1 \tag{2.9}$$

where at $x = \lambda = 0$

$$p_1 = \frac{\partial p_1}{\partial x} = \frac{\partial^2 p_1}{\partial x^2} = \frac{\partial p_1}{\partial \lambda} = 0.$$

The remainder p_1 may be expanded

$$p_1 = x^3 q_1(x,\lambda) + \lambda x\, q_2(x,\lambda) + \lambda^2 q_3(x,\lambda),$$

so p_1 admits the representation (2.7). In fact we can do a little better by returning to (2.6). Observe that

$$x = g_\lambda \tag{2.10a}$$

$$\lambda = -3x^2 + g_x . \tag{2.10b}$$

We may use (2.10a) to express the term bx in (2.9) in the form (2.6), and we may use (2.10b) to express $d\lambda$ in terms of x^2 plus a remainder which admits the representation (2.6). In other words, for any smooth $p(x,\lambda)$ there exist constants α and β

such that

$$p(x,\lambda) = \alpha + \beta x^2 + p_2(x,\lambda) \tag{2.11}$$

where p_2 admits the representation (2.6). The two functions 1 and x^2 which remain in (2.11) cannot be eliminated; indeed these are the two functions which remain in the universal unfolding of Theorem 2.1.

We have summarized the calculations needed to determine the universal unfolding of $x^3 + \lambda x$. As stated above, the proof that these calculations suffice is given in Golubitsky and Schaeffer [1979a].

The specific bifurcation problem $x^3 + \lambda x$ which we have discussed here is fairly representative of the general case. For reference we give a more general theorem now. Suppose that g is a bifurcation problem such that there are k functions $\pi_i(x,\lambda)$ with the property that

$$\mathcal{E}_{x,\lambda} = \mathbb{R}\{\pi_1,\ldots,\pi_k\} + \mathcal{E}_{x\lambda}\{g,g_x\} + \mathcal{E}_\lambda\{g_\lambda\} .$$

This means that any function p can be represented in the form

$$p = a_1\pi_1 + \ldots + a_k\pi_k + s(x,\lambda)g + g_x X(x,\lambda) + g_\lambda \Lambda(\lambda)$$

where $a_i \in \mathbb{R}$. For example $x^3 + \lambda x$ satisfies this hypothesis with $k = 2$, $\pi_1 = 1$, $\pi_2 = x^2$. Then a universal unfolding of g is provided by

$$G(x,\lambda,\alpha) = g(x,\lambda) + \sum_{i=1}^{k} \alpha_i \pi_i(x,\lambda) .$$

This is still not the most general result--in particular the theory covers finite dimensional systems of equations--but it indicates the direction.

We now turn to Theorem 2.2. The following concept is the basis for the proof of this result. A bifurcation problem g is called <u>persistent</u> if for every perturbation $p, g + \varepsilon p$ is contact equivalent to g for small ε. (In singularity theory "stable" is often used for this concept, but this seems confusing for bifurcation theory.) There are three obvious ways in which a bifurcation problem can fail to be persistent; these possibilities are illustrated on the left in Figure 2.11. Typical perturbed diagrams with different structure are sketched on the right. It was proved in Golubitsky and Schaeffer [1979a] that a bifurcation problem is persistent unless it has one of these three degeneracies.

We want to apply the result just quoted to the universal un-

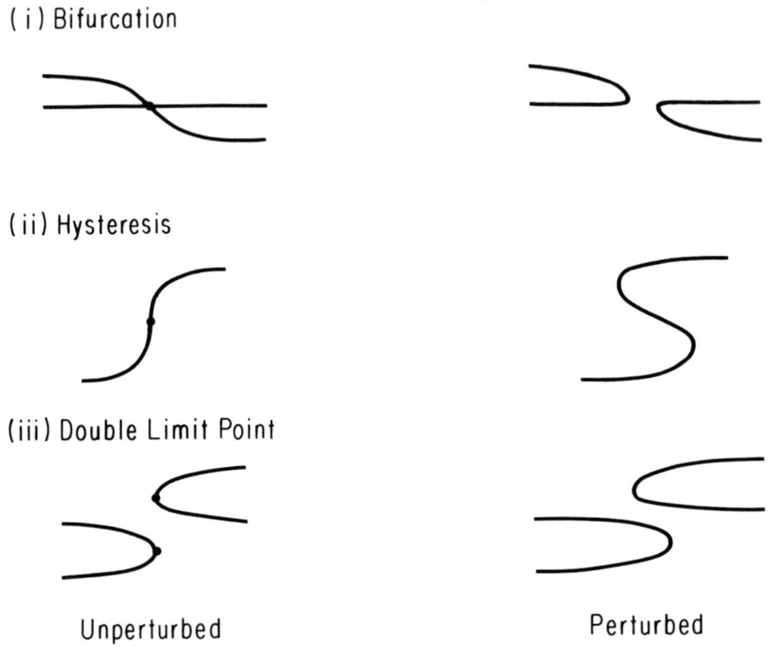

Figure 2.11: Degeneracies causing nonpersistence

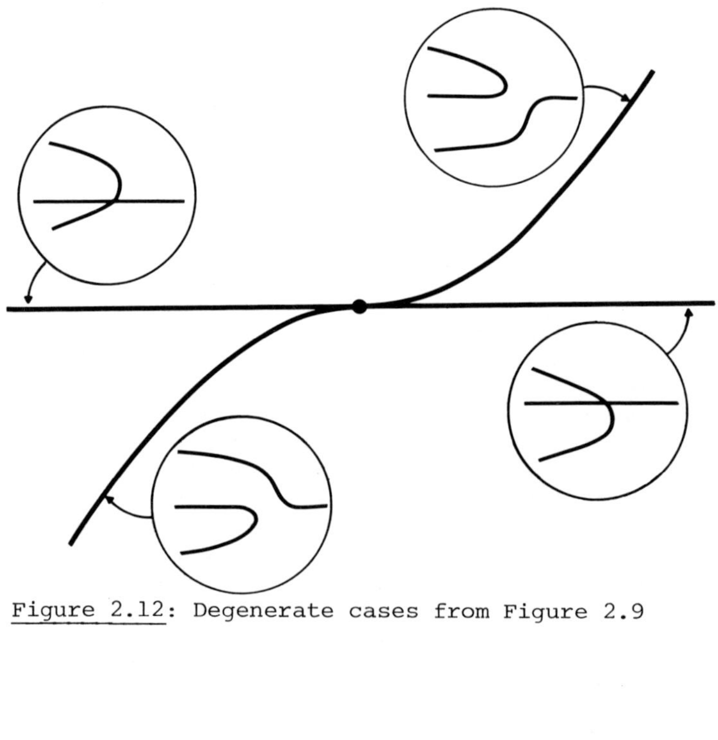

Figure 2.12: Degenerate cases from Figure 2.9

folding of $x^3 + \lambda x$. Specifically we want to know for what values of α,β does $G(\cdot,\cdot;\alpha,\beta)$ exhibit one of these three degeneracies. It turns out that there are no double limit points, so we only consider the first two degeneracies.

We claim that given a pair of parameter values α,β, the bifurcation diagram $\{(x,\lambda): G(x,\lambda,\alpha,\beta) = 0\}$, can exhibit a bifurcation degeneracy only if there exist x and λ satisfying the equations

$$G = G_x = G_\lambda = 0. \tag{2.12}$$

This follows from the implicit function theorem: at a bifurcation point the solution set of $G = 0$ is not a manifold, and if either derivative in (2.12) were nonzero one could solve the equation $G = 0$ for one of the variables as a function of the other. Similarly we claim that at a hysteresis point the equations

$$G = G_x = G_{xx} = 0 \tag{2.13}$$

must be satisfied. This expresses the condition that $\lambda = O(x^3)$ at a hysteresis point. In both (2.12) and (2.13) other derivatives might well vanish--the conditions are merely necessary. This fact is not important for us.

We now substitute the universal unfolding $G(x,\lambda,\alpha,\beta) = x^3 + \lambda x + \alpha + \beta x^2$ into (2.12) and (2.13). In each case we use two of the three equations to solve for x and λ, and the third equation then yields a relation between α and β that must be satisfied. We find

Bifurcation: $\alpha = 0$

Hysteresis : $\alpha = \beta^3/27$.

These are the two curves in Figure 2.9. The result quoted above implies that all bifurcation diagrams within one connected component of the complement must be equivalent. One can draw the bifurcation diagrams of Figure 2.9 by graphing one choice of parameters from each of the four regions. There are also useful checks on internal consistency. In particular the bifurcation diagrams on either side of the bifurcation curve $\{\alpha = 0\}$ or the hysteresis curve $\{\alpha = \beta^3/27\}$ must come from perturbations of a diagram with a bifurcation point or a hysteresis point. Appropriate diagrams for the case at hand are shown in Figure 2.12.

We close with an indication of the general result. Let $G(x,\lambda;\alpha_1,\ldots,\alpha_k)$ be the universal unfolding of a bifurcation problem g. Consider the following three subsets of the parameter space $\mathbb{R}^k$.

$$B = \{\alpha \in \mathbb{R}^k : (\exists x,\lambda) \text{ s.t. } G = G_x = G_\lambda = 0\}$$

$$H = \{\alpha \in \mathbb{R}^k : (\exists x,\lambda) \text{ s.t. } G = G_x = G_{xx} = 0\}$$

$$D = \{\alpha \in \mathbb{R}^k : (\exists x,y,\lambda) \text{ s.t. } x \neq y,\ G = G_x = G = G_y = 0\}$$

Then B, H, and D are semialgebraic varieties of codimension one in $\mathbb{R}^k$. (Note for example that the equation count for B is right--two of the three equations may be solved for x and λ, leaving one relation between the α's.) The finitely many components of $\mathbb{R}^k \sim (B \cup H \cup D)$ enumerate the qualitatively different bifurcation diagrams which can be obtained from small perturbations of g.

In the next lecture we turn to bifurcation problems with symmetry. In that context Theorem 2.1 has a complete analogue but Theorem 2.2 does not, at least not yet.

LECTURE 3: SYMMETRY IN BIFURCATION THEORY

David G. Schaeffer

Duke University, Durham, NC 27706

In this lecture we consider the effects of symmetry on bifurcation problems. Even a simple reflectional symmetry can have important consequences. For example, recall the two problems with the elastica from Lectures 1 and 2 which exhibited pitchfork bifurcations. Why should the pitchfork ever be seen? After all, we saw in the last lecture that a bifurcation diagram which contains either a bifurcation point (i.e., a solution of (2.12)) or a hysteresis point (i.e., a solution of (2.13)) is nonpersistent, and the pitchfork has both. How does it happen that a natural mathematical description of these problems predicts this apparently very unstable phenomenon, a pitchfork? The answer lies in the fact that both these problems possessed a nontrivial reflectional symmetry: $u \to -u$ for the compression problem of Lecture 1, $u(s) \to u(\pi - s)$ for the shallow arch of Lecture 2. It turns out that the pitchfork _is_ persistent within the class of functions admitting a nontrivial action of the group Z_2. (This fact indicates some of the problems of generalizing Theorem 2.2 to cases where a symmetry group acts.) See Golubitsky and Schaeffer [1979b] for a proof of persistence; here we do not pursue it further.

Our approach to bifurcation in the presence of symmetry is through examples rather than general theory. Specifically we consider two examples, the Rivlin cube and the planar Bénard problem (Cf. Lecture 1). The first admits a finite group of symmetries, the latter a continuous group. The examples are representative of the general theory.

The Rivlin cube (Rivlin [1948,1974]) is a homogeneous, isotropic elastic material subjected to equal tensions in all three coordinate directions. (See Figure 3.1.) The tensions are dead

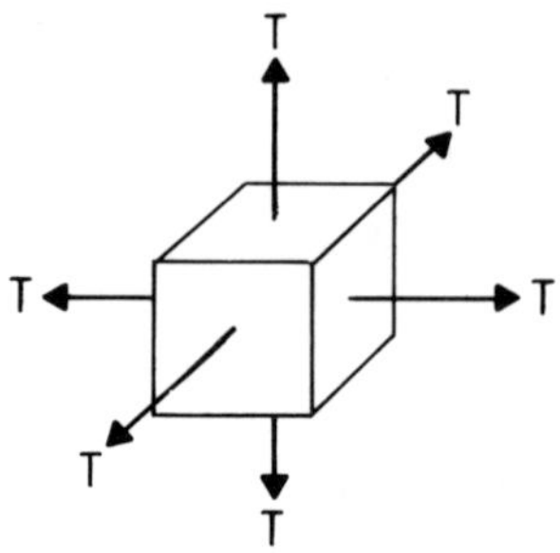

Figure 3.1: The Rivlin cube

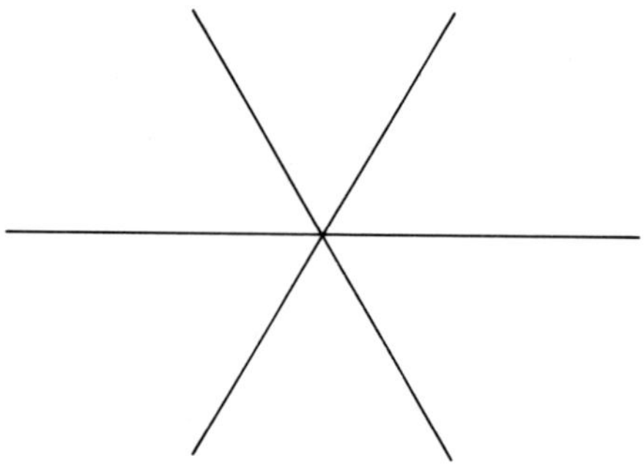

Figure 3.2: Fixed points of the action of S_3

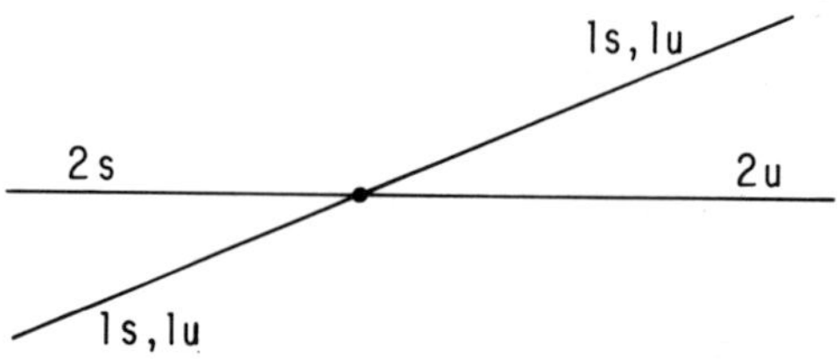

Figure 3.3: Bifurcation diagram for Theorem 3.2

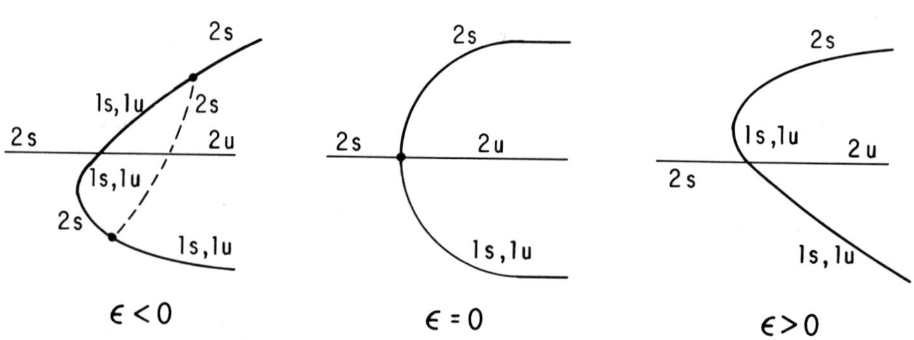

Figure 3.4: Bifurcation diagram for Theorem 3.3

loads, which means that the same total force acts, uniformly distributed over the face per unit area in the reference configuration, regardless of whatever deformations the body may undergo. We consider only homogeneous deformations of the form

$$(x,y,z) \to (v_1 x, v_2 y, v_3 z) \tag{3.1}$$

where $v_i > 0$. (Remark: In Ball and Schaeffer [1982], the source for this part of the lecture, more general homogeneous deformations are considered. Also it is shown there that the same analysis applies to a noncubical body.) Let $W(v_1,v_2,v_3)$ be the stored energy function for this material. We shall assume that this body is incompressible. Thus W need only be defined on the surface $\{v_1v_2v_3 = 1\}$, but we will regard W as defined for all positive v_i. Two natural choices for W are:

(i) Neo-Hookean $\quad W = \mu \Sigma v_i^2$

(ii) Mooney-Rivlin $\quad W = \mu \Sigma v_i^2 + \nu \Sigma v_i^{-2}$.

Here μ and ν are constants.

This setup gives rise to the following mathematical problem. Given a traction T,

$$\text{minimize} \quad W(v_1,v_2,v_3) - T(v_1+v_2+v_3) \quad \text{on} \quad \{v_1v_2v_3\} = 1. \tag{3.2}$$

The second term in (3.2) represents the work done by the external forces.

It is convenient in (3.2) to make the substitution $w_i = \log v_i$, because this reduces the constraint to a linear one:

$$w_1 + w_2 + w_3 = 0. \tag{3.3}$$

The permutation group S_3 acts on this problem. The expression (3.2) is unchanged by any permutation of the w's. This is clear for the second term; for the first it follows from isotropy. (Remark: A much larger group acts on this problem, the group of symmetries of the cube. However many elements act trivially, leaving only the quotient group S_3.) Our primary interest in this lecture is symmetry, so we separate the specific application from the discussion as much as practical.

Let us introduce complex coordinates $z = x + iy$ on the plane (3.3) as follows.

$$w_1 = x, \quad w_2 = -\frac{x}{2} + \frac{\sqrt{3}}{2} y, \quad w_3 = -\frac{x}{2} - \frac{\sqrt{3}}{2} y .$$

In these coordinates the cyclic permutation $w_1 \to w_2 \to w_3 \to w_1$ is represented as rotation through 120°,

$$z \to e^{\frac{2\pi i}{3}} z \tag{3.3a}$$

and the interchange permutation $w_2 \leftrightarrow w_3$ is represented as complex conjugation,

$$z \to \bar{z} \tag{3.3b}$$

In other words, the action of S_3 on the plane (3.3) is the standard two (real) dimensional, irreducible representation of S_3.

It is useful to ask whether any points in the plane (3.3) are fixed by any nontrivial group elements. The rotation (3.3a) fixes only the origin, while complex conjugation (3.3b) fixes the real axis. In terms of the w's, points on the real axis have the symmetry $w_2 = w_3$. The other interchange permutations also have lines of fixed points as illustrated in Figure 3.2. Points not lying on one of these three lines have six distinct images under the action of S_3, while points on them (other than the origin) have three.

In the singularity theory approach to bifurcation one considers not the potential function as in (3.2), but rather the derivative equation which expresses the usual necessary condition for a minimum. This gives an equation

$$g(z,T) = 0 \tag{3.4}$$

where $g: \mathbb{C} \times \mathbb{R} \to \mathbb{C}$ is equivariant, defined as follows.

A mapping $g: \mathbb{C} \times \mathbb{R} \to \mathbb{C}$ is called <u>equivariant</u> if $g(\gamma \cdot z,T) = \gamma \cdot g(z,T)$ for all $\gamma \in S_3$. Here $\gamma \cdot z$ of course indicates the action of this group element on z. Two such mappings g and h are <u>equivariantly contact equivalent</u>, or simply <u>equivalent</u>, if (1.14) is satisfied with coordinate transformations s, X, and Λ which respect the symmetry. (See Golubitsky and Schaeffer [1979b] for the precise requirements on s, X, and Λ.)

<u>Proposition 3.1</u>: If $g: \mathbb{C} \times \mathbb{R} \to \mathbb{C}$ is a smooth equivariant map, there exist smooth, real coefficients a and b such that

$$g(z,\lambda) = a(|z|^2, R\ell\, z^3, \lambda)z + b(|z|^2, R\ell\, z^3, \lambda)\bar{z}^2 . \tag{3.5}$$

Here we have switched to $\lambda = T - T_0$ as the bifurcation parameter (T_0 defined below). Any (scalar valued) function on $\mathbb{C}$ invariant under S_3 can be expressed as a function of $|z|^2$ and $R\ell\, z^3$. The proposition states that the most general equivariant mapping can be expressed as a linear combination, with invariant functions as coefficients, of the two basic equivariant mappings z and $\bar{z}^2$. See Golubitsky and Schaeffer [1982] for the proof.

The following theorem provides a normal form for the equation of the simplest bifurcation problem with S_3 symmetry.

Theorem 3.2: Let $g: \mathbb{C} \times \mathbb{R} \to \mathbb{C}$ be an equivariant map, with a and b defined by (3.5). Suppose that at $(z,\lambda) = (0,0)$ we have

$$\text{(i)} \quad a = 0, \qquad \text{(ii)} \quad a_\lambda \neq 0, \qquad \text{(iii)} \quad b \neq 0. \tag{3.6}$$

Then g is equivalent to $\bar{z}^2 \pm \lambda z$. This problem is persistent; i.e., any small perturbation of g is equivalent to g.

In other words, the three hypotheses in (3.6) state the following: (i) Bifurcation occurs--the equation is nonsingular if $a \neq 0$. (ii) The eigenvalue moves across zero with nonzero speed. (iii) There are quadratic terms present in g. Since this bifurcation problem is persistent, no universal unfolding is required.

In the Rivlin cube it turns out that for certain parameter values the quadratic terms in g vanish; i.e., $b(0,0) = 0$. The next theorem gives an appropriate normal form for that case. We expand the coefficients to first order in their arguments,

$$a = A|z|^2 + B \,\mathrm{R}\ell\, z^3 + \alpha\lambda + \text{hot} \tag{3.7a}$$

$$b = C|z|^2 + D \,\mathrm{R}\ell\, z^3 + \beta\lambda + \text{hot}\ . \tag{3.7b}$$

Note that the notation in (3.7) incorporates the hypotheses

$$a(0,0) = 0, \qquad b(0,0) = 0.$$

We shall make the following assumptions insuring nondegeneracy.

$$A \neq 0, \quad \alpha \neq 0, \quad A\beta - C\alpha \neq 0, \quad AD - BC \neq 0.$$

Theorem 3.3: Given the above hypotheses, g is equivalent to

$$(|z|^2 \pm \lambda)z + (\pm|z|^2 + \Delta \,\mathrm{R}\ell\, z^3)\bar{z}^2 \tag{3.8}$$

where $\Delta = (AD - BC)\alpha^2(A\beta - C\alpha)^{-2}$ and the two unspecified signs in (3.8) equal $\mathrm{sign}(A\alpha)$ and $\mathrm{sign}\{(C\alpha - A\beta)A\alpha\}$ respectively. A universal unfolding of (3.8) is provided by

$$G(z,\lambda;\varepsilon,\delta)=(|z|^2\pm\lambda)z+(\pm|z|^2+(\Delta+\delta)\mathrm{R}\ell\, z^3+\varepsilon)\bar{z}^2\ . \tag{3.9}$$

The universal unfolding (3.9) contains two parameters. However δ plays an entirely subordinate role; indeed if we were to consider C^0 equivalence rather than C^∞ equivalence δ would not be needed at all. Any two versions of (3.8) with different values of Δ of the same sign may be related as in (1.14) where s, X, and Λ are Lipschitz continuous. However the C^∞ structure is finer, and one needs to allow for changes in Δ to guarantee the existence of a C^∞ coordinate transformation. Unfortunately it seems that C^0 equivalence cannot be studied directly--one must work through C^∞ equivalence. (Remark: In singularity theory terminology, δ is called a modal parameter.)

Let us now consider the solution set of an equation

$$g(z,\lambda) = 0 \tag{3.10}$$

when g is equivariant. Recalling (3.5), we see that there are three different circumstances under which (3.10) has solutions:

(i) $z = 0$

(ii) $z \neq 0$, the complex numbers z and $\bar{z}^2$ are parallel, and the two terms in (3.5) add up to zero

(iii) z and $\bar{z}^2$ are not parallel, but $a = b = 0$.

In case (ii), z and $\bar{z}^2$ are parallel (or antiparallel) if and only z^3 is real; i.e., iff z lies on one of the three lines of Figure 3.2. In other words, the solutions which arise in case (ii) possess some extra symmetry. In technical language the isotropy group of such solutions is Z_2. By contrast the solutions arising in cases (i) and (iii) have isotropy groups S_3 and $\{I\}$ respectively.

We determine the solution set of the canonical form of Theorem 3.2. Case (iii) above does not arise here, since $b \equiv 1$. For case (ii) it suffices to consider z real, say $z = x$, as the other two cases may be obtained by letting the group act. Thus we have the equation

$$x^2 \pm \lambda x = 0.$$

This has the solution $x = 0$, the trivial solution of case (i), and $x = \mp\lambda$. These are graphed in Figure 3.3. This figure also gives the stability characteristics of each branch; 2s for both eigenvalues stable, etc. See Golubitsky and Schaeffer [1982] for proofs.

It is noteworthy that the only bifurcating solutions here have the partial symmetry $w_i = w_j$ for some $i \neq j$. On generic grounds one might suspect that bifurcating solutions would usually <u>avoid</u> the three distinguished lines in Figure 3.2, but in fact the exact opposite is true. It seems that in situations where symmetry is broken, the system yields its symmetry reluctantly, losing as little as possible at each stage.

Now we consider the zero set of (3.9) in Theorem 3.3. We shall ignore δ, in keeping with the discussion above. We fix the two unspecified signs in (3.9) as minus and plus, respectively, and take $\Delta > 0$, as this is the situation which occurs for the Rivlin cube. For solutions of type (ii) we may again write $z = x$, which leads to solution branches $x = 0$ and $\lambda = x^2 + x^3 + \Delta x^4 + \varepsilon x$. For case (iii) we have

$$\text{(a) } \lambda = |z|^2 \quad \text{and} \quad \text{(b) } |z|^2 + \Delta \,\mathrm{R}\ell\, z^3 + \varepsilon = 0. \tag{3.11}$$

If $\varepsilon > 0$, (3.11b) has no solutions locally near $z = 0$. (Since our theory is a local one we only consider solutions close to the trivial solution.) If $\varepsilon = 0$, (3.11b) only has the solution $z = 0$, already found above. If $\varepsilon < 0$, then (3.11b) has a topological circle of solutions close to $z = 0$; (3.11a) tells how λ varies along this solution branch. These solutions are indicated schematically in Figure 3.4. The solutions arising from case (iii) are shown as dotted. It should be recalled that the one curve of solutions arising in case (ii) actually represents an orbit under the group action of three such curves, and the topological circle of type (iii) solutions actually connects all three type (ii) branches. Other choices of sign lead to similar diagrams.

In an important paper Bauer, Keller, and Reiss [1975] showed that perturbation of a bifurcation problem with a multiple eigenvalue often leads to secondary bifurcation. The diagrams of Figure 3.4 are an illustration of this principal, or perhaps an extended principle "Degeneracy leads to secondary bifurcation."

Finally we relate the theoretical results above to the Rivlin cube example. We quote the results of Ball and Schaeffer [1982], referring to that paper for calculations. First consider a Neo-Hookean material, as did Rivlin. In this case the system bifurcates at the load $T_0 = 2\mu$. As illustrated in Figure 3.5 the bifurcation is of the "hard" type; i.e., there are no stable equilibria near the trivial solution for $T > T_0$, so the system must jump to a state far from the trivial solution.

The parameter μ in the Neo-Hookean stored energy function is an inessential one--it just defines the scale of the problem. With the Mooney-Rivlin stored energy function the bifurcation occurs for $T_0 = 2(\nu + \mu)$, which defines the scale. However, the ratio ν/μ is an essential parameter, and in fact the interesting behavior exhibited in Figure 3.6 occurs as this parameter is varied.

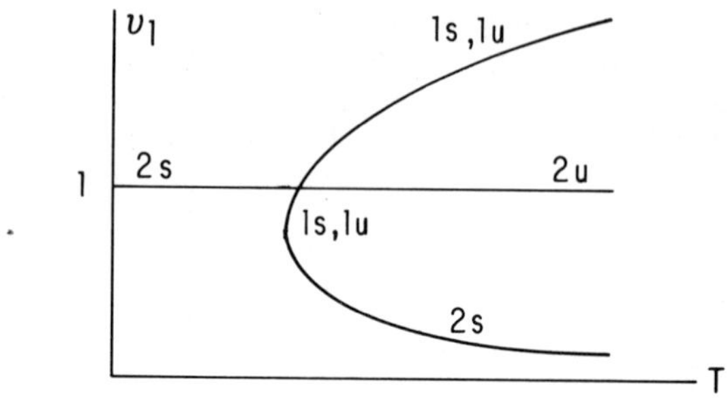

Figure 3.5: Bifurcation diagrams for the Rivlin cube

Note: $v_2 = v_3$ determined from $v_1 v_2 v_3 = 1$

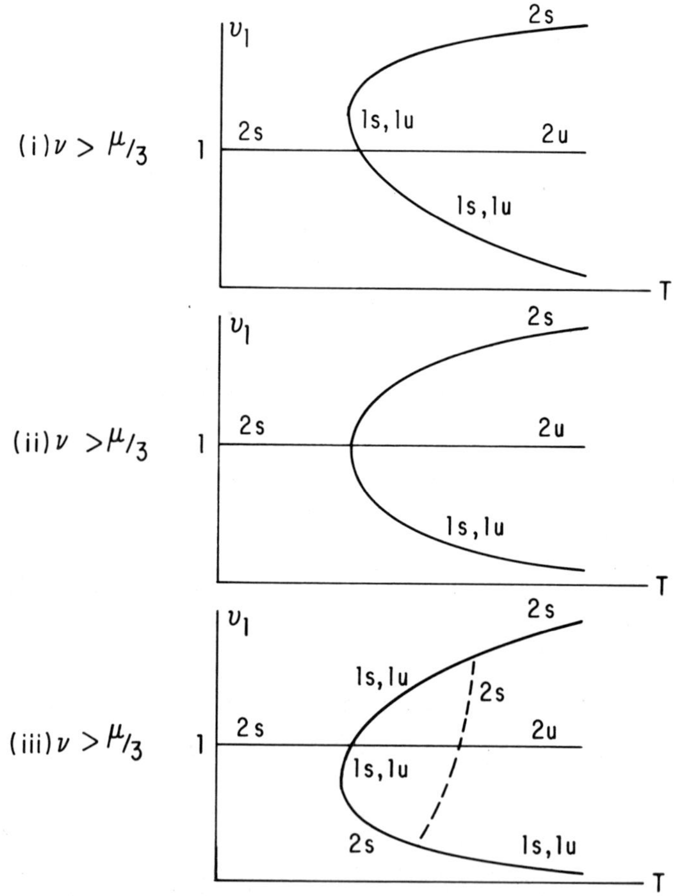

Figure 3.6: Bifurcation diagram for Mooney-Rivlin material

For small ν the bifurcation diagram is apparently just a perturbation of the Neo-Hookean case. None the less, as $T \to \infty$ the main solution branch loses stability through a secondary bifurcation. This secondary bifurcation moves into the primary bifurcation point as ν is increased further, merging with the primary bifurcation when $\nu = \mu/3$ and going off into the complex plane for $\nu > \mu/3$. The degenerate bifurcation (3.8) occurs for $\nu = \mu/3$, and ν may be related to the unfolding parameter ε.

Singularity theory made three contributions to this problem. In the first place it provided the motivation to consider a more general material than Rivlin did. The canonical forms of Theorems 3.2 and 3.3 were available from analysis of other problems, and they suggested that more complex behavior could occur in a problem with extra parameters. Thus we considered Mooney-Rivlin material and found exactly the complication we anticipated. Secondly, understanding the degenerate case when $\nu = \mu/3$ was aided by knowing the canonical forms. Finally we were able to use the persistence results to show that similar results would continue to hold if the equations were perturbed somewhat, even in cases where the equations could not be solved explicitly. For example, introducing a slight degree of compressibility into the system is such a perturbation.

We now turn to our second example of bifurcation with symmetry, the planar Bénard problem which was introduced in Lecture 1. We are reporting on the very recent results of Buzano and Golubitsky [1982] and Golubitsky and Swift [1982]. Indeed the latter work is not yet complete; some of the results we quote below are based on calculations that have not yet been checked and therefore might be subject to small perturbations. (And unfoldings?) Our treatment here is necessarily rather sketchy--we refer to the original papers for greater detail.

Recall that in Lecture 1 we restricted our attention to flows in a certain space (say X) of spatially periodic functions admitting the double Fourier series representation (1.28); this was done to get bifurcation from an eigenvalue of finite multiplicity. We claim the the group

$$\Gamma = D_6 + \mathbb{T}^2 + Z_2 \tag{3.12}$$

acts on X and that the equation

$$\Phi(u,R) = 0 \tag{3.13}$$

is equivariant with respect to this action. Here D_6 denotes the group of symmetries of the hexagon, a group of order 12--although the original problem was equivariant under the full rotation group O(2), the specification of definite coordinate axes in (1.28) cuts the symmetry down to D_6. (Indeed a less careful choice of the period lattice would lead to an even smaller group.) The second term in (3.12) is the 2-torus--the original problem was invariant under $\mathbb{R}^2$, the group of all translations; because X contains only periodic functions the effective group acting is $\mathbb{R}^2$ mod the period lattice. Rotations and translations do <u>not</u> commute; in writing plus in (3.12) we do not imply that the group is a direct sum. The third factor in (3.12) is a reflectional symmetry, $z \to \pi - z$--the equations of motion (1.15) are equivariant with respect to this symmetry, even though this is far from obvious on physical grounds.

As indicated in Lecture 1, the Liapunov-Schmidt reduction of (3.13) at the lowest eigenvalue leads to a 6×6 system of equations $g(x,R) = 0$, where $g: \mathbb{R}^6 \times \mathbb{R} \to \mathbb{R}^6$. Sattinger [1979] showed that this reduction may be performed so as to maintain the equivariance of the reduced problem under the group Γ. There is not enough time to give complete normal forms for g in this problem, as was done in Theorems 3.2 and 3.3 for the previous example. It turns out that no bifurcation problem equivariant under Γ is persistent (Cf. Theorem 2.2 for S_3.); i.e., any such problem requires at least one additional parameter to describe the effect of an arbitrary perturbation (up to C^∞ equivalence). However there are bifurcation problems equivariant under Γ whose unfoldings have only modal parameters; such problems would in fact be persistent if one accepted Lipschitz equivalence transformations. We consider only this kind of problem, but even with this restriction there are still several cases. In other words, symmetry considerations and genericity alone do not suffice to determine the qualitative behavior of the Bénard problem--one must insert some further information specific to the problem. Schluter, Lortz, and Busse [1965] have shown that rolls are the only stable bifurcating solution. Of the several cases mentioned above, only one is compatible with this conclusion; below we restrict our attention exclusively to this case.

For the bifurcation problem singled out by the above considerations four different kinds of nontrivial solutions appear at the bifurcation point. We distinguish these by their isotropy groups; i.e., the subgroup of Γ which leaves the solution invariant. (Remark: In the situation of Theorem 3.2, the closest analogue of what we are considering here, all bifurcating solutions had the isotropy group Z_2.) The four isotropy subgroups are listed in Figure 3.7. Here V denotes the <u>vier gruppe</u>, the subgroup of D_6 generated by the two reflections $x \to -x$ and $y \to -y$; this group is of course isomorphic to $Z_2 \oplus Z_2$. $\mathbb{T}^1$ denotes the subgroup of translations along one axis only. The Z_2 groups in Figure 3.7 are generated by the reflection $z \to \pi - z$ composed with some operation in x and y of order 2, translation by half a period or a reflection; the latter operation is

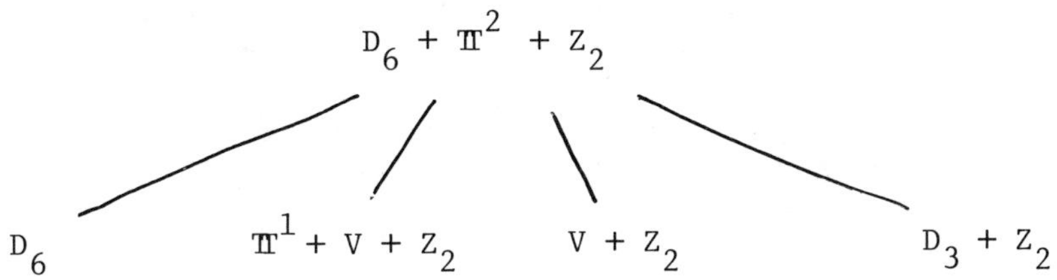

Hexagons	Rolls	Rectangles	Triangles
$\sum_{k=1}^{3} \cos(\vec{k}_i \cdot (x,y))$	$\cos \vec{k}_1 \cdot (x,y)$	$\sum_{i=2}^{3} \cos \vec{k}_i \cdot (x,y)$	$\sum_{i=1}^{3} \sin \vec{k}_i \cdot (x,y)$

Figure 3.7: Isotropy subgroups

different in the various cases. Names associated to the different kinds of flow are listed below the isotropy group. Below these names the x,y spatial dependence of a typical element from each class is listed. This may be regarded as an approximating the z-component of the velocity in the midplane $z = \pi/2$. The wave number vectors $\vec{k}_i$ appearing here are

$$\vec{k}_1 = k^*(0,1), \quad \vec{k}_2 = k^*(\frac{\sqrt{3}}{2},-\frac{1}{2}), \quad \vec{k}_3 = k^*(-\frac{\sqrt{3}}{2},-\frac{1}{2}). \tag{3.14}$$

The planforms of the several types of solutions are shown in Figure 3.8. The x and y-axes of (3.14) are horizontal and vertical respectively in Figure 3.8; the minimum distance between points on the period lattice is $2\pi/k^*$. Areas where the flow is upwards are shown hatched. In Figure 3.8a the hexagonal boundaries between cells are shown as dotted lines; these indicate planes across which no fluid flows. They are included for historical reasons.

The information discussed above was obtained by the very laborious process of finding the normal forms of the equations and analyzing their solutions. Included as steps in this process are characterizing all invariant functions and all equivariant mappings under the group action. It would be extremely valuable to be able to short cut this process by using more group theory. As a hint of how this might be done we mention the following very interesting fact. Consider the lattice of isotropy subgroups of Γ; i.e., subgroups which are isotropy subgroups of some function in the six

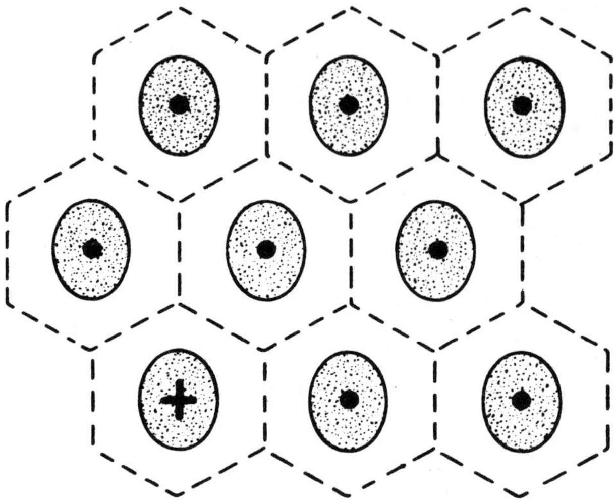

Figure 3.8a: Hexagons

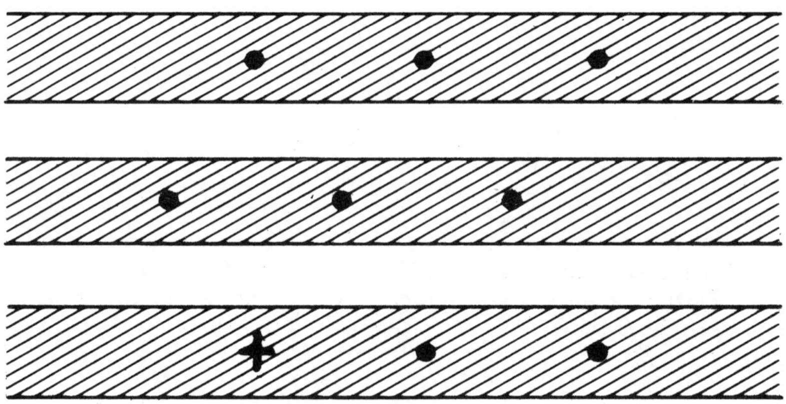

Figure 3.8b: Rolls

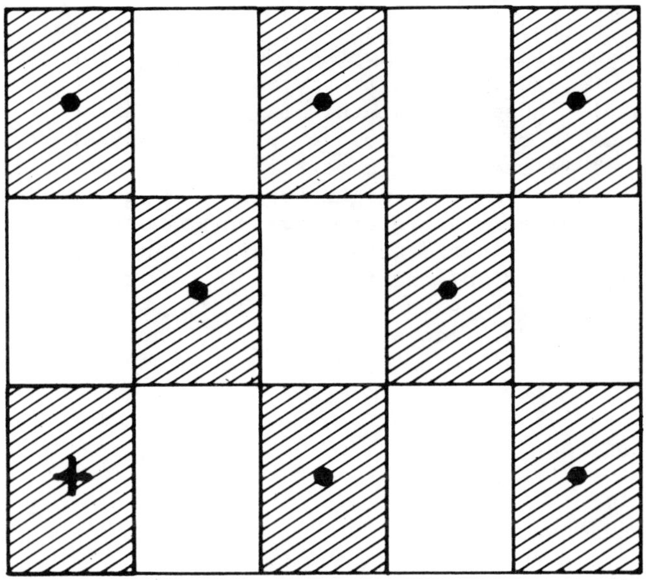

Figure 3.8c: Rectangles

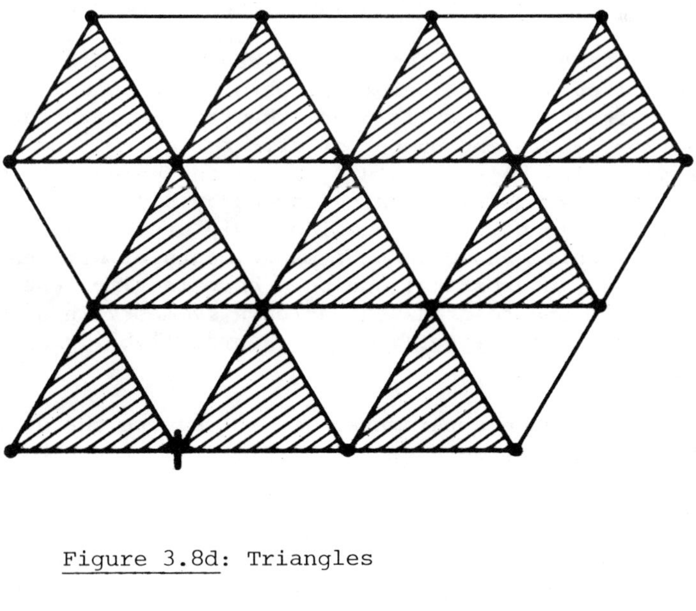

Figure 3.8d: Triangles

dimensional space spanned by $R\ell$ and $\text{Im}\exp\{i\,\vec{k}_j(x,y)\}$, $j = 1,2,3$. The four subgroups which occur in Figure 3.7 are precisely the maximal ones. (Cf. the remark above that when symmetry breaking occurs, as little symmetry as possible is given up.) Several examples suggest that this might be indicative of a general phenomenon, but as yet not even precise conjectures have been formulated. This seems like a fertile area for further work. (Remark: At first sight it might appear that the isotropy subgroup of rectangles is a proper subgroup of that of rolls; however this is false. Both rectangles and rolls admit an up/down reflectional symmetry of the form $z \to \pi - z$ composed with a translation; for rectangles the translation is through $\vec{e}_1/2$, for rolls $\vec{e}_2/2$ (Notation of Lecture 1). Thus these group elements are slightly different, and both groups are in fact maximal.)

In Figure 3.9 we have drawn a bifurcation diagram for this problem using the following abbreviations:

R: rolls
H: hexagons
P: rectangles (patchwork quilt)
T: triangles.

This diagram is highly schematic--each point on one of the curves represents a whole orbit of distinct solutions, either one dimensional (for rolls) or two dimensional, and the orbits have several components. The number of stable and unstable eigenvalues along each branch is also shown. The one or two zero eigenvalues associated with directions tangent to the orbit are not indicated.

The mathematical problem we have been discussing so far is a highly idealized description of the physical problem. In an actual experiment distant side walls break the translational and rotational symmetries, while any number of effects, among them the dependence of viscosity on temperature, break the up/down symmetry. Analysis of the symmetry breaking caused by side walls has hardly gotten off the ground, so we consider only breaking of the up/down symmetry. In this connection the experiments of Krishnamurti [1968] are most interesting; she used certain time dependent boundary conditions to break the up/down symmetry in a controllable way, analogous to the experiments of Roorda [1965] discussed in Lecture 2.

Let us introduce a parameter ε to measure the degree of symmetry breaking. Thus we have an equation

$$\Phi(u,R,\varepsilon) = 0 \tag{3.15}$$

to solve. For $\varepsilon = 0$ (3.15) is equivariant with respect to the full group $D_6 + \mathbb{T}^2 + Z_2$, while for $\varepsilon \neq 0$ (3.15) is only equi-

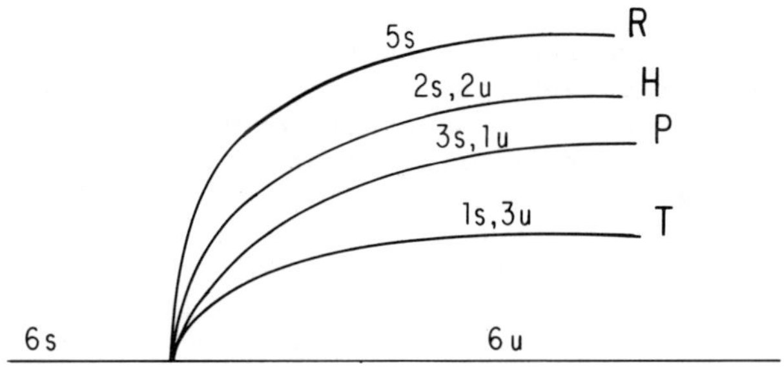

Figure 3.9: Bifurcation diagram

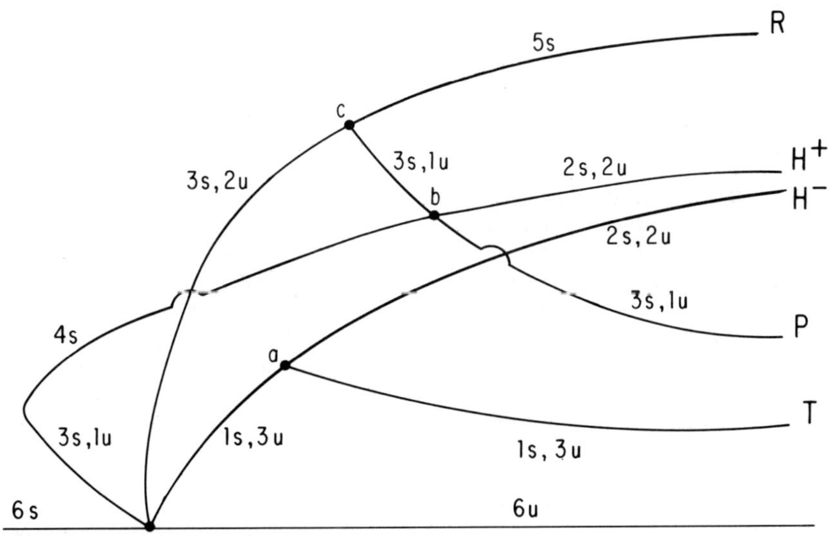

Figure 3.10: Bifurcation diagram after breaking symmetry

variant with respect to $D_6 + \mathbb{T}^2$. Repetition of the Liapunov-Schmidt process leads to a reduced set of equations $g(x,\lambda,\varepsilon) = 0$ with the same symmetry properties. It appears that there is essentially only one way to break the symmetry; i.e., there is a certain monomial in the reduced equations which dominates the situation if its coefficient is nonzero, and all nonzero choices lead to equivalent problems. In other words, here symmetry and genericity considerations <u>are</u> sufficient to determine the qualitative behavior being studied.

For this distinguished problem, the bifurcation diagram which occurs for $\varepsilon > 0$ is shown in Figure 3.10. There are two primary bifurcating solutions--hexagons and rolls--and two secondary branches with three secondary bifurcating points. As shown in Figure 3.11, there are also two maximal isotropy subgroups of

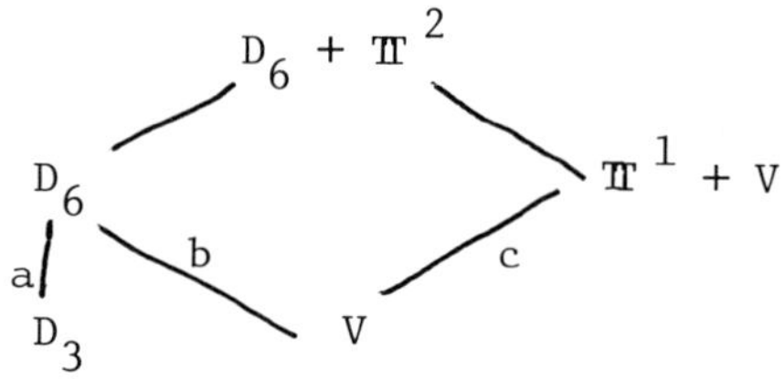

<u>Figure 3.11</u>: Isotropy subgroups

$D_6 + \mathbb{T}^2$, precisely the isotropy subgroups of hexagons and rolls. Still more suggestive, there are two maximal isotropy subgroups at the next level, those of the secondary branches. There are also three containments at this level, corresponding to the three points of secondary bifurcation as indicated by the letters. It even appears that for each containment the symmetry which is broken at the associated secondary bifurcation point is somehow related to the quotient of the two isotropy subgroups involved. This close correspondence can hardly be pure coincidence, but the general picture is far from clear.

Some comments about Figure 3.10 may be helpful. The hexagonal solutions, the only ones in Figure 3.9 not admitting an up/down symmetry, are split into two branches when $\varepsilon > 0$; one has the planform of Figure 3.8a, the other its negative. The bifurcation of the hexagonal solutions from the trivial branch is transcritical, analogous to Figure 3.3 above. The secondary branches in Figure 3.10 (labeled P and T) have smaller isotropy subgroups than the corresponding branches in Figure 3.9--the up/down symmetry is missing. As a consequence the planforms of these solutions is somewhat different from that illustrated in Figures 3.8c and d. In the limit $R \to \infty$ this distinction disappears.

Our second example of bifurcation with symmetry has perhaps raised more questions than it has answered. However, I hope it illustrates an exciting, active area of research where pure and applied mathematics make significant contact.

REFERENCES

[1969] S. Antman and J. B. Keller, Bifurcation Theory and Nonlinear Eigenvalue Problems. Benjamin, New York.

[1982] J. Ball and D. Schaeffer, "Bifurcation and stability of homogeneous equilibrium configurations of an elastic body under dead-load tractions," in preparation.

[1975] L. Bauer, H. Keller, and E. Reiss, "Multiple eigenvalues lead to secondary bifurcation," SIAM J. Appl. Math. 17, pp. 101-122.

[1982] E. Buzano and M. Golubitsky, "Bifurcation involving the hexagonal lattice," submitted.

[1961] S. Chandrasekhar, Hydrodynamic and Hydromagnetic Stability. Clarendon Press, Oxford.

[1973] M. Golubitsky and V. Guillemin, "Stable maps and their singularities," Springer, New York.

[1979a] M. Golubitsky and D. Schaeffer, "A theory for imperfect bifurcation theory via singularity theory," Comm. Pure Appl. Math. 32, pp. 21-98.

[1979b] M. Golubitsky and D. Schaeffer, "Imperfect bifurcation in the presence of symmetry," Comm. Math. Phys. 67, pp. 205-232.

[1982] M. Golubitsky and D. Schaeffer, "Bifurcations with O(3) symmetry including applications to the Bénard problem," Comm. Pure Appl. Math. 35, pp. 81-111.

[1974] E. Koschmieder, "Bénard connection," Adv. Chem. Phys. 26, pp. 177-212.

[1968] R. Krishnamurti, "Finite amplitude convection with changing mean temperature. Parts 1 and 2," JFM 33, pp. 445-455 and pp. 457-463.

[1982] J. Maddocks, "Stability of the elastica," these Proceedings.

[1948] R. Rivlin, "Large elastic deformations of isotropic materials II," Phil. Trans. Roy. Soc., Ser. A 240, pp. 491-508.

[1974] R. Rivlin, "Stability of pure homogeneous deformations of an elastic cube under dead loading," Quart. Appl. Math. 32, pp. 265-271.

[1975] J. Roorda, "Stability of structures with small imperfections," J. Engg. Mech. Div. Am. Soc. Civ. Engrs. 91, p. 87.

[1979] D. Sattinger, _Group Theoretic Methods in Bifurcation Theory_. Springer, New York.

[1965] A. Schlüter, D. Lortz, and F. Busse, "On the stability of finite amplitude convection," JFM 23, pp. 129-144.

[1959] M. Stein, "The phenomenon of change of buckle patterns in elastic structures," NASA technical report R-39.

[1973] J. M. Thompson and G. W. Hunt, _General Theory of Elastic Stability_, Wiley, New York.

THE COMPENSATED COMPACTNESS METHOD APPLIED TO SYSTEMS OF CONSERVATION LAWS

Luc Tartar
17 allée Blaise Pascal
78460 Chevreuse
France

One of the main difficulties in solving nonlinear partial differential equations lies in the following fact: after introducing a suitable sequence of approximations one needs enough a priori estimates to ensure the convergence of a subsequence to a solution; this argument is based on compactness results and in a nonlinear case one needs more estimates than in the linear case where weak continuity results can be used.

This lack of continuity in nonlinear problems has long restricted the use of weak convergence to linear problems; actually a careful analysis can render it more useful and this is the essence of the compensated compactness method. One of its most interesting applications is the recent work of R. DiPerna of global existence for some 2×2 systems of conservation laws with bounded data.

Although the reader may share my feeling that the method should be improved in order to simplify and extend the existing proofs, I think that DiPerna's result has shown that the method has given the possibility of tackling some of the important problems in nonlinear partial differential equations.

The method has not been exclusively designed for hyperbolic systems (it has come out of joint, mostly unpublished, work with F. Murat on homogenization, and fruitful interaction with J. Ball's work on nonlinear elasticity).

I want to thank the organizing committee for having given me the opportunity of lecturing on this topic.

I apologize for the fact that these written notes will not

J. M. Ball (ed.), Systems of Nonlinear Partial Differential Equations, 263–285.

convey the same enthusiasm I had giving these lectures, the support of so many of my friends in the audience being so helpful then; I want to thank them all.

I. WEAK CONVERGENCE

Weak convergence is a classical tool in functional analysis and it has been early stressed that it is not well suited for nonlinear problems.

I want to consider here weak convergence from a different point of view. First I want to point out its natural relationship with the concept of microscopic versus macroscopic quantities; then, and this is one of the goals of the compensated compactness method, I want to show that it can be efficient in nonlinear problems when used carefully.

1. The Electrostatics Model

The simplest model for describing weak convergence and compensated compactness is electrostatics, which is described by the following equations

$$\circledast\begin{cases} E = -\operatorname{grad} U \\ \operatorname{div} D = \rho \\ D = aE \\ e = \frac{1}{2}E.D \end{cases}$$

(U is the electrostatic potential, E and D the electric and electric induction fields, ρ and e the densities of charge and of electrostatic energy, a the dielectric permittivity).

In the equation $\operatorname{div} D = \rho$ we must remember that a distribution of point charges has been replaced by a density of charge and this is the first case of a weak convergence: a distribution of point charges cannot converge strongly in the space of measures to an integrable density and weak* convergence is then useful.

In the equation $E = -\operatorname{grad} U$ we must remember that through pointwise measurement of the potential one can have a good knowlege of U but that this does not give a good description of E as small oscillations of U can give big oscillations for E; but if one uses weak convergence for E (in $L^2(\Omega)^N$, which is a natural space for E) then this difficulty disappears.

Understanding the role of a in the equation $D = aE$ is the subject of homogenization; the last equation $e = \frac{1}{2}E.D$ will give the basic example of compensated compactness.

When a sequence of fields E_n converges weakly to E_∞ we can understand the oscillations in E_n as a microscopic description of the field but E_∞, where oscillations have disappeared, will give a macroscopic description.

The compensated compactness result applies to E.D and says that a macroscopic description of E and D gives a macroscopic description of $e = \frac{1}{2}E.D$.

2. Abstract Functional Analysis

Let us survey quickly the classical results of functional analysis on weak convergence.

(a) On any Banach space X, with dual X', one defines the weak topology $\sigma(X,X')$ as the one defined by the family of seminorms $(p_{x'},\ x' \in X')$ given by $p_{x'}(x) = |(x,x')|$. If X' is separable (i.e. has a countable dense set) then on bounded sets of X this topology is metrizable, and x_n converges weakly to x_∞ means that

$$(x_n,x') \to (x_\infty,x') \qquad \forall\ x' \in X'.$$

(b) On the dual X' of a Banach space X one defines the weak * topology $\sigma(X',X)$ as the one defined by the family of seminorms $(q_x,\ x \in X)$ given by $q_x(x') = |(x,x')|$. If X is separable then on bounded sets of X' this topology is metrizable and x'_n converges weakly * to x'_∞ means that

$$(x,x'_n) \to (x,x'_\infty) \qquad \forall\ x \in X.$$

The crucial property of this topology is that the unit ball of X' is weakly * compact (consequence of Ascoli's theorem).

(c) Closed convex sets are weakly closed and convex functions that are strongly lower semicontinuous are also weakly lower semicontinuous (consequence of the Hahn-Banach theorem).

3. Applied Functional Analysis

We look at particular spaces: $L^p(\Omega)$, $1 \leqslant p \leqslant +\infty$, $\mathcal{M}(\Omega)$ the space of Radon measures, where Ω is an open set of $\mathbb{R}^N$ equipped with the Lebesgue measure.

(a) The dual space of $L^p(\Omega)$ is isometric to $L^{p'}(\Omega)$, $1/p + 1/p' = 1$. From a bounded sequence f_n in $L^p(\Omega)$ one can extract a subsequence f_m converging weakly to f_∞ if $1 < p < +\infty$ (weakly * if $p = +\infty$)
i.e.

$$\int_\Omega f_m g\,dx \to \int_\Omega f_\infty g\,dx \qquad \forall\, g \in L^{p'}(\Omega).$$

(b) $L^1(\Omega)$ is isometrically imbedded into $\mathcal{M}_b(\Omega)$ (the space of measures with finite total mass) which is the dual of $C_b(\Omega)$ (the space of continuous bounded functions with the sup norm). From a bounded sequence f_n in $L^2(\Omega)$ one can extract a subsequence f_m weakly * converging to a measure μ i.e.

$$\int_\Omega f_m g\,dx \to \langle \mu, g\rangle \quad \forall\, g \in C_b(\Omega).$$

(c) The space of measures $\mathcal{M}(\Omega)$ is not a Banach space but for each compact K of Ω, if $X_K = C^0_K(\Omega)$ denotes the space of continous functions with support in K with the sup norm, then $\mathcal{M}(\Omega)$ is a subspace of X_K' and we can use the weak * topology on X_K'. When dealing with measures and extracting weakly * converging subsequences it is understood that we have a bounded sequence in each X_K' and that the extracted subsequence is weakly * converging in X_K' for a countable family of compact sets whose union is Ω.

(d) An instructing example of a weakly converging sequence is given by oscillating functions with periodic structure. Let $y_1, \ldots, y_N$ be independent vectors of $\mathbb{R}^N$ and let

$$Y = \{\sum_i \theta_i y_i \;,\quad 0 \leqslant \theta_i \leqslant 1 \quad \forall_i\} \;;$$

let $F(x,y)$ be a continuous bounded function on $\Omega \times \mathbb{R}^N$ which is Y-periodic (i.e. $F(x, y+y_j) = F(x,y)$ $\forall\, j = 1,\ldots,N$); then as n goes to ∞ the sequence $f_n(x) = F(x,nx)$ converges weakly * in $L^\infty(\Omega)$ to

$$f_\infty(x) = \frac{1}{\text{meas } Y} \int_Y F(x,y)\,dy.$$

This shows why the weak * limit of $\phi(f_n)$ is not $\phi(f_\infty)$ for non-affine ϕ.

(e) Actually the weak (or weak *) limits of the sequences g_n and $\phi(g_n)$ are not independent: for example if g_n converges weakly * in $L^\infty(\Omega)$ to g_∞ and $\phi(g_n)$ converges to h_∞ then if ϕ is convex one has $h_\infty(x) \geqslant \phi(g_\infty(x))$ a.e. .

If one wants to compare limits for a small number of different functions the following result will be useful:

<u>Proposition 1:</u>

Let K be a bounded set of $\mathbb{R}^p$ and let u^n be a sequence of vector valued functions from Ω to $\mathbb{R}^p$ such that

(i) $u^{(n)}$ converges to u^∞ in $(L^\infty(\Omega))^p$ weak *

(ii) $u^{(n)}(x) \in K$ a.e. .

Then

(iii) $u^\infty(x) \in \overline{\text{conv}\, K}$ a.e. (closed convex hull of K).

Conversely if a function u^∞ satisfies (iii) one can find a sequence $u^{(n)}$ satisfying (i) and (ii). □

<u>Outline of Proof.</u> To show that (i) + (ii) ⇒ (iii) one must notice that $\overline{\text{conv}\, K}$ is the intersection of half spaces containing K and that, if for an affine function L one has $L(u^{(n)}) \geqslant 0$ a.e. then $L(u^\infty) \geqslant 0$ a.e. .

To construct a sequence $u^{(n)}$ one first restricts attention to the case where $u^{(\infty)}$ is a step function and in each set where u^∞ is constant use a construction as in (d). □

4. L.C. Young's Generalized Functions

A more detailed description of the oscillations of the sequence $u^{(n)}$ considered previously can be given by the knowledge of all weak * limits of functions of $u^{(n)}$.

(a) To a function $u^{(n)} \in (L^\infty(\Omega))^p$ we associated a Radon measure $\mu^{(n)}$ defined on $\Omega \times \mathbb{R}^p$ by the formula

$$\left.\begin{array}{l} \langle \mu^{(n)}, F \rangle = \int_\Omega F(x, u^{(n)}(x))\,dx. \\ \\ F(x,\lambda) \text{ being continuous in } (x,\lambda) \text{ with} \\ \text{compact support in } x \end{array}\right\} \qquad (1)$$

(as $u^{(n)}$ is bounded a.e. it is not necessary that F has compact support in λ).

Working on $\mu^{(n)}$ instead of $u^{(n)}$ leads to the fundamental

<u>Theorem 1</u>

Assume $u^{(n)}(x) \in K$ a.e. where K is a bounded set of $\mathbb{R}^p$; then one can extract a subsequence $\mu^{(m)}$ (of the sequence defined by (1)) weakly * converging to a measure μ^∞: i.e.

$$\left.\begin{array}{l} \langle \mu^\infty, F \rangle = \lim \int_\Omega F(x, u^{(m)}(x))\,dx \text{ for every } F \text{ continuous} \\ \text{in } (x,\lambda) \text{ with compact support in } x \end{array}\right\} (2)$$

satisfying

$$\mu^\infty \geqslant 0 \quad \text{and support} \quad \mu^\infty \subset \Omega \times \overline{K} \tag{3}$$

If $F(x,\lambda) = G(x)$ with G continuous with compact support,

$$\langle \mu^\infty, F \rangle = \int_\Omega G(x)\,dx. \tag{4}$$

Conversely if μ^∞ is a measure satisfying (3)(4) there exists a sequence $u^{(m)}(x) \in K$ a.e. and (2) holds. □

A measure like μ^∞ will be called a generalized function; actually it is useful to split μ^∞ with respect to x and use the following

Corollary 1

There exists a (measurable) family ν_x of probability measures on $\mathbb{R}^p$ such that

$$\text{support } \nu_x \subset \overline{K} \quad \text{a.e.} \tag{5}$$

$\langle \mu^\infty, F \rangle = \int_\Omega \langle \nu_x, F(x,\cdot) \rangle dx$ for every F continuous in (x,λ) with compact support in x. (6)

For every continuous function ϕ on $\mathbb{R}^p$, $\phi(u^{(m)})$ converges in $L^\infty(\Omega)$ weak * to a function ψ with $\psi(x) = \langle \nu_x, \phi \rangle$ a.e. . (7)

□

Outline of Proof. The existence of μ^∞ satisfying (2)(3)(4) follows from quoted results on $\mathcal{M}(\Omega \times \mathbb{R}^p)$ and then the lemma follows from results of measure theory; actually one can construct directly the family ν_x in the following way: first extract a subsequence $u^{(m)}$ such that for every polynomial with rational coefficients P, $P(u^{(m)})$ converges in L^∞ weak * to p. As

$$\| P(u^{(m)}) \|_{L^\infty} \leqslant \max_{k \in \overline{K}} |P(k)|$$

we have

$$|p(x)| \leqslant \max_{k \in \overline{K}} |P(k)| \quad \text{a.e. .}$$

Outside a set of measure zero all these inequalities are true.

For these x the map $P \to p(x)$ is linear and defines a unique probability measure ν_x with support in $\overline{K}$. Approximation of continuous functions by polynomials will complete the proof of the corollary.

(b) Without going into the proof that any (measurable) family of probability measures ν_x with support in $\overline{K}$ can be obtained as a limit of a sequence $u^{(m)}$ it is important to notice that theorem 1 gives the more general information about the oscillations of the sequence $u^{(m)}$. Intuitively $\nu_x(A)$ measures the probability that $u^{(m)}(y) \in A$ for y "near" x. The theorem says that the oscillations are arbitrary (all probability measures with support in $\overline{K}$ can be obtained) and independent at different points (only measurability in x).

When we add information on derivatives of $u^{(n)}$ we will see that this is no longer true but we will not be able to characterize the proper ν_x and their dependence on x: in some situations propagation of oscillations (that one should not mistake for propagation of singularities of the linear theory) can be observed; unfortunately very little is known in this direction.

It is important to know when no oscillations take place in some part of Ω: this is related to the fact that ν_x is a Dirac measure:

<u>Proposition 2</u>. $u^{(m)} \to u^{\infty}$ in $(L^r_{loc}(\omega))$ $\forall\, r < +\infty$ if and only if almost every ν_x is a Dirac measure for $x \in \omega$ (necessarily then $\nu_x = \delta_{u(x)}$ a.e.). □

Indeed one should notice that $u_j^{(m)} \to u_j^{\infty}$ in $L^r_{loc}(\omega)$ strongly equivalent to $u_j^{(m)} \rightharpoonup u_j^{\infty}$ and $(u_j^{(m)})^2 \rightharpoonup (u_j^{\infty})^2$ in $L^{\infty}(\omega)$ weak * and this is equivalent to ν_x supported in $\lambda_j = u_j^{\infty}(x)$ a.e. in ω.

II. COMPENSATED COMPACTNESS

1. We now consider the case where we have some information on derivatives of $u^{(n)}$. For technical reasons (quadratic functionals will be involved and Fourier transforms will be used) we will now work in an $(L^2(\Omega))^p$ setting: we assume that

$$u^{(n)} \text{ converges to } u^{\infty} \text{ in } (L^2(\Omega))^p \text{ weakly} \tag{8}$$

$$\sum_{jk} a_{ijk} \frac{\partial u_j^{(n)}}{\partial x_k} \in \text{compact of } H^{-1}_{loc}(\Omega) \quad i = 1,\dots,q. \tag{9}$$

In (9) the coefficients a_{ijk} are real constants (the functions $u^{(n)}$ are also real, but the complex case is easy to deduce); $H^{-1}_{loc}(\Omega)$ is the space of distributions of the form $f_0 + \sum_j \partial f_j/\partial x_j$ with $f_k \in L^2_{loc}(\Omega)$, $k = 0,\ldots,N$. In order to verify condition (9) one should remember that

(i) bounded sets of $L^2_{loc}(\Omega)$ are relatively compact in $H^{-1}_{loc}(\Omega)$

(ii) If g^n converges strongly in $L^2_{loc}(\Omega)$ then $\partial g^n/\partial x_j$ converges strongly in $H^{-1}_{loc}(\Omega)$

(iii) Murat's lemma, which we will state later, can imply (9) in some technical situations that we will encounter.

To the list of information on derivatives (9) we associate the following characteristic set

$$\Lambda = \{\lambda \in \mathbb{R}^p : \exists\, \xi \in \mathbb{R}^N\setminus\{0\} : \forall\, i = 1,\ldots,q \quad \sum_{jk} a_{ijk}\lambda_j\xi_k = 0\} \tag{10}$$

Actually for more precise information one needs the characteristic set

$$\mathcal{V} = \{(\lambda,\xi) \in \mathbb{R}^p \times (\mathbb{R}^N\setminus\{0\} : \forall\, i = 1,\ldots,q \ \sum_{jk} a_{ijk}\lambda_j\xi_k = \{0\}). \tag{11}$$

The basic theorem of compensated compactness is the following

<u>Theorem 2</u>

Let Q be a quadratic form on $\mathbb{R}^p$ satisfying

$$Q(\lambda) \geqslant 0 \qquad \forall\, \lambda \in \Lambda. \tag{12}$$

Assume that $u^{(n)}$ satisfies (8),(9) and that

$$Q(u^{(n)}) \text{ converges weakly * to } \mu \text{ in } (\Omega); \tag{13}$$

then we have

$$\mu \geqslant Q(u^{\infty}) \text{ in the sense of } \mathcal{M}(\Omega). \qquad \square \tag{14}$$

<u>Corollary 2</u>

Let Q be a quadratic form on $\mathbb{R}^p$ satisfying

$$Q(\lambda) = 0 \qquad \forall\, \lambda \in \Lambda. \tag{15}$$

Assume that $u^{(n)}$ satisfies (8) and (9); then we have

$$Q(u^{(n)}) \text{ converges weakly* to } Q(u^{\infty}) \text{ in } \mathcal{M}(\Omega). \tag{16}$$

□

Corollary 3

If $u^{(n)}$ is bounded in $(L^{\infty}(\Omega))^p$ and $u^{(m)}$ converges to the generalized function μ^{∞} defined by probability measures ν_x then for almost every x, ν_x satisfies the condition

$$\left.\begin{array}{l} \langle \nu, Q\rangle \geqslant Q(\bar{\lambda}) \text{ where } \bar{\lambda} \text{ is the centre of mass of} \\ \nu = \langle \nu, Id\rangle \text{ for every } Q \text{ satisfying (12).} \end{array}\right\} \tag{17}$$

□

Remark 1. Conditions (12) or (15) are optimal in the sense that:

(i) If (8)(9)(13) implies (14) then Q must satisfy (12).

(ii) If (8)(9) implies (16) then Q must satisfy (15). □

Remark 2. (17) is not always a characterization of the probability measures ν_x. But if one is only interested in first and second moments of ν, it is then optimal. □

Remark 3. If one knows that $Q(u^{(n)})$ is bounded in $L^p(\Omega)$ with $1 < p \leqslant +\infty$ then weakly* in $\mathcal{M}(\Omega)$ can be replaced by weakly in $L^p(\Omega)$ (weakly* if $p = +\infty$). □

Outline of Proof. By translation one can assume that $u^{\infty} = 0$ and then by localization assume that one works on $\mathbb{R}^N$ with functions having their support in a fixed compact set C. Then we need only to prove that $\liminf \int_{\mathbb{R}^N} Q(u^{(n)})dx \geqslant 0$. By using Fourier transforms and the Plancherel formula this is the same as $\liminf \int_{\mathbb{R}^N} \mathcal{R}e\, Q(\hat{u}^{(n)})d\xi \geqslant 0$. As $\hat{u}^{(n)}$ converges to 0 uniformly on bounded sets the difficulty lies at infinity. Condition (9) shows that $\frac{1}{1+|\xi|} \sum_{jk} a_{ijk}\xi_k \hat{u}_j^{(n)}(\xi)$ converges strongly to 0 in L^2 and this intuitively says that near infinity $\hat{u}^{(n)}(\xi)$ wants to be in $\Lambda + i\Lambda$ and the condition we need is $\mathcal{R}e\, Q(\lambda) \geqslant 0$ if $\lambda \in \Lambda + i\Lambda$ which is (12). (Of course some technical inequality is needed to make this precise).

Example 1. (Compactness) $\Lambda = \{0\}$. In this case every quadratic function Q satisfies (15) and every ν satisfying (17) is a Dirac measure. So a subsequence $u^{(m)}$ converges in $(L^2(\Omega))^p$ strong and a.e. . In this case the system of linear differential equation given by (9) is elliptic. □

Example 2. (Convexity) $\Lambda = \mathbb{R}^p$. In this case only convex Q satisfy (12) and ν can be any probability measure. The list of information on derivatives has not improved the knowledge of the generalized function; actually the improvement lies in the fact that the ν at different points are no longer independent. □

Example 3. (Elastostatics) Here $u^{(n)} = (E^{(n)}, D^{(n)}) \in (L^2(\Omega))^{2N}$; the list (9) is given by

$$\frac{\partial E_i^{(n)}}{\partial x_j} - \frac{\partial E_j^{(n)}}{\partial x_i} \quad \forall\, i,j \quad \text{and} \quad \sum_j \frac{\partial D_j^{(n)}}{\partial x_j} \quad ;$$

the set Λ is $\{(E,D) : \xi \neq 0 : \xi_j E_i - \xi_i E_j = 0 \quad \forall\, i,j$ and $\sum_j \xi_j D_j = 0\}$ which is $\{(E,D) : E.D = 0\}$. Clearly $Q(u) = E.D$ is quadratic and 0 on Λ so $E^{(n)}.D^{(n)}$ converges to $E^\infty.D^\infty$ weakly * in $\mathcal{M}(\Omega)$. It is useful to notice that this property is also the basis of the monotonicity method. □

Example 4. (Maxwell's system). We work here in 3 space variables and time, $u^{(n)} = (E^{(n)}, D^{(n)}, H^{(n)}, B^{(n)}) \in (L^2(\Omega))^{12}$ and the list (9) is given by

$$\frac{\partial B^{(n)}}{\partial t} + \operatorname{curl} E^{(n)}, \quad \operatorname{div} D^{(n)}, \quad \frac{\partial D^{(n)}}{\partial t} - \operatorname{curl} H^{(n)}, \quad \operatorname{div} B^{(n)}.$$

If all these quantities are in a compact set of $(H^{-1}_{loc}(\Omega))^m$ and $u^{(n)}$ converges weakly to u^∞ in $(L^2(\Omega))^{12}$ then we have

$E^{(n)}.B^{(n)}$ converges to $E^\infty.B^\infty$ weakly * in $\mathcal{M}(\Omega)$

$D^{(n)}.H^{(n)}$ converges to $D^\infty.H^\infty$ weakly * in $\mathcal{M}(\Omega)$

$E^{(n)}.D^{(n)} - B^{(n)}.H^{(n)}$ converges to $E^\infty.D^\infty - B^\infty.H^\infty$ weakly * in $\mathcal{M}(\Omega)$.

It is worth noticing that the density of electromagnetic energy $\frac{1}{2}(E.D + B.H)$ is not in the list; energy can propagate in the form of oscillations which cannot be distinguished through weak convergence. □

Other useful examples arise in situations connected with continuum mechanics and physics. For our application to hyperbolic systems in one space variable and time we will only use the following simple case:

Example 5. Ω is an open set in the (x,t) plane.

If $v_j^{(n)}$ converges to v_j weakly * in $L^\infty(\Omega)$ $j = 1,\dots,4$ and if

$$\frac{\partial v_1^{(n)}}{\partial t} + \frac{\partial v_2^{(n)}}{\partial x} \quad \text{and} \quad \frac{\partial v_3^{(n)}}{\partial t} + \frac{\partial v_4^{(n)}}{\partial x}$$

are in compact sets of $H^{-1}_{loc}(\Omega)$ then $v_1^{(n)}v_4^{(n)} - v_2^{(n)}v_3^{(n)}$ converges to $v_1v_4 - v_2v_3$ weakly * in $L^\infty(\Omega)$. Indeed a quick calculation shows that $\Lambda = \{u = (u_1,\dots,u_4): u_1u_4 - u_2u_3 = 0\}$

□

2. Now let us come back to our general setting: we have a sequence $u^{(n)}$ satisfying some linear partial differential equation like (9) (the components of $u^{(n)}$ will usually have a physical meaning) and we have relations between components of $u^{(n)}$ expressed by $u^{(n)}(x) \in K$ a.e. (the constitutive relations are usually involved here. We do not consider here the case where K depends on x which will be the case in the equation $D = a(x)E$: oscillations in a will produce homogenization effects and the adequate convergence for a is not a weak convergence). The probability measures ν_x involved in the generalized function are restricted by (17) and (5); it is evident that we have not used at all the interaction of both hypotheses which will arise in geometric properties of $\overline{K}$ with respect to the characteristic set $\mathcal{V}$ given by (11) (the dimension N plays a role in $\mathcal{V}$); by this lack we are only able to study the interplay of (17) and (5) (we will do a more careful analysis in the next chapter) and to conjecture what happens in general:

Conjecture 1. Assume that K and Λ satisfy the following compatibility condition:

$$a,b \in \overline{K} \text{ and } b - a \in \Lambda \text{ implies } b = a. \tag{18}$$

Then the conjecture is that ν_x is a Dirac measure a.e.

□

Remark 4. If $a,b \in K$ with $b - a \in \Lambda$ then there are oscillating $u^{(n)}$ jumping from a to b (if $b \neq a$) and ν_x can be a combination of δ_a and δ_b. Indeed there is a $\xi \neq 0$ such that $\sum_{jk} a_{ijk}(b-a)_j\xi_k = 0$ for $i = 1,\dots,q$. Let $u^{(n)} = a + (b-a)f^{(n)}(\xi.x)$ with $f^{(n)}$ smooth; then

$$\sum_{jk} a_{ijk}\frac{\partial u_j^{(n)}}{\partial x_k} = 0 \quad \text{for } i = 1,\dots,q$$

and this will remain true for a discontinuous $f^{(n)}$ taking values 0 or 1: but then $u^{(n)}$ only takes values a or b. If $f^{(n)}$ converges to $\frac{1}{2}$ in $L^{\infty}(\mathbb{R})$ weak* then $\nu_x = \frac{1}{2}(\delta_a + \delta_b)$. □

Remark 5. The conjecture is obvious for $N = 1$. In this case Λ is a subspace $\Lambda = \{\lambda : \sum_j a_{ij}\lambda_j = 0 \quad i = 1,\dots,q\}$. But if Λ is a subspace (17) implies that ν_x has its support in an affine subspace parallel to Λ (this is general and not related to $N = 1$): this is seen by taking $Q(\lambda) = \mathrm{dist}(\lambda,\Lambda)^2$ in (17). As condition (18) says that the intersection of K with a subspace parallel to Λ is at most one point then ν_x is a Dirac measure. □

The conjecture says that if one-dimensional oscillations are forbidden then no oscillation can take place. Although I made the conjecture I agree that my understanding of the geometry of $\mathcal{V}$ is so poor that I may be wrong: I do believe the conjecture to be true for $N = 2$. A local form of the conjecture can be proved:

Theorem 3

Assume that

(a) K is a p_0-dimensional C^1 manifold imbedded in $\mathbb{R}^p$.

(b) For every $a \in K$ the tangent space $T_a K$ (imbedded in $\mathbb{R}^p$) intersects $a + \Lambda$ only at a

Then if a probability measure ν has support in K, satisfies (17) and has a small support then it is a Dirac measure.

Proof. The size of the allowed support will be described in the proof. We first need the support of ν to be in the image of a chart G defined on a convex open set $\mathcal{O}$ of $\mathbb{R}^{p_0}$ and the size will be measured on the image ν_0 of ν by G^{-1}. Let a_0 be the centre of mass of ν_0. The tangent space at $a = G(a_0)$ is $a + \mathrm{Im}\, G'(a_0)$ so by (b) $\Lambda \cap \mathrm{Im}\, G'(a_0) = \{0\}$. So there exists $\alpha > 0$ such that

(i) $\quad \mathrm{dist}(\lambda, \mathrm{Im}(G'(a_0)))^2 \geq \alpha \|\lambda\|^2 \qquad \forall\, \lambda \in \Lambda.$

(We have chosen a euclidian structure on $\mathbb{R}^p$). The precise size of the support of ν_0 is that

(ii) $\begin{cases} \|G(b) - G(a_0) - G'(a_0)(b-a_0)\| \leqslant \beta \|G(b) - G(a_0)\| & \text{for every} \\ b \text{ in the support of } \nu_0 \text{ and } \beta < \sqrt{\dfrac{\alpha}{1+\alpha}} \end{cases}$

($G'(a_0)$ is one to one so the inequality is true in a neighbourhood of a_0).

Let

(iii) $Q(\lambda) = \mathrm{dist}(\lambda, \mathrm{Im}\, G'(a_0))^2 - \alpha \|\lambda\|^2$.

So Q is quadratic and $\geqslant 0$ on Λ by (i) so by (17)

$$L = \langle \nu, Q(\cdot - G(a_0)) \rangle \geqslant Q(\langle \nu, \cdot - G(a_0) \rangle) = R.$$

We bound the right hand side R from below by noticing that

$$\langle \nu, \cdot - G(a_0) \rangle = \langle \nu_0, G(\cdot) - G(a_0) \rangle = \langle \nu_0, G'(a_0)(\cdot - a_0) + r(\cdot) \rangle$$
$$= \langle \nu_0, r(\cdot) \rangle$$

as a_0 is the centre of mass of ν_0.

(We have noted that

$$r(b) = G(b) - G(a_0) - G'(a_0)(b-a_0).)$$

As

$$Q(\lambda) \geqslant -\alpha \|\lambda\|^2$$

we obtain

$$R \geqslant -\alpha \| \langle \nu_0, r(\cdot) \rangle \|^2.$$

We bound the left hand side L from above by noticing that

$$\langle \nu, Q(\cdot - G(a_0)) \rangle = \langle \nu_0, Q(G(\cdot) - G(a_0)) \rangle$$

and that, as

$$\mathrm{dist}(G(b) - G(a_0), \mathrm{Im}\, G'(a_0)) \leqslant \|r(b)\|,$$

we have

$$Q(G(b) - G(a_0)) \leqslant \|r(b)\|^2 - \alpha \|G(b) - G(a_0)\|^2 ;$$

so

$$L \leqslant \langle \nu_0 \|r(\cdot)\|^2 \rangle - \alpha \langle \nu_0, \|G(\cdot) - G(a_0)\|^2 \rangle.$$

Finally

$$\alpha < \nu_0, \|G(\cdot) - G(a_0)\|^2 > \ \leqslant \ < \nu_0, \|r\|^2 > \ + \alpha \| < \nu_0, r > \|^2$$

$$\leqslant (1+\alpha) < \nu_0, \|r\|^2 > .$$

Using now (ii) we have

$$\|r(b)\|^2 \leqslant \beta^2 \|G(b) - G(a_0)\|^2$$

on the support of ν_0 and as $(1+\alpha)\beta^2 < \alpha$ we conclude that

$$< \nu_0, \|G(\cdot) - G(a_0)\|^2 > \ = 0$$

So $G(b) = G(a_0)$ on support ν_0 and this implies $b = a_0$ so $\nu_0 = \delta_{a_o}$ and ν is a Dirac measure. □

Remark 5. The smallness of the support is necessary here because conditions (a)(b) do not imply (18) in general. □

This theorem contains applications to nonlinear elliptic systems (condition (b) is the ellipticity of the linearized problem).

III. APPLICATIONS TO HYPERBOLIC SYSTEMS OF CONSERVATION LAWS

We are interested here in systems of conservation laws

$$\frac{\partial U}{\partial t} + \frac{\partial}{\partial x} F(U) = 0 \tag{19}$$

where U is an m-vector. Although the method does not need hyperbolicity we will assume that this system is strictly hyperbolic i.e. $F'(a)$ has m real distinct eigenvalues $\lambda_1(a) < \ldots < \lambda_m(a)$ with right eigenvector $r_j(a)$ for a in some domain of $\mathbb{R}^m$. The first thing to notice is that there are always oscillating solutions of (19) (so (18) is not true): indeed let $a,b \in \mathbb{R}^m$ be such that they can be connected by a shock, i.e. they satisfy the Rankine-Hugoniot condition

$$F(b) - F(a) = s(b-a) \qquad s \in \mathbb{R} \qquad a \neq b.$$

Let

$$\left.\begin{array}{l} U^{(n)}(x,t) = a + b\, f^{(n)}(x-st) \\[1em] f^{(n)} \text{ a characteristic function} \end{array}\right\} . \tag{20}$$

Then $U^{(n)}$ satisfies (19) and oscillates between a and b.

(There are, in contrast with the elliptic setting of theorem 3, small oscillations: the set of b satisfying (20) consists in the neighbourhood of a of m distinct curves, in the directions of $r_j(a)$, $j = 1,\dots,m$).

The second thing to notice is that, except in the case of linear degeneracy, one of the shocks from a to b or from b to a is forbidden by entropy conditions and thus we do not want to consider such sequences of solutions $U^{(n)}$ but want to restrict our attention only to solutions satisfying entropy conditions.

1. Entropy conditions

To do this we will use Lax's framework and consider the artificial viscosity method

$$\frac{\partial U_\varepsilon}{\partial t} + \frac{\partial F(U_\varepsilon)}{\partial x} - \varepsilon \frac{\partial^2 U_\varepsilon}{\partial x^2} = 0 \tag{21}$$

and assume that a sequence $(\varepsilon_n \to 0)$ of solutions of (21) is given satisfying the following estimates

$$\left.\begin{array}{l} U_\varepsilon \text{ is bounded in } (L^\infty(\Omega))^m \\ \qquad\qquad\qquad\qquad (\Omega \text{ is an open set in } x,t) \\ \sqrt{\varepsilon}\,\dfrac{\partial U_\varepsilon}{\partial x} \text{ is bounded in } (L^2(\Omega))^m \end{array}\right\} \tag{22}$$

An entropy ϕ of system (19) is a real function on $\mathbb{R}^m$, associated to a flux ψ satisfying

$$\phi'(a)F'(a) = \psi'(a) \quad \forall\ a \text{ in domain of definition of } F. \tag{23}$$

Multiplying (21) by $\phi'(U_\varepsilon)$ one gets easily

$$\frac{\partial}{\partial t}\phi(U_\varepsilon) + \frac{\partial}{\partial x}\psi(U_\varepsilon) - \varepsilon\frac{\partial^2}{\partial x^2}\phi(U_\varepsilon) + \varepsilon\,\phi''(U_\varepsilon)\left(\frac{\partial U_\varepsilon}{\partial x}, \frac{\partial U_\varepsilon}{\partial x}\right) = 0 . \tag{24}$$

If U_ε was converging a.e. to U then for a convex ϕ we would know the sign of the last term and know the limit of the three first ones and obtain the following property for U:

$$\frac{\partial}{\partial t}\phi(U) + \frac{\partial}{\partial x}\psi(U) \leqslant 0 \quad \text{for all convex entropies } \phi. \tag{25}$$

This constitutes the entropy conditions.

2. The method

We want to use (24) as an equation of type (9) (and use an infinite number of them); by (22) $\phi(U_\varepsilon)$ and $\psi(U_\varepsilon)$ are bounded in $L^\infty(\Omega)$ and $\varepsilon \frac{\partial^2}{\partial x^2} \phi(U_\varepsilon)$ converges strongly to 0 in $H^{-1}_{loc}(\Omega)$ because it is $\sqrt{\varepsilon}\frac{\partial}{\partial x}\left(\phi'(U_\varepsilon)\sqrt{\varepsilon}\frac{\partial U_\varepsilon}{\partial x}\right)$ but the last term $\varepsilon\,\phi''(U_\varepsilon)\left(\frac{\partial U_\varepsilon}{\partial x}, \frac{\partial U_\varepsilon}{\partial x}\right)$ is only bounded in $L^1(\Omega)$ (for ϕ a C^2 function) and we need the following lemma of F. Murat in order to know the desired compactness in $H^{-1}_{loc}(\Omega)$.

Murat's Lemma: Let Ω be an open set of $\mathbb{R}^N$ and let a sequence of distributions $f^{(n)}$ satisfy

(i) $f^{(n)}$ is bounded in $W^{-1,r}_{loc}(\Omega)$ for some $r > 2$

(ii) $\begin{cases} f^{(n)} = g^{(n)} + h^{(n)} \text{ with } g^{(n)} \in \text{compact set of } H^{-1}_{loc}(\Omega) \\ \text{and } h^{(n)} \in \text{bounded set of } \mathcal{M}(\Omega). \end{cases}$

Then we have

(iii) $f^{(n)} \in$ compact set of $H^{-1}_{loc}(\Omega)$. □

With this in hand we can extract a subsequence U_{ε_m} giving a generalized function defined by a family $\nu_{x,t}$ of probability measures on $\mathbb{R}^m$ and apply the example 5 with $v_1 = \phi_1(U_\varepsilon)$, $v_2 = \psi_1(U_\varepsilon)$, $v_3 = \phi_2(U_2)$, $v_4 = \psi_2(U_\varepsilon)$ where (ϕ_1,ψ_1) and (ϕ_2,ψ_2) are 2 (entropy, flux) pairs. Corollary 1 of theorem 1 with example 5 shows that for almost every $(x,t) \in \Omega$ the probability measure $\nu_{x,t}$ satisfies:

$$\left.\begin{array}{l} \langle\nu,\phi_1\psi_2 - \phi_2\psi_1\rangle = \langle\nu,\phi_1\rangle\langle\nu,\psi_2\rangle - \langle\nu,\phi_2\rangle\langle\nu_1,\psi_1\rangle \\ \text{for all pairs } (\phi_1,\psi_1)(\phi_2,\psi_2) \text{ satisfying (23).} \end{array}\right\} \quad (26)$$

The method is then to characterize such probability measures ν and eventually try to prove that only Dirac measures satisfy (26). The case $m = 1$ can be completely solved because every function ϕ is an entropy; one finds that the support of ν lies in an interval (which can be reduced to a point) where F is affine, so oscillations can only happen if some linear degeneracy holds.

For $m = 2$ entropies are solutions of a linear second order hyperbolic equation and we will need to look at (23) more carefully.

3. Lax's family of entropies

If we write (23) in the basis of eigenvectors of $F'(a)$ we obtain

$$\psi'(a).r_j(a) = \lambda_j(a)\phi'(a).r_j(a) \qquad j = 1,\ldots,m . \tag{27}$$

For $m = 2$ it is natural to change coordinates with the help of Riemann-invariants

$$w_j \text{ is a j-Riemann-invariant if } w_j'(a).r_j(a) = 0 \quad \forall\, a . \tag{28}$$

We will assume that w_1, w_2 have been chosen (nonsingular) so that they define a global change of coordinates on the U domain used. Then (27) becomes ($\lambda_1 < \lambda_2$ are now functions of w_1, w_2)

$$\left.\begin{aligned} \frac{\partial\psi}{\partial w_1} &= \lambda_2 \frac{\partial\phi}{\partial w_1} \\ \frac{\partial\psi}{\partial w_2} &= \lambda_1 \frac{\partial\phi}{\partial w_2} \end{aligned}\right\} . \tag{29}$$

We now look for solutions having the asymptotic form

$$\left.\begin{aligned} \phi_k(w_1,w_2) &= e^{kw_1}\left[A_0 + \frac{A_1}{k} + O\left(\frac{1}{k^2}\right)\right] \\ & \qquad\qquad\qquad k \to \pm\infty \\ \psi_k(w_1,w_2) &= e^{kw_1}\left[B_0 + \frac{B_1}{k} + O\left(\frac{1}{k^2}\right)\right] \end{aligned}\right\} . \tag{30}$$

Such solutions exist (estimates are uniform on bounded sets) if the formal compatibility conditions are true

$$\left.\begin{aligned} & B_0 = \lambda_2 A_0 \\ & B_{j+1} + \frac{\partial B_j}{\partial w_1} = \lambda_2\left(A_{j+1} + \frac{\partial A_j}{\partial w_1}\right) \quad j = 0,1,\ldots . \\ & \frac{\partial B_j}{\partial w_2} = \lambda_1 \frac{\partial A_j}{\partial w_2} \quad j = 0,1,\ldots . \end{aligned}\right\} \tag{31}$$

This gives

$$\left.\begin{aligned} & (\lambda_2 - \lambda_1)\frac{\partial A_0}{\partial w_2} + \frac{\partial \lambda_2}{\partial w_2} A_0 = 0 \\ & B_0 = \lambda_2 A_0 \end{aligned}\right\} . \tag{32}$$

$$\left.\begin{aligned}&(\lambda_2-\lambda_1)\frac{\partial A_1}{\partial w_2}+\frac{\partial \lambda_2}{\partial w_2}A_1=\frac{\partial}{\partial w_2}\left(\frac{\partial \lambda_2}{\partial w_1}A_0\right)\\ &B_1=\lambda_2 A_1-\frac{\partial \lambda_2}{\partial w_1}A_0 .\end{aligned}\right\} \quad (33)$$

The important fact we will use from (32) is that $A_0 > 0$ and $B_0 = \lambda_2 A_0$; from (33) we will use the fact that $B_1 - \lambda_2 A_1$ has the opposite sign to $\partial\lambda_2/\partial w_1$.

The fact that $\partial\lambda_2/\partial w_1 \neq 0$ is related to genuine nonlinearity, defined in the general case for the j-th characteristic field as $\lambda_j'(a).r_j(a) \neq 0$.

Of course a similar family with an exponential e^{kw_2} and permutation of the roles of λ_1,λ_2 exists.

4. DiPerna's result

We assume that one can use w_1,w_2 as a global change of coordinates of the U domain. Let ν satisfy (26) and let

$$\left.\begin{array}{l}R \text{ be the smallest rectangle } = \{(w_1,w_2): w_j^- \leqslant w_j \leqslant w_j^+\}\\ \text{with sides parallel to the w-axes containing the}\\ \text{support of } \nu\end{array}\right\} (34) .$$

Theorem 4

If $w_1^- < w_1^+$ then each closed vertical side of R must contain a point where $\dfrac{\partial\lambda_2}{\partial w_1} = 0$

Proof: First Step. We assume that $w_1^- < w_1^+$. DiPerna's argument relies on the introduction of boundary probability measures $\mu^{\pm}$ defined as follows: for $|k|$ large enough define the probability measure μ_k by

$$<\mu_k, f> = \frac{<\nu,\phi_k f>}{<\nu,\phi_k>} \quad (35)$$

where (ϕ_k,ψ_k) is a family of (entropy, flux) pairs having the form (3) with $A_0 > 0$. Because $\phi_k e^{-kw_1} = A_0 + O\left(\frac{1}{k}\right)$ is > 0 for large k, (35) has a meaning and defines a probability measure with support in R. Now we extract weakly * converging subsequences such that

$$\text{For } k_n \to +\infty \quad \left.\begin{array}{l} \mu_{k_n} \text{ converges weakly * to } \mu^+ \\ \mu_{-k_n} \text{ converges weakly * to } \mu^- \end{array}\right\} \tag{36}$$

Because as k goes to infinity e^{kw_1} gives most of the weight to a neighbourhood of w_1^- or w_1^+ it is seen, as R is the smallest rectangle containing the support, that

$$\left.\begin{array}{l} \text{support } \mu^- \subset R \cap \{w_1 = w_1^-\} \\ \text{support } \mu^+ \subset R \cap \{w_1 = w_1^+\} \end{array}\right\}$$

Denote

$$\lambda_2^{\pm} = \langle \mu^{\pm}, \lambda_2 \rangle . \tag{37}$$

Second Step. We want to show that

$$\langle \nu, \psi - \lambda_2^{\pm}\phi \rangle = \langle \mu^{\pm}, \psi - \lambda_2\phi \rangle \text{ For every (entropy, flux).} \tag{38}$$

For this we use in (26) the pairs (ϕ,ψ) and (ϕ_k,ψ_k), divide by $\langle \nu,\phi_k \rangle$ and let k go to $\pm\infty$ along the sequence $\pm k_n$. We must notice that $\dfrac{\langle \nu,\psi_k \rangle}{\langle \nu,\phi_k \rangle}$ can be written, using (30),(32) as

$$\frac{\langle \nu,\lambda_2\phi_k \rangle}{\langle \nu,\phi_k \rangle} + \frac{\langle \nu,\phi_k \cdot O(\frac{1}{k}) \rangle}{\langle \nu,\phi_k \rangle} = \langle \mu_k,\lambda_2 \rangle + O(\tfrac{1}{k})$$

so it has limit $\lambda_2^{\pm} = \langle \mu^{\pm},\lambda_2 \rangle$.

Similarly $\dfrac{\langle \nu,\psi_k\phi \rangle}{\langle \nu,\phi_k \rangle}$ will have limit $\langle \mu^{\pm},\lambda_2\phi \rangle$. So the equality

$$\frac{\langle \nu,\phi_k\psi - \phi\psi_k \rangle}{\langle \nu,\phi_k \rangle} = \langle \nu,\psi \rangle - \langle \nu,\phi \rangle \frac{\langle \nu,\psi_k \rangle}{\langle \nu,\phi_k \rangle} \quad \text{gives (38).}$$

Third Step. We want to show that

$$\lambda_2^- = \lambda_2^+ \tag{39}$$

For this we use in (26) the pairs (ϕ_k,ψ_k) and (ϕ_{-k},ψ_{-k}), divide by $<\nu,\phi_k><\nu,\phi_{-k}>$ and let k go to $+\infty$ along the sequence k_n. Thus

$$\frac{<\nu,\phi_k\psi_{-k}-\phi_{-k}\psi_k>}{<\nu,\phi_k><\nu,\phi_{-k}>} = \frac{<\nu,\psi_{-k}>}{<\nu,\phi_{-k}>} - \frac{<\nu,\psi_k>}{<\nu,\phi_k>} .$$

The right hand side converges to $\lambda_2^- - \lambda_2^+$.

If we notice that $\phi_k\psi_{-k} - \phi_{-k}\psi_k$ is $O(\frac{1}{k})$ and that $<\nu,\phi_k><\nu,\phi_{-k}>$ goes to $+\infty$ faster than $e^{k(w_1^+-w_1^- -\varepsilon)}$ (by the definition of R) we see that the left hand side goes to 0.

Fourth Step. We want to show that

$$<\mu^\pm , A_o\frac{\partial\lambda_2}{\partial w_1}> = 0 \tag{40}$$

Using (38) and (39) we see that

$$<\mu^+,\psi-\lambda_2\phi> = <\mu^-,\psi-\lambda_2\phi> \tag{41}$$

for every (entropy, flux). Again use (ϕ_k,ψ_k). Thus

$$<\mu^*,\psi - \lambda_2\phi> = e^{kw_1^+}<\mu^+,\frac{B_1 - \lambda_2A_1}{k} + O(\frac{1}{k^2})>$$

and

$$<\mu^-,\psi-\lambda_2\phi> = e^{kw_1^-}(\mu^-,\frac{B_1-\lambda_2A_1}{k} + O(\frac{1}{k^2})>$$

As $w_1^- < w_1^+$ we must have $<\mu^\pm, B_1-\lambda_2A_1> = 0$ (and also $<\mu^\pm,B_n-\lambda_2A_n> = 0$ if we use a longer asymptotic expansion); by (33) this gives (40).

The theorem is proved because $A_o > 0$ and $\mu^\pm$ being probability measures $A_o\frac{\partial\lambda_2}{\partial w_1}$ must vanish on each vertical sides of R.

Corollary 4

If both characteristic fields are genuinely nonlinear ($\frac{\partial \lambda_1}{\partial w_2} \neq 0$, $\frac{\partial \lambda_2}{\partial w_1} \neq 0$ everywhere) then ν is a Dirac measure.

Corollary 5

If both characteristic fields are genuinely nonlinear outside a curve $w_2 = f(w_1)$ with f strictly monotone then ν is a Dirac measure.

Proof: By Theorem 4 each closed side of the rectangle R should intersect the line; necessarily the line should go through opposite vertices P and Q. μ^- and μ^+ should be Dirac measures at these points because by (40) they give 0 against a function vanishing only at the end of the interval supporting the measure. By (38), (39) we obtain that

$$\psi(Q) - \lambda_2(Q)\ \phi(Q) = \psi(P) - \lambda_2(P)\ \phi(P)$$

Similarly we have the same remark by replacing λ_2 by λ_1 and this gives $(\lambda_2(Q) - \lambda_1(Q))\phi(Q) = (\lambda_2(P) - \lambda_1(P))\phi(P)$ for every entropy ϕ, in particular for every affine function ϕ in the U variable and this implies $P = Q$ so the rectangle is a point.

Remark 6. The averages of λ_2 at both sides against $\mu^{\pm}$ are the same and a stronger property for λ_2 is conjectured in general.

Conjecture 2. If R is the smallest rectangle with sides parallel to w_1, w_2-axes and ν satisfies (26) then λ_1 and λ_2 are constant on R and ν is a tensor product. So, as for $m = 1$, it is conjectured that only linear degeneracy can be responsible for oscillations.

Example. We consider the system

$$\frac{\partial}{\partial t}\begin{pmatrix} u_1 \\ u_2 \end{pmatrix} - \frac{\partial}{\partial x}\begin{pmatrix} u_2 \\ \sigma(u_1) \end{pmatrix} = 0 \qquad (42)$$

which is strictly hyperbolic if

$$\sigma'(u_1) > 0 \qquad (43)$$

The eigenvalues are $\pm\sqrt{\sigma'(u_1)}$ and the Riemann invariants are easily computed.

$$\begin{cases} w_1 = u_2 - \int^{u_1} \sqrt{\sigma'(\nu)}\,d\nu \\ w_2 = u_2 + \int^{u_1} \sqrt{\sigma'(\nu)}\,d\nu \end{cases} \tag{44}$$

The characteristic fields are genuinely nonlinear where $\sigma''(u_1) \neq 0$. Corollary 4 or 5 apply if σ'' has at most one zero: indeed u_1 = constant is equivalent to $w_2 - w_1$ = constant.

Remark 7. All this depends upon the estimates (22). It is indeed true in the example (41) if σ is concave and then convex; under this constraint some rectangles (which are convex in the U plane) in the w plane are invariant and this gives the crucial L^∞ estimates, and this proves global existence of a bounded solution satisfying the entropy conditions if the data are bounded.

IV. COMMENTS ON IDEAS

The idea of generalised functions, due to L.C. Young, has been mostly used for $N = 1$ in relation with non-existence theorems in optimisation. It is the same idea that leads to Pontryagin's maximum principle: writing optimality conditions for a functional only defined on characteristic functions is not always easy and the introduction of generalised functions often leads to an extension of the functional defined on a convex set where optimality conditions are obvious; the same problem for $N > 1$ (related to optimal design) is much more difficult and to go beyond Hadamard's necessary conditions it is useful to understand homogenization.

It was the study of this problem by Murat and myself in 1973-74 which gave some of the basic results of homogenization and the first example (example 3) of compensated compactness. Only after meeting J. Ball in the fall of 1975 in Marseille and learning about his results on nonlinear elasticity did I foresee the power of the forecoming theory.

The first general statement, in a slightly less general setting, analogous to corollary 2 was given by Murat and soon after I could, in my Cours Peccot on homogenization in the spring of 1977, add theorem 2 and suggest some relations between homogenization and compensated compactness. Only in the fall, while discussing with DiPerna in Madison, did I find the way to handle entropy conditions by using Murat's lemma and solve the scalar case.

After that time all the ideas are due to R. DiPerna. Special mention should be made concerning theorem 3 which I proved while in Oxford: actually this was only an attempt to simplify one of DiPerna's results on nonlinear elasticity; so the underlying idea should be regarded as his although the technical proof is mine.

Added in proof: As noticed by D. Serre, in the linear degenerate case

$$\left(\frac{\partial\lambda_1}{\partial w_2} \equiv \frac{\partial\lambda_2}{\partial w_1} \equiv 0\right)$$

if ν satisfies (26) then one can only conclude that $\frac{\nu}{\lambda_2 - \lambda_1}$ is a tensor product.

Actually, given two probability measures μ_1 and μ_2 on $\mathbb{R}$, one can then construct oscillating solutions corresponding to $\nu = c(\lambda_2 - \lambda_1)\mu_1 \otimes \mu_2$. This construction can be used when linear degeneracy only holds on the sides of a rectangle to have such a ν supported by the four vertices. After such a bad guess I will leave it to the reader to make and prove the right conjecture.

BIBLIOGRAPHY

The details missing in these lectures will be found in

[1] Dacorogna, B.: 1982, '*Weak semicontinuity and weak lower semicontinuity of nonlinear functionals*', Lecture Notes in Mathematics, No. 922, Springer.

[2] DiPerna, R.J.: *Convergence of approximate solutions to conservation laws*, to appear.

[3] Tartar, L.C.: 1979, '*Compensated compactness and applications to partial differential equations*', in Nonlinear Analysis and Mechanics, Heriot-Watt Symposium, IV, pp.136-192. Research Notes in Mathematics, Pitman.

PART II

SPECIAL SESSIONS

COERCIVITY CONDITIONS IN NONLINEAR ELASTICITY

Stuart S. Antman

Department of Mathematics and
Institute for Physical Science and Technology
University of Maryland
College Park, Maryland 20742, U.S.A.

Abstract. The simplest theories for quasilinear elliptic equations for an unknown scalar-valued function u are based upon coercivity conditions that require some energy-like term to exceed const$|u|^p$ where $p > 1$. There are many physical applications of such theories. Theories for quasilinear elliptic systems for a vector-valued $\underset{\sim}{u}$ can likewise be based on analogous coercivity conditions. If, however, the system is to consist of the equilibrium equations of nonlinear elasticity with $\underset{\sim}{u}(\underset{\sim}{z})$ denoting the position of a material point $\underset{\sim}{z}$, then this kind of coercivity condition is singularly inadequate on both physical and mathematical grounds. This paper first describes the mathematical structure of the quasilinear equations of elastostatics, focusing on the far more intricate kinds of coercivity conditions that are needed to meet reasonable physical restrictions and to promote mathematical analysis. The role of these new coercivity conditions is then illustrated by an analysis of the existence, regularity, and qualitative behavior of solutions of specific boundary value problems.

From the viewpoint of the theory of quasilinear elliptic systems, elasticity theory may seem like a collection of unpleasant pathologies. From the viewpoint of science, however, elasticity theory is perhaps the most accessible source of such systems. In this paper we first describe the mathematical structure of the quasilinear, elliptic system of partial differential equations governing the equilibrium of nonlinearly elastic bodies. We show how peculiarities of the theory that reflect the underlying physics lead to a rich and mathematically novel set of coercivity conditions. We finally indicate how these coercivity conditions play an essential role in the analysis of specific boundary value

J. M. Ball (ed.), Systems of Nonlinear Partial Differential Equations, 289–297.

problems.

We identify a material point of a body with its position $\mathbf{z}$ in some reference configuration. We may accordingly identify the body with a domain Ω in $\mathbb{E}^3$. Let $\mathbf{p}(\mathbf{z})$ be the position of $\mathbf{z}$ in a deformed configuration. An elastic material is defined by prescribing the first Piola-Kirchhoff stress tensor as a function $\mathbf{T}$ of the gradient $\mathbf{F} \equiv \partial\mathbf{p}/\partial\mathbf{z}$ and of $\mathbf{z}$. (This stress tensor measures the intensity of force per unit reference area over material surfaces. $\mathbf{T}$ must be invariant under rigid motions of the body and must satisfy a certain symmetry condition that ensures that the angular momentum principle reduces to an identity.) The equilibrium equations, expressing the requirement that the resultant force on each part of the body vanishes, have the classical form

$$\operatorname{Div} \mathbf{T}(\partial\mathbf{p}/\partial\mathbf{z},\mathbf{z}) + \mathbf{f} = \mathbf{0} \qquad (\mathbf{z} \in \Omega) \tag{1}$$

where the body force $\mathbf{f}$ is a given function of $\mathbf{z}$ and $\mathbf{p}$, say. (A detailed definition of how the differential operator $\widetilde{\operatorname{Div}}$ acts on the tensor-valued function $\mathbf{T}$ is not needed for our analysis.) For simplicity, we henceforth assume that $\mathbf{f} = \mathbf{0}$.

It is clear that the analysis of (1) devolves on the form of $\mathbf{T}$, which characterizes the material response. We could choose a particular form of $\mathbf{T}$ and seek solutions of suitable boundary value problem for (1). (This could be done numerically even when we do not know whether the problem possesses solutions.) On the other hand, if we leave the form of $\mathbf{T}$ relatively unrestricted, then any analysis we perform will illuminate the nature of material response and will guide, rather than follow, physical inquiry. (Reliable experimental data on nonlinear material response can at best be characterized as sparse.) Such a strategy is perfectly suited to functional analysis.

What features distinguish (1) from the kinds of quasilinear elliptic systems treated in the extant theory, such as it is? There are two:

(i) Since $\mathbf{p}$ defines a deformation of a material body we require that $\mathbf{p}$ be one-to-one, so that two parts of the body cannot simultaneously occupy the same region of space. This is a global restriction on the function $\mathbf{p}$, which is to be determined by the local equation (1) and by boundary conditions. Since the treatment of this injectivity seems beyond our present resources, we adopt the more modest requirement that $\mathbf{p}$ preserve orientation:

$$\det \partial\mathbf{p}(\mathbf{z})/\partial\mathbf{z} > 0, \tag{2}$$

at least almost everywhere. (The serious technical problem of defining this Jacobian when $\underset{\sim}{p}$ belongs to a Sobolev space has been resolved by Ball (1977a,b).) Consequently the domain of $\underset{\sim}{T}$ is not the (convex) set of all second-order tensors, but rather $\{\underset{\sim}{F} : \det \underset{\sim}{F} > 0\}$, which is not convex. (Indeed $\{\underset{\sim}{p} \in C^1 : \partial\underset{\sim}{p}/\partial\underset{\sim}{z} > 0\}$ may have a countable infinity of non-convex components. That $\underset{\sim}{p}$ is a geometrical entity is also the source of another serious difficulty, namely that there is an ambiguity in the specification of mixed boundary conditions. Cf. Alexander & Antman (1982).)

(ii) It is reasonable to require that pulling a specimen in a given direction causes it to elongate in that direction. A <u>mathematical</u> generalization of this notion is that $\underset{\sim}{T}$ be strictly monotone. If $\underset{\sim}{T}$ is differentiable, this requirement is equivalent to

$$\underset{\sim}{A} : (\partial\underset{\sim}{T}/\partial\underset{\sim}{F}) : \underset{\sim}{A} > 0 \tag{3}$$

for all non-zero second order tensors $\underset{\sim}{A}$. (The left side of (3) is a general quadratic form for the fourth-order tensor $\partial\underset{\sim}{T}/\partial\underset{\sim}{F}$.) When (3) holds, one can do a great deal of analysis with (1). Unfortunately (3) implies a universal uniqueness that says that a rod however thin never buckles under a compressive (dead) load however large. (One of the chief motivations for suffering the complexities of nonlinear elasticity is to describe just such buckling.) Moreover, a monotone $\underset{\sim}{T}$ cannot become unbounded at the extremes of a non-convex set such as $\{\underset{\sim}{F} : \det \underset{\sim}{F} > 0\}$.

The problem of giving a physically reasonable statement that some mapping from deformations to stresses should have some order-preserving properties is quite delicate. There are many ways to measure strains and stresses. Which are appropriate for this purpose? Most materials when pulled in one direction not only lengthen in that direction but also contract in the transverse directions. How should this property be expressed?

A convenient mathematical answer is that (1) should be <u>strongly elliptic</u>, i.e., (3) should hold for all non-zero $\underset{\sim}{A}$'s of rank 1, which are dyadic (or tensor) products $\underset{\sim}{a}\underset{\sim}{b}$ (also denoted $\underset{\sim}{a} \otimes \underset{\sim}{b}$) of vectors $\underset{\sim}{a}$ and $\underset{\sim}{b}$. This assumption ensures that the dynamical equations corresponding to (1) are totally hyperbolic and have the richest variety of wave-like behavior. Moreover, if $\underset{\sim}{a}$ and $\underset{\sim}{b}$ are independent of $\underset{\sim}{F}$, then the strong ellipticity condition reduces to

$$\partial(\underset{\sim}{a} \cdot \underset{\sim}{T} \cdot \underset{\sim}{b})/\partial(\underset{\sim}{a} \cdot \underset{\sim}{F} \cdot \underset{\sim}{b}) > 0 \tag{4}$$

for $\underset{\sim}{a}\underset{\sim}{b} \neq \underset{\sim}{0}$. (Here $\underset{\sim}{a} \cdot \underset{\sim}{T} \cdot \underset{\sim}{b}$ is the component of $\underset{\sim}{T}$ in the $\underset{\sim}{a}\underset{\sim}{b}$ direction.) Inequality (4) says that a given component of stress is an increasing function of the corresponding component of deformation gradient. Thus the strong ellipticity condition captures our primitive physical notion of order-preservation. Moreover, the domain of the increasing function $\underset{\sim}{a}\cdot\underset{\sim}{F}\cdot\underset{\sim}{b} \mapsto \underset{\sim}{a}\cdot\underset{\sim}{T}\cdot\underset{\sim}{b}$ is a half-line, or a whole line, or the empty set (because a determinant is an affine function of any one of its entries).

Hughes, Kato, & Marsden (1977) have used the strong ellipticity condition to obtain a short-time solution of the Cauchy problem for the quasilinear hyperbolic dynamical equations. Ball (1977a) introduced the notion of polyconvexity, a refinement of the strong ellipticity condition, to prove that the functional whose Euler-Lagrange equation is (1) has a minimizer satisfying (2). Ball (1980) also treated regularity questions in light of these inequalities. These are the central papers in the analysis of the equations of three-dimensional nonlinear elasticity.

That one can obtain a rich and physically reasonable theory with the strong ellipticity condition does not mean that is is universally valid. Ericksen, in his article in this volume, describes a number of crystalline materials for which (4) is explicitly violated. In short, monotonicity of $\underset{\sim}{T}$ (or equivalently convexity of the corresponding strain energy for hyperelastic materials) is never a reasonable assumption, whereas the strong ellipticity condition seems to be quite reasonable, but is not a universally valid condition for three-dimensional theory. (In this connection, see the article of Dafermos in this volume.)

Our basic coercivity condition is that an infinite stress must accompany an extreme strain, i.e., that

$$|\underset{\sim}{T}| \to \infty \quad \text{as any eigenvalue of} \quad \underset{\sim}{F}^*\cdot\underset{\sim}{F} \quad \text{aproaches 0 or } \infty. \tag{5}$$

Here the dot denotes tensor multiplication and the $*$ denotes transpose. We refine this by requiring that

$$a\cdot T\cdot b \to \pm\infty \text{ as } a\cdot F\cdot b \text{ approaches an extreme of its interval,} \tag{6}$$

the sign being dictated by (4). Our final set of coercivity conditions are related to armchair experiments in which a specimen is pulled in one direction by an infinitely large stress. Not only should the specimen suffer an infinite elongation in the direction of pull, but it should also suffer a total contraction in the transverse directions or else be prevented from doing so by infinite tensile stresses in the transverse directions. There are a number of analogous conditions, which we do not spell out. These conditions can all be stated rather easily in terms of principal axes, but the algebraic complications in finding these axes makes such statements practically useless. The appropriate versions relative to arbitrary axes are complicated, delicate, and lengthy, and will accordingly be omitted. (Cf. Antman (1983).) The heuristic principle underlying all these conditions is that extremes in behavior always involve at least two directions.

We now wish to see the effects of these conditions by studying a class of problems simple enough to be tractable, complicated enough to reflect the full complexity of the theory, and interesting enough to be worthwhile. The problems we choose are a class of semi-inverse problems describing deformations of tubes, wedges, and cylinders of inner radius R and outer radius 1. Here the deformation is described in cylindrical material coordinates (ρ,ϕ,z) by

$$(\rho,\phi,z) \to (f(\rho),g(\rho)+\alpha\phi+\beta z,h(\rho)+\gamma\phi+\delta z) \tag{7}$$

where α, β, γ, δ are constants. If the material properties do not vary with ϕ and z, then the equilibrium problem reduces to a system of ordinary differential equations and side conditions for the three functions f, g, h and the four numbers α, β, γ, δ. Condition (2) reduces to

$$(\alpha\delta-\beta\gamma)(f/\rho)f' > 0. \tag{8}$$

We study just the case that

$$\alpha\delta-\beta\gamma > 0, \qquad f > 0, \quad f' > 0. \tag{9}$$

We also find that

$$|\underset{\sim}{F}|^2 = (f')^2+(fg')^2+(h')^2+(\alpha f/\rho)^2+(\beta f)^2+(\gamma/\rho)^2+\delta^2. \tag{10}$$

Let

$$w = (f,g,h,\alpha,\beta,\gamma,\delta). \tag{11}$$

Then the problem has a weak formulation corresponding to the Principle of Virtual Work in the bilinear form

$$\langle M(w),w^{\#}\rangle \equiv \int_R^1 \rho\{T_{11}(w,\rho)f'_{\#}+\cdots\}d\rho+\cdots = 0 \qquad \forall \text{ admissible test functions } w^{\#}. \tag{12}$$

Here T_{11} is the radial component of stress on the cylindrical face and $M(w)$ describes the stress field corresponding to the strain field w. We seek w satisfying (9).

We construct such solutions from approximate solutions w_n satisfying

$$f_n(\rho)/\rho \geq 1/n, \quad f'_n(\rho) \geq 1/n \text{ a.e.}, \quad \alpha\delta-\beta\gamma \geq 1/n \tag{13}$$

and the variational inequality

$$\langle M(w_n)w_n - \overline{w}\rangle \leq 0 \quad \forall \text{ admissible } \overline{w} \text{ satisfying (13).} \tag{14}$$

If we assume for simplicity that there is a w_0 and a constant $C > 0$ such that

$$T_{11}(w,\rho)(f'-f_0') + \cdots \geq C|\underset{\sim}{F}|^p, \tag{15}$$

then we may take w_n to lie in the weighted Sobolev space consisting of all w for which

$$\int_R^1 \rho\,|\underset{\sim}{F}(\rho)|^p d\rho < \infty. \tag{16}$$

(Actually the construction of natural function spaces for problems in which (15) is replaced by a more general inequality can be quite delicate and can require the machinery of Orlicz space theory.)

The specialization of the strong ellipticity condition to the problem at hand implies that the leading terms of the governing ordinary differential equations are monotone in their highest derivatives. We impose an inequality complementary to (15) that has the effect that M takes functions satisfying (13), (16), and the specified side conditions into the dual space of that defined by (16). This means that M can be shown to be pseudomonotone. Since the domain of M, defined essentially by (13), is convex we can invoke a theorem of Brezis (1968) to show that there is a w_n satisfying (13), (14), (16).

Our objective now is to use $\{w_n\}$ to construct a solution to our original problem. For this purpose we need sharp estimates on w_n that are independent of n. We get these by making judicious choices of $\overline{w}$ in (14) and by exploiting our coercivity conditions. The last such condition is especially important because it enables us to control stresses or strains in one direction if we can control the stresses and strains in transverse directions.

We now exploit (15) and (14) in a standard way to show that

$$\|w_n\| \leq C \tag{17}$$

where here and below C represents a positive constant independent of n. A fortiori

$$\int_R^1 \rho (f_n')^p d\rho < C. \tag{18}$$

A straightforward embedding theorem shows that f_n is continuous on $[R,1] \cap (0,1]$ while the positivity of f_n' ensures that f_n is continuous on $[R,1]$ itself when $R = 0$.

Let us suppose that T_{11} at $\rho = 1$ is prescribed to equal $\overline{T}_{11}$ (and that $f(1)$ is free). We then set

$$\overline{f}(\rho) = f_n(\rho) + \begin{cases} 0 & \text{if } \rho \leq \omega - \varepsilon, \\ \varepsilon^{-1}[\rho - \omega + \varepsilon] & \text{if } \omega - \varepsilon \leq \rho \leq \omega, \\ 1 & \text{if } \omega \leq \rho, \end{cases} \tag{19}$$

where ε is a small positive number and $(\omega-\varepsilon,\omega) \subset (R,1)$. This $\overline{f}$ satisfies (13). By substituting it into (14), letting $\varepsilon \to 0$, and replacing ω with ρ we get among other results an inequality of the form

$$\rho T_{11}(w_n(\rho),\rho) \geq \overline{T}_{11} - \int_\rho^1 K(w_n(\sigma),\sigma)d\sigma. \tag{20}$$

Here K is essentially the circumferential stress acting across radial planes. As a function of the circumferential stretch f/ρ, it has the form shown in Fig. 1.

The inequality complementary to (15), which ensures that M takes functions satisfying (13), (16) into the dual of the space defined by (16), implies that the integral of $K(w_n,\rho)$ over the set where $K(w_n,\rho)$ is positive has an upper bound independent of n. Consequently (20) yields

$$\rho T_{11}(w_n(\rho),\rho) \geq - C. \tag{21}$$

Now we show that the $f_n(\rho)/\rho$ are bounded below by a function that is positive on $(R,1]$ and independent of n. If not, there would be a $\sigma > R$ and a subsequence such that $f_n(\sigma)/\sigma \to 0$, so that (13) would imply that $f_n(\rho) \to 0$ for $\rho \in (R,\sigma]$ and a further subsequence of f_n would have the property that $f_n'(\rho) \to 0$ for $\rho \in (R,\sigma]$. Our last class of coercivity conditions would imply that these simultaneous extreme compressions in two directions on a set of positive measure must be accompanied by an extreme behavior of stress or strain in complementary directions. But it is easy to show that for our problem this last possibility is impossible. A much subtler measure-theoretic argument relying on (21) is needed to show that the f_n' are bounded below by a function that is positive on $[R,1] \cap (0,1]$ and is independent of n.

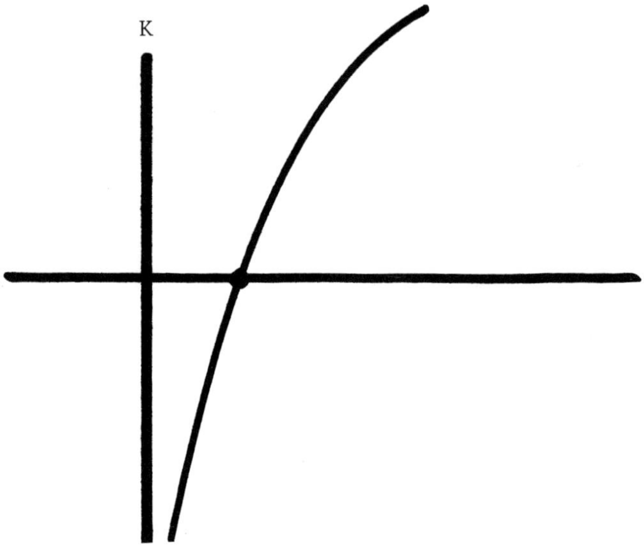

Figure 1. Typical form of K.

Now suppose that $R > 0$. Then there is a $c > 0$ such that

$$f_n'(\rho) \geq c \quad \text{whence} \quad f_n(\rho) \geq R/n + c(\rho - R). \tag{22}$$

If we now fix n such that $nc > 2$, then w_n satisfies (13) and we can take

$$\bar{f} = f_n \pm k \text{ with } k(\rho) = 0 \text{ for } \rho \in [R, R+\varepsilon],\ 2|k'(\rho)| \leq c. \tag{23}$$

This $\bar{f}$ also satisfies (13). Then by standard methods of the Calculus of Variations we obtain from (14) that (20) holds with equality replacing the inequality. A standard bootstrap procedure enables us to obtain from such equations that w_n is a classical solution of the boundary value problem.

If $R = 0$ and if $p \geq 2$, then (10) and (16) imply that γ and $\alpha f(0)$ must vanish. To handle the singularity at $\rho = 0$ we use the above methods to show that for each integer m, w_n is regular on $[1/m, 1]$. To get a solution with the right behavior at $\rho = 0$ we take a diagonal subsequence of the restrictions of w_n to $[1/m, 1]$ and exploit the available compactness.

Now consider the case in which Ω has the shape of a wedge.

Inequalities (10) and (16) imply that h is Hölder continuous if $p > 2$. Thus if the fiber $\rho = 0$ of the wedge is subjected to a concentrated vertical load, it will undergo a finite vertical displacement $h(0)$ for $p > 2$. Otherwise $h(0)$ could be infinite. A more delicate analysis is needed to determine the amount of rotation $g(0)$ when the fiber $\rho = 0$ is subjected to a concentrated torque about its axis. (In this case $|g'|^p$ has weight $\rho f(\rho)^p$ in (16).) Finally if $p < 2$ and if the transverse response to a longitudinal stress or strain is appropriate, then a solid cylinder can undergo a cavitation (in which the fiber $\rho = 0$ opens into a hollow cylinder). This observation complements Ball's (1983) definitive treatment of cavitation. Finally our analysis indicates that included in (7) is a rich collection of dislocations.

Many of the results stated in this paper are based on those of Antman (1983a,b).

Acknowledgment. The research reported here was partially supported by National Science Foundation Grant MCS-80-01844.

REFERENCES

Alexander, J. C., and Antman, S. S.: 1982, The ambiguous twist of Love. Quart. Appl. Math. 40, pp. 83-92.

Antman, S. S.: 1983a, Regular and singular problems for large elastic deformations of tubes, wedges, and cylinders. Arch. Rational Mech. Anal. To appear.

Antman, S. S.: 1983b, Constitutive restrictions permitting cavitation in nonlinearly elastic cylinders. In preparation.

Ball, J. M.: 1977a, Convexity conditions and existence theorems in nonlinear elasticity. Arch. Rational Mech. Anal. 63, pp. 337-403.

Ball, J. M.: 1977b, Constitutive inequalities and existence theorems in nonlinear elastostatics, in *Nonlinear Analysis and Mechanics: Heriot-Watt Symposium*, Vol. I, Pitman Research Notes in Math. 17, London, pp. 187-241.

Ball, J. M.: 1980, Strict convexity, strong ellipticity, and regularity in the calculus of variations. Math. Proc. Camb. Phil. Soc. 87, pp. 501-513.

Ball, J. M.: 1982, Discontinuous equilibrium solutions and cavitation in nonlinear elasticity. Proc. Roy. Soc. London. To appear.

Brezis, H.: 1968, Équations et inéquations non linéaires dans les espaces vectoriels en dualité. Ann. Inst. Fourier, 18, pp. 115-175.

Hughes, T. J. R., Kato, T., and Marsden, J. E.: 1977, Well-posed quasilinear second-order hyperbolic systems with applications to nonlinear elastodynamics and general relativity. Arch. Rational Mech. Anal. 63, pp. 273-294.

CONSTITUTIVE INEQUALITIES AND DYNAMIC STABILITY IN THE LINEAR THEORIES OF ELASTICITY, THERMOELASTICITY AND VISCOELASTICITY

Louis BRUN (C.E.A., Sevran - France)

Michel POTIER-FERRY (Paris VI)

In linear elasticity theory, one usually assumes that the strain energy function is non-negative. Our purpose is to show that this assumption is necessary. This is an old question that has already been discussed by Kirchhoff and Gibbs during the last century. Recently, the problem was raised again by Truesdell [7] and Ericksen [4].
In Section I we establish that this assumption of a positive strain energy follows from a physically meaningful definition of the solid state. Next, we show that the classical Energy Criterion remains a good stability test for materials which are thermoelastic (Section II) and viscoelastic of the rate type (Section III).

1. CONSTITUTIVE INEQUALITIES AND ELASTIC STABILITY

The constitutive law of linear elasticity is a linear relation between the Cauchy stress tensor $\vec{\sigma}$ and the small strain tensor $\vec{\varepsilon}$:

$$\vec{\sigma} = \vec{A}\,\vec{\varepsilon} \tag{1}$$

where $\vec{A} = (a_{ijkl})$ is a fourth order tensor which is symmetric with respect to the exchanges of the indices i j, of k l and of the couples (ij) (kl). The assumption of a positive strain energy function is :

$$\exists C_1 > 0 \;:\; \forall \vec{\varepsilon} = \vec{\varepsilon}^T \,,\quad \vec{A}\,\vec{\varepsilon} : \vec{\varepsilon} \geqslant C_1\, \vec{\varepsilon} : \vec{\varepsilon} \tag{M}$$

J. M. Ball (ed.), Systems of Nonlinear Partial Differential Equations, 299–304.

We denote this condition by (M), because it implies that the stress-strain relation is monotonous.

The body occupies a bounded domain Ω of R^3. This displacement $\vec{u}(\vec{x})$ is prescribed on a part of the boundary

$$\vec{u}\,(\vec{x}) = 0 \quad \text{for } \vec{x} \text{ on } \partial\,\Omega_u\,. \tag{2}$$

The remainder of the boundary is stress free. The vector fields which satisfy (2) are said to be admissible. Obviously, Condition (M) implies that the elastic energy is positive :

$$a\,(u,u) = \int_\Omega \vec{A}\,\vec{\varepsilon}\,(u) : \vec{\varepsilon}\,(u)\,dx > 0 \tag{3}$$

for any admissible non zero $u(x)$. The notation $\vec{\varepsilon}\,(u)$ denotes the symmetric part of the gradient. The property (3) is called stability in the sense of the Energy Criterion or <u>static stability</u>. When the infimum of the elastic energy is zero or negative, we shall say that there is <u>neutral stability</u> or <u>static instability</u>. The aim of the following theorem is to establish a kind of equivalence between Condition (M) and static stability.

<u>THEOREM 1</u>

Assume that there exists a symmetric tensor $\vec{\varepsilon}_o$ such that :

$$\vec{A}\,\vec{\varepsilon}_o : \vec{\varepsilon}_o < 0 \tag{4}$$

The body is homogeneous. The part $\partial\Omega_u$ of the boundary is contained in a ball of radius η.

Then, if η is sufficiently small, the equilibrium is statically unstable.

Theorem 1 can be explained in terms of a definition of the solid state. The motion of a rigid body is prohibited when the body is fastened at three points. We should wish to require something similar for a deformable solid body. This can be done if one replaces pointwise ties by quasipointwise ones. <u>By definition</u>, <u>a homogeneous body is said to be solid if it is stable once fastened by a small part of its boundary</u>. With this definition, Theorem 1 states that <u>the strain energy function of an elastic solid body must be non-negative</u>. The proof of Theorem 1 will be found in [1] [2].

As said previously, Condition (M) is sufficient for static stability, while it is well known that the following strong ellipticity condition is necessary :

$$\exists\, c_2 > 0 \quad : \forall\, \vec{\eta}, \vec{\xi} \qquad \vec{A}\,\vec{\xi} \otimes \vec{\eta} : \vec{\xi} \otimes \vec{\eta} > c_2\, |\vec{\xi}|^2 |\vec{\eta}|^2 \quad \text{(SE)}$$

When (S.E.) holds, but (M) does not, the stability depends on boundary conditions and on the homogeneity of the body. Several examples are given in [1], as well as in previous papers quoted therein.

It remains to show how the static stability condition is related to properties of dynamic stability. This has been done in [1] when the motion is governed by elastodynamics. The equations are the law (1) and the principle of virtual power :

$$\int_\Omega \left\{\rho \frac{\partial^2 \vec{u}}{\partial t^2} \cdot \vec{v} + \vec{\sigma} : \vec{\varepsilon}(v)\right\} dx \equiv 0 \qquad (5)$$

for any admissible $\vec{v}(\vec{x})$. We assume that the tensor $\vec{A}$ is strongly elliptic to ensure the existence of the dynamical process.

> <u>THEOREM 2</u> (cf. [1])
>
> The equations of the motion are (1) and (5) and the tensor $\vec{A}$ is strongly elliptic.
>
> Then, if one excludes the case of neutral stability, the static definition of stability is equivalent to dynamic stability with respect to the norm :

$$\|u\|_{H^1(\Omega)} + \left\|\frac{\partial u}{\partial t}\right\|_{L^2(\Omega)}.$$

In the following sections, we establish similar equivalence theorems when the body is not purely elastic.

2. ENERGY CRITERION IN LINEAR THERMOELASTICITY

We show that thermal couplings have no effect on the Energy Criterion. The stress-strain law (1) is modified in order to take the temperature increment θ into account and a heat equation is introduced so that we have :

$$\vec{\sigma} = \vec{A}\,\vec{\varepsilon} - \theta\,\vec{b} \qquad (6)$$

$$\rho\left(c\frac{\partial \theta}{\partial t} + b_{ij}\frac{\partial^2 u_i}{\partial x_j\,\partial t}\right) = \frac{\partial}{\partial x_j}\left(k_{ij}\frac{\partial \theta}{\partial x_j}\right) \qquad (7)$$

where $\vec{b} = (b_{ij})$ is the dilatation tensor, the conductivity tensor k_{ij} is positive definite, C measures the specific heat and the mass density $\rho(\vec{x})$ is greater than a positive constant. The mechanical boundary conditions are the same as those in Section I. As to θ, we assume that it must be zero on a part of the boundary and there is no heat flux through the remainder of the boundary. In order that the initial value problem be well posed, we require that :

$$C(\vec{x}) \geqslant C_3 > 0 \tag{8}$$

$$\vec{A} \text{ is strongly elliptic} \tag{9}$$

THEOREM 3 (cf. [1])

The equations of the motion are (5), (6) and (7). The assumptions, especially (8) (9), have been set above.
Then, if one excludes the case of neutral stability, the static definition of stability is equivalent to dynamic stability with respect to the norm

$$\left\|\frac{\partial u}{\partial t}\right\|_{L^2(\Omega)} + \|u\|_{H^1(\Omega)} + \|\theta\|_{L^2(\Omega)} .$$

The stability theorem follows from the decay of the available energy

$$\frac{1}{2}\frac{d}{dt}\left\{\int_\Omega \rho\left[\left|\frac{\partial \vec{u}}{\partial t}\right|^2 + C\,\theta^2\right] dx + a(u,u)\right\} = - \int_\Omega k_{ij}\frac{\partial \theta}{\partial x_i}\frac{\partial \theta}{\partial x_j} \leqslant 0 \tag{10}$$

The instability theorem relies heavily on Hale's invariance principle [1] and on the idea of limit process. In thermoelasticity, for such a process, the right hand-side of (10) becomes zero : hence a limit process is thermodynamically reversible and isothermal.

Other applications of the invariance principle in thermoelasticity have been given in [3] [6].

3. ENERGY CRITERION IN LINEAR VISCOELASTICITY

We study the effect of viscous damping on the Energy Criterion. It will be found that this criterion remains a good stability test when expressed in terms of equilibrium elasticity.

In the present viscoelasticity theory, the volume free energy w is a function of the strain tensor $\vec{\varepsilon}$ and of "internal variables" $\vec{\alpha} = (\alpha_1, \dots \alpha_n)$:

$$2\,w\,(\vec{\varepsilon},\vec{\alpha}) = \vec{A}\,\vec{\varepsilon} : \vec{\varepsilon} + 2\,\vec{B}\,\vec{\alpha} : \vec{\varepsilon} + \vec{C}\,\vec{\alpha}.\vec{\alpha} \quad (11)$$

where $\vec{A}$, $\vec{B}$, $\vec{C}$ are some matrices. The behaviour of the material is governed by :

$$\vec{\sigma} = \frac{\partial w}{\partial \vec{\varepsilon}} = \vec{A}\,\vec{\varepsilon} + \vec{B}\,\vec{\alpha} \quad (12)$$

$$\vec{L}\,\frac{\partial \vec{\alpha}}{\partial t} = -\frac{\partial w}{\partial \vec{\alpha}} = -\vec{B}^{*}\,\vec{\varepsilon} - \vec{C}\,\vec{\alpha} \quad (13)$$

where $\vec{B}^{*}$ is the adjoint matrix of $\vec{B}$. The matrix $\vec{L}$ is positive definite, as required by the second law of thermodynamics. The boundary conditions are the same as those in Section II.

When a shock is prescribed, the instantaneous response of the material is governed by the moduli $\vec{A}$. Hence it is consistent to assume that $\vec{A}$ is strongly elliptic to ensure the existence of the process. In this case, we obtain a hyperbolic system coupled with the O . D . E . (13).

As in Section II, the available energy decays with time :

$$\frac{1}{2}\frac{d}{dt}\int_{\Omega}\{\rho\,(\frac{\partial \vec{u}}{\partial t})^2 + w\,(\vec{\varepsilon},\,\vec{\alpha})\}\,dx = -\int_{\Omega}\vec{L}\,\frac{\partial \vec{\alpha}}{\partial t}\cdot\frac{\partial \vec{\alpha}}{\partial t}\,dx \leqslant 0 \quad (14)$$

This suggests that the assumption of a positive strain energy can be formulated as follows :

$$\exists\, c_4 > 0 \;:\; w\,(\vec{\varepsilon},\vec{\alpha}) \geqslant c_4\,(|\vec{\varepsilon}|^2 + |\vec{\alpha}|^2) \quad (15)$$

To find the mechanical significance of Property (15), we look at the equilibrium behaviour, in the case of a positive definite matrix $\vec{C}$. Equations (12) and (13) lead to :

$$\vec{\sigma} = \vec{A}_{\infty}\,\vec{\varepsilon} \quad \text{where} \quad \vec{A}_{\infty} = \vec{A} - \vec{B}.\vec{C}^{-1}.\vec{B}^{*} \quad (16)$$

We have defined the tensor of equilibrium elasticity $\vec{A}_{\infty}$. Furthermore, Property (15) is equivalent to the positive definiteness of $\vec{A}_{\infty}$. Contrary to the cases of elasticity and thermoelasticity, the instantaneous and equilibrium behaviours are not identical.

The following results show that the tensor of equilibrium elasticity $\vec{A}_\infty$ must be non-negative in order that the body be stable when fastened by a small part of its boundary.

Theorem 1 can be applied : then if the sign of $\vec{A}_\infty$ is indefinite, the "equilibrium energy" $a_\infty(u,u)$ -whose definition is obvious- has an indefinite sign when the body is fastened along a small part of its boundary. Lastly, the following theorem shows that the Energy Criterion remains a good stability test in viscoelasticity theory.

THEOREM 4

The equations of the motion are (5), (12) and (13). The tensor $\vec{A}$ is strongly elliptic ; $\vec{C}$ and $\vec{L}$ are positive definite. Then, if one excludes the case of neutral stability, the static definition of stability is equivalent to dynamic stability with respect to the norm

$$\left\|\frac{\partial u}{\partial t}\right\|_{L^2(\Omega)} + \| u \|_{H^1(\Omega)} + \| \alpha \|_{L^2(\Omega)}$$

(static stability is understood in terms of the "equilibrium energy" a_∞ (.,.)).

The proof is similar to that of Theorem 3. Equivalence theorems like this one can also be established even if $\vec{C}$ is not positive definite.

REFERENCES

[1] BRUN L., POTIER-FERRY M. : submitted to J. Thermal Stresses.

[2] BRUN L., POTIER-FERRY M. : submitted to C. Rend. Acad. Sci. Paris, série B.

[3] DAFERMOS C. : 1976, in Lecture Notes in Math. 503, pp. 295-306, Springer-Verlag.

[4] ERICKSEN J.L. : 1978, J. Elasticity 8, pp. 439-446.

[5] HALE J.K. : 1969, J. Math. Anal. Appl. 26, pp. 39-59.

[6] SLEMROD M., INFANTE E.F. : 1971, in Instability of continuous systems, pp. 215-221, Springer-Verlag.

[7] TRUESDELL C. : 1978, in Contemporary developments in continuum mechanics, North-Holland.

GENERALIZED SOLUTIONS TO CONSERVATION LAWS

Ronald J. DiPerna

Duke University, Durham, NC 27706

Abstract: We introduce a framework for studying degenerate limits for conservation laws of hyperbolic type and initial value problems for conservation laws of mixed hyperbolic-elliptic type.

The purpose of this note is to suggest a framework for analyzing certain aspects of singular limits for conservation laws of hyperbolic type, in particular the zero diffusion limit vs the zero dispersion limit and for studying the Cauchy problem for conservation laws of mixed type. The general setting is provided by a system of n equations in one space dimension,

$$u_t + f(u)_x = 0 \tag{1}$$

where the flux function f is a smooth nonlinear map from R^n to R^n. First let us assume that f is strictly hyperbolic in the sense that its Jacobian ∇f has n real and district eigenvalues. With regard to singular limits we are interested in diffusion processes generated by parabolic regularization,

$$u_t + f(u)_x = \varepsilon u_{xx} , \tag{2}$$

and in dispersion processes generated by third order regularization,

$$u_t + f(u)_x = \delta u_{xxx} .$$

One may of course introduce a more general operator on the right hand side depending upon several parameters and entertain the corresponding process.

Basic problems with the singular limit of vanishing parameters

J. M. Ball (ed.), Systems of Nonlinear Partial Differential Equations, 305–309.

arise even at the level of the scalar equation. We recall that the solution $u(x,t,\varepsilon)$ of Burger's equation

$$u_t + (u^2/2)_x = \varepsilon\, u_{xx}$$

converges as the diffusion parameter ε vanishes to a solution $u(x,t)$ of the hyperbolic equation

$$u_t + (u^2/2)_x = 0.$$

Verification of the zero diffusion limit for general scalar equations in one space dimension can be achieve with either the classical theory of compactness or with the recently developed theory of compensated compactness. In the latter connection we refer the reader to the article of L. Tartar in this volume.

On the other hand, the zero dispersion limit for the KdV equation

$$u_t + (u^2/2)_x = \delta\, u_{xxx}, \qquad (3)$$

is not governed by the hyperbolic law (2), due to the persistence of oscillations as the dispersion parameter δ vanishes. For a class of smooth data, Lax and Levermore [2] have shown that the solution $u(x,t;\delta)$ of (3) converges strongly to a solution $u(x,t)$ of (1) before shock waves develop. In the presence of shocks convergence takes place in the weak topology with a limiting solution governed by modulation equations.

The problem of describing singular limits of solutions to conservation laws is one motivation for introducing the following notion of solution. A measurable map

$$(x,t) \to \nu_{(x,t)}$$

from the physical space to the space of probability measures over the state space R^n is a generalized solution of system (1) if the expected value of the vector field (u,f) has zero divergence, i.e. if

$$\bar{u}_t + \bar{f}_x = 0$$

in the sense of distributions. Here the average value of a continuous map $g\colon R^n \to R^n$ is denoted by

$$\bar{g} = \bar{g}(x,t) \equiv \int_{R^n} g(\lambda)\, d\nu_{(x,t)} \, .$$

A classical distribution solution $u(x,t)$ of system (1) is identified with the map which sends (x,t) to a point mass at $u(x,t)$:

$$(x,t) \to \delta_{u(x,t)} \ .$$

We remark that the existence problem for generalized solutions is rather simple. Any conservative approximation process for system (1) which is stable in L^∞ yields a generalized solution: one merely associates with an approximating sequence the corresponding family of representing measures. In this regard we recall that if

$$w_j(y): R^m \to R^n$$

is an arbitrary sequence of measurable functions uniformly bounded in L^∞, there exists a subsequence w_k and a family of representing probability measures $\{\nu_y\}$ over the range space R^n with the following property: for any continuous real-valued map g on R^n, the sequence $g(w_k)$ converges in the weak-star topology of L^∞ to a function $\bar{g}$ which coincides almost everywhere with the expected value of g relative to ν_y:

$$\bar{g}(y) = \lim_{k\to\infty} g(w_k(y))$$

$$\bar{g}(y) = \int_{R^n} g(\lambda)\, d\nu_y(\lambda) \ .$$

The main problem is one of classification. It remains to characterize the generalized measures ν in terms of the regularizing operator. At the basic level of the diffusion process one is lead to the following conjecture. The only generalized solution which dissipates all convex entropy fields is a point mass. To be specific, let (η,q) denote an entropy pair for system (1), [1]. Assume that

$$\bar{\eta}_t + \bar{q}_x \leq 0 \tag{4}$$

for all convex entropies η and show that

$$\nu_{(x,t)} = \delta_{\bar{u}(x,t)}$$

where

$$\bar{u}(x,t) = \int \lambda \, d\nu_{(x,t)}$$

We recall that for a scalar equation an arbitrary function may serve as an entropy η provided that the entropy flux q is choosen appropriately. Thus, in the setting of a scalar problem, one would postulate the dissipation inequality (4) for all convex functions η. For systems of two equations there exists a broad class of convex entropies [1], perhaps sufficiently rich to force the reduction of the representing measure to a point mass.

For systems of three or more equations the existence of an entropy is a rare event, confined mainly to the equations of continuum mechanics, so that an alternative formulation is needed.

We note that the generalized version of the Lax entropy inequality formulated in (4) necessarily holds for all weak limits of a diagonal diffusion process (2). If $u(x,t;\varepsilon)$ is a sequence of solutions to (2) which is uniformly bounded in L^{∞} then (4) is easily verified for all convex entropy η provided that $\nu_{(x,t)}$ is the representing measure associated with an appropriate subsequence. Here we are concerned with the converse problem of showing that (4) implies that ν is a point mass and that the point mass corresponds to the zero diffusion limit.

We remark that in the setting of several space dimensions a more refined notion of generalized notion may be required for the purpose of distinguishing various singular limits.

We conclude with some brief remarks concerning the Cauchy problem for equations of mixed type. It has been suggested that certain aspects of phase transitions may be modeled by the quasi-linear system

$$u_t - \sigma(v)_x = 0$$

$$v_t - u_x = 0$$

with σ given by a nonmonotone function. This system provides an interesting example of a general class of equations which change type from hyperbolic to elliptic as the state variable crosses a parabolic manifold of codimension one. It is an open problem to determine the extent to which the Cauchy problem is meaningful for such systems. One may anticipate the existence of solutions within an appropriately enlarged class of solutions, but a lack of uniqueness. A reasonable approach to the existence question would be to introduce a parabolic regularization of the form

$$u_t - \sigma(v)_x = \varepsilon\, u_{xx}$$

$$v_t - u_x = \varepsilon\, v_{xx} ,$$

and attempt to establish a uniform L^{∞} estimate. Establishing such an estimate leads to existence of a generalized solution in the sense mentioned above.

References

[1] Lax, P.D., Shock waves and entropy, in Contributions to Nonlinear Functional Analysis, ed. E.A. Zarantonello, Academic

Press, 1971, pp. 603-604.

[2] Lax, P. D. and C. D. Levermore, The zero dispersion limit for the Korteweg de Vries equation, Proc. Nat. Acad. Sci. U.S.A., 76 (1979), no. 8, pp. 3602-3606.

STABILITY OF THE ELASTICA

J.H. Maddocks

Mathematics Department
Stanford University
Stanford, Ca. 94305

The equilibrium configurations of a nonlinearly elastic rod in three-space are here regarded as extremals of a constrained variational principle. A new technique, applying to the second variation in constrained problems of the calculus of variations, is then exploited to ascertain stability properties of various planar buckled equilibria of a rod subjected to end-thrust. Perturbations both in, and out of, the plane are considered. Marked contrasts between in and out of plane stability are discovered. Secondary bifurcation from the first branch of buckled solutions, with associated loss of stability, is also found.

1. INTRODUCTION

This lecture describes stability properties of various buckled equilibria of a nonlinearly-elastic rod which is subject to a compressive end-load. Euler's elastica (e.g. Love, p.401) is a particular case of the model here treated, and new information about this classic problem is obtained.

The mathematical setting is a constrained variational principle, the main objective of the analysis being to determine which constrained extremals (i.e. stationary points) are actually constrained local minima. The presence of constraints is a novel feature of the inquiry as standard tests for minima are not applicable. A new technique is introduced to overcome this difficulty.

The following customary notion of stability is assumed

J. M. Ball (ed.), Systems of Nonlinear Partial Differential Equations, 311–322.

throughout: an equilibrium configuration will be called <u>stable</u> if the second variation about the corresponding extremal is positive for all admissible non-trivial perturbations, and will be called <u>unstable</u> if there is an admissible perturbation making the second variation negative. Whether a perturbation is admissible is dependent both upon boundary conditions and upon constraints; the required properties are stated later. Connexions between this static criterion for stability and the concept of Liapunov stability in the underlying dynamical system are not discussed here, but are currently under investigation (Caflisch and Maddocks).

The equilibria here treated are buckled only in the plane; however, they are actually extremals of a variational principle in three-space. Accordingly, stability is investigated both to variations in, and out of, the plane of the equilibrium state.

The physical assumptions leading to the model particularized in this presentation are discussed by Maddocks (a), which paper also contains details of the calculations outlined here. In addition, various other boundary-value problems are analysed there, and more general models describing, for example, effects of physical twist are treated.

2. THE MODEL

The model assumed is a line that resists bending according to a nonlinearly elastic law. The arc-length of the line is taken to be unity, so the strain energy can be expressed as

$$\int_0^1 W(\kappa)\, ds.$$

Here s represents arc-length, $\kappa(s)$ is the curvature of the line, and the strain energy density function $W(\kappa)$ is assumed to be smooth, and to satisfy

$$W(\kappa) = W(-\kappa) \quad , \quad W(0) = 0 \, , \qquad W_{\kappa\kappa}(\kappa) > 0 \, . \tag{2.1}$$

An additional convexity hypothesis is also made, namely

$$W_{\kappa\kappa}(\kappa) \geqslant W_\kappa(\kappa)/\kappa \, . \tag{2.2}$$

(The subscript denotes differentiation with respect to κ). In the linearly elastic case, corresponding to Euler's elastica model,

$$W(\kappa) = \frac{1}{2} B\kappa^2 \quad , \qquad B > 0$$

which function satisfies (2.1) and (2.2).

The coordinates chosen to describe the line are s, and the two Euler angles $\theta(s)$ and $\phi(s)$ of the tangent vector at the point corresponding to arc-length s (cf. Figure 1). The curvature κ is related to the Euler angles by

$$\kappa^2(s) = \theta'(s)^2 + \phi'(s)^2 \sin^2\theta(s)$$

(The prime signifies differentiation with respect to s.)

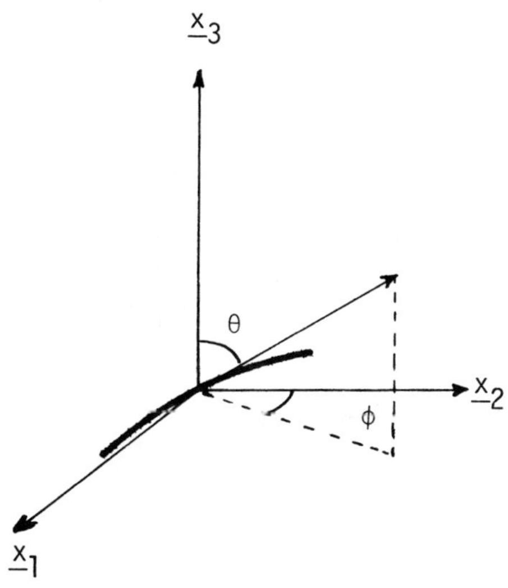

Figure 1. The coordinate system.

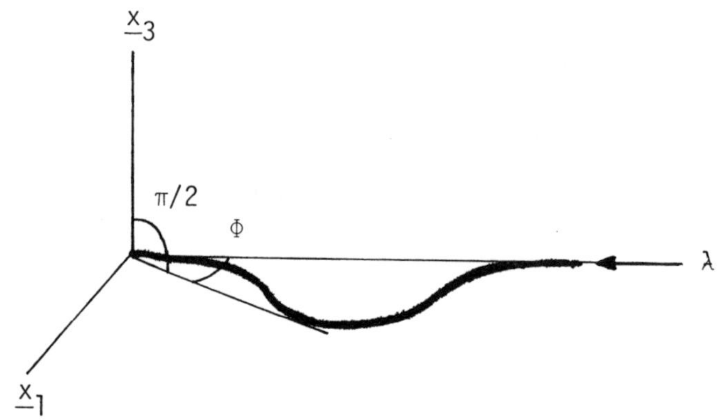

Figure 2. The boundary-value problem.

3. VARIATIONAL PRINCIPLE AND ITS EXTREMALS

The boundary-value problem to be considered is the following (cf. Figure 2). One end of the rod is fixed at the origin; the other is constrained to lie in the $\underline{x}_2$-axis, but may slide under the action of a compressive dead load λ; and the tangents to the rod at its end points are required to coincide with the $\underline{x}_2$-axis. The mathematical formulation is to make stationary the functional

$$V(\phi,\theta,\lambda) = \int_0^1 \{W(\kappa) + \lambda\cos\phi\sin\theta\}\, ds \tag{3.1}$$

subject to constraints

$$\int_0^1 \cos\theta\, ds = 0 \ , \tag{3.2}$$

$$\int_0^1 \sin\phi\, \sin\theta\, ds = 0 \ , \tag{3.3}$$

and boundary conditions

$$\phi(s) = 0 \ , \quad \theta(s) = \frac{\pi}{2} \Bigg|_{s\,=\,0,1} \ . \tag{3.4}$$

This system has planar extremals

$$(\theta,\phi) = (\frac{\pi}{2}, \Phi) \tag{3.5}$$

where Φ satisfies

$$\frac{d}{ds} W_{\kappa}(\Phi') + \lambda \sin\Phi + \nu\cos\Phi = 0 \quad . \tag{3.6}$$

Here ν is a Lagrange multiplier corresponding to constraint (3.3), and the Lagrange multiplier for constraint (3.2) is zero. Of course, in the plane $\theta = \frac{\pi}{2}$ we have that

$$\kappa(s) = \phi'(s) \, .$$

Equation (3.6) has been studied extensively by Antman and Rosenfeld, and a bifurcation diagram for it is shown in Figure 3. The multiplier ν vanishes on the solution $\Phi(s) \equiv 0$, and on solution branches bifurcating from the trivial solution at the first and all odd eigenvalues. The properties crucial to determination of stability are that $\Phi'(s-\frac{1}{2})$ is an even function, and that the nth bifurcating mode has n interior zeros of Φ, and $(n+1)$ interior zeros of Φ'. The configuration shown in Figure 2 is actually a sketch of the first buckled mode.

4. ANALYSIS OF THE SECOND VARIATIONS

At the solutions (3.5), variations in ϕ describe perturbations of the equilibrium in the plane, and variations in θ measure spatial perturbations, that is perturbations out of the plane. For the sake of brevity, only buckled solutions with $\nu = 0$ will be considered. Then the two corresponding second variations are

$$\frac{\delta^2}{\delta\phi^2} V(\Phi,\frac{\pi}{2},\lambda)\,[g] = \int_0^1 \{W_{\kappa\kappa}(\Phi')g'^2 - \lambda\cos\Phi\, g^2\}\, ds \, ,$$

and

$$\frac{\delta^2}{\delta\theta^2} V(\Phi,\frac{\pi}{2},\lambda)\,[h] = \int_0^1 \{W_{\kappa}(\Phi')/\Phi'\; h'^2 - [W_{\kappa}(\Phi')\Phi' + \lambda\cos\Phi]h^2\}\, ds.$$

Because of boundary conditions (3.4), admissible variations g and h must satisfy

$$g(s) = h(s) = 0 \,\Big|_{s=0,1} \quad . \tag{4.1}$$

Integration by parts then demonstrates that the planar and

spatial second variations can be written as, respectively,

$$\frac{\delta^2}{\delta\phi^2} V(\Phi, \frac{\pi}{2}, \lambda)[g] = \langle g, Pg \rangle , \tag{4.2}$$

and

$$\frac{\delta^2}{\delta\theta^2} V(\Phi, \frac{\pi}{2}, \lambda)[h] = \langle h, Sh \rangle . \tag{4.3}$$

Here P and S are linear second-order self-adjoint differential operators defined by

$$Pg \equiv - \frac{d}{ds} \{W_{KK}(\Phi')g'\} - \lambda\cos\Phi\, g$$

and

$$Sh \equiv - \frac{d}{ds} \{W_K(\Phi')/(\Phi' h'\} - \{W_K(\Phi')\Phi' + \lambda\cos\Phi\}h$$

and $\langle\ ,\ \rangle$ denotes the (formal) L^2-inner product.

Hypotheses (2.1) and (2.2) imply that Sturm Separation and Comparison Theorems apply to the operators P and S. In particular, solutions of

$$Su = 0$$

oscillate faster than solutions of

$$Pu = 0$$

A combination of Sturm-Liouville theory for eigenvalues, the number of interior zeroes of Φ and Φ', and the observations that

$$S\{\sin\Phi\} = 0 \tag{4.4}$$

and

$$P\{\Phi'\} = 0 \tag{4.5}$$

then allow us to determine the number of nonpositive eigenvalues of S and P when subject to boundary conditions (4.1). Identity (4.4) arises from rotational symmetry about the $\underline{x}_2$-axis and corresponds to a rigid-body rotation of the equilibrium;

identity (4.5) is obtained from differentiation of equation (3.6) with $\nu = 0$.

The conclusions finally reached are that on the nth mode the operators P and S subject to boundary conditions (4.1) each have n negative eigenvalues. Moreover, the kernel or null-space of P, denoted ker P, is empty, and ker S is spanned by $\sin\Phi$.

5. STABILITY CRITERION

The results of §4 demonstrate that neither (4.2) nor (4.3) is positive definite; however, because of the presence of constraints this is not the correct criterion for stability. According to standard theory (e.g. Hestenes), the correct criterion is that the quadratic forms (4.2) and (4.3) need only be positive for variations satisfying, in addition to boundary conditions, the linearization of constraints (3.2) and (3.3). Thus the criterion for stability in the plane is that

$$\langle g,Pg\rangle > 0,\ \forall\ g \text{ satisfying } g(0)=g(1)=0, \text{ and the constraint}$$

$$\langle g,\cos\Phi\rangle = 0\ , \tag{5.1}$$

and the criterion for stability out of the plane is that

$$\langle h,Sh\rangle > 0,\ \forall\ h \text{ satisfying } h(0)=h(1)=0, \text{ and constraints}$$

$$\langle h,1\rangle = 0 \quad \text{and} \quad \langle h,\sin\Phi\rangle = 0. \tag{5.2}$$

The second constraint in (5.2) does not arise from the linearization of (3.2) and (3.3), but is added to exclude the effect of a rigid-body rotation of the rod, which is not of interest.

Consider any linear self-adjoint operator L that has at least two negative eigenvalues μ_1 and μ_2, with corresponding orthonormal eigenfunctions ξ_1 and ξ_2. Then α and $\beta \in \mathbb{R}$ can be chosen such that

$$u = \alpha\xi_1 + \beta\xi_2$$

satisfies any linear constraint

$$\langle u,T\rangle = 0.$$

Furthermore,

$$\langle u,Lu\rangle = \mu_1\alpha^2 + \mu_2\beta^2 < 0\ .$$

Consequently, it can be concluded that condition (5.1) fails on all branches other than the first; equivalently, higher buckled modes are unstable in the plane. The same conclusion holds for condition (5.2) because the negative eigenfunctions of S are necessarily orthogonal to ker S, and hence to $\sin\Phi$. Therefore higher buckled modes are unstable both out of, and in the plane.

The situation on the first buckled mode is not as straightforward. We invoke the following theorem, which is a simple case of a general theory presented by Maddocks (b).

Theorem

Let L be a linear self-adjoint operator with precisely one negative eigenvalue. Consider the following subspace of the domain of definition for L,

$$U = \{u: \ u \text{ is orthogonal to ker } L, \text{ and } \langle u,T\rangle = 0\}.$$

Here T is some vector itself orthogonal to ker L, so that a vector η can be defined as the solution of

$$L\eta = T\ .$$

Then the quadratic form $\langle u,Lu\rangle$ is positive $\forall\ u \in U\backslash\{o\}$, if and only if

$$\langle \eta,L\eta\rangle < 0\ .$$

Indication of Proof

Suppose that the vector v and $\alpha \in \mathbb{R}$ are such that

$$\langle v+\alpha\eta,T\rangle = 0\ , \tag{5.3}$$

Then,

$$\langle v+\alpha\eta,L(v+\alpha\eta)\rangle \ = \ \langle v,Lv\rangle + 2\alpha\langle v,T\rangle + \alpha^2\langle\eta,L\eta\rangle\ ,$$

which, by (5.3), equals

$$\langle v,Lv\rangle - \alpha^2\langle\eta,L\eta\rangle\ . \tag{5.4}$$

In particular, no cross-terms arise. The proof then proceeds straightforwardly. For example, if $\langle\eta,L\eta\rangle$ is positive, then (5.4) is made negative whenever v is taken to be a multiple of the 'negative' eigenfunction, and α is chosen such that (5.3) is satisfied.

Of course, the Theorem applies to criteria (5.1) and (5.2). Moreover, the solution to

$$P\eta_1 = \cos\Phi \quad , \quad \eta_1(0) = \eta_1(1) = 0 ,$$

is

$$\eta_1 = c\Phi' - 1/\lambda ,$$

where $c \in \mathbb{R}$ can, by symmetry of Φ', be chosen so that the boundary conditions are satisfied.

Now

$$<\eta_1, P\eta_1> = \int_0^1 \{c\Phi' - 1/\lambda\}\cos\Phi \, ds = c\sin\Phi\Big|_0^1 - 1/\lambda\int_0^1 \cos\Phi \, ds ,$$

which may be written

$$<\eta_1, P\eta_1> = -\ell/\lambda , \tag{5.5}$$

where $\ell = \int_0^1 \cos\Phi \, ds$ is the distance between the end-points. As λ is positive the conclusion is that the first mode is stable in the plane when ℓ is positive, but is unstable in the plane when ℓ is negative (cf. Figure 4).

The solution to

$$S\eta_2 = 1 \quad , \quad n_2(0) = \eta_2(1) = 0$$

can be written as

$$\eta_2 = y - 1/\lambda \cos\Phi$$

where y is a solution to

$$Sy = 0 \quad , \quad y(0) = y(1) = 1/\lambda .$$

Accordingly,

$$<\eta_2, S\eta_2> = <y, S\eta_2> - 1/\lambda <\cos\Phi, 1>$$

which equals, after integration by parts on the first term,

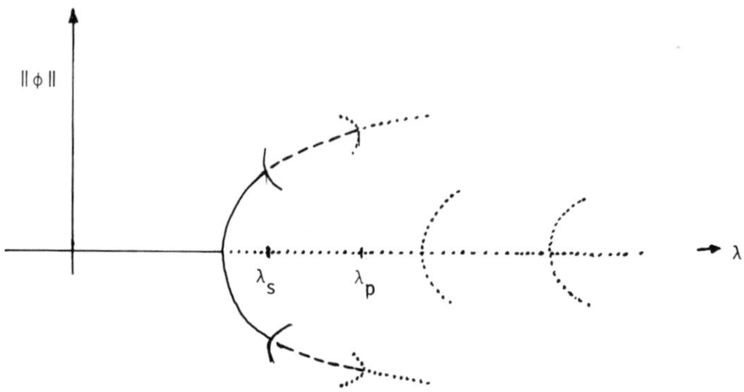

Figure 3. Bifurcation diagram for equation (3.6).

There is an infinite set of buckled modes bifurcating from the trivial solution. Solid lines denote equilibria that are stable; dashed lines denote equilibria that are stable to perturbations in the plane, but unstable out of the plane; and dotted lines denote equilibria that are unstable both in and out of the plane. Secondary bifurcation branches are not discussed here, but are illustrated for completeness. The critical load λ_p corresponds to the ends of the rod touching. The load λ_s is strictly less than λ_p.

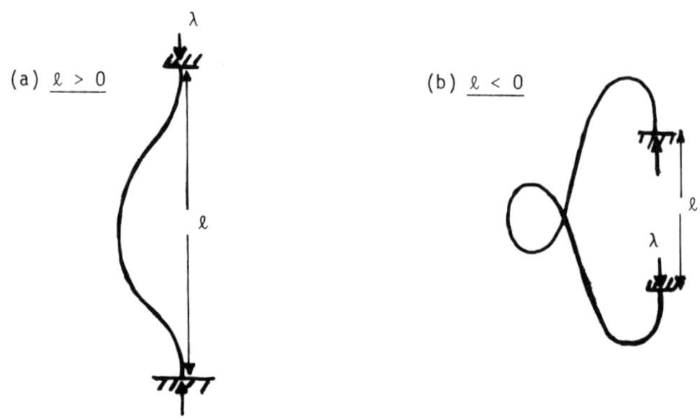

Figure 4. Illustrations of equilibria on the first bifurcating branch

The equilibrium sketched in (a) is stable in the plane, but is unstable out of the plane provided ℓ is sufficiently small. The equilibrium sketched in (b) is unstable both in, and out of the plane.

$$-y \frac{W_\kappa(\Phi')}{\Phi'} \eta_2' \Big|_0^1 + \langle Sy, \eta_2 \rangle - 1/\lambda \langle \cos\Phi, 1 \rangle ,$$

or

$$\langle \eta_2, S\eta_2 \rangle = - 1/\lambda \frac{W_\kappa(\Phi')}{\Phi'} y' \Big|_0^1 - 1/\lambda \, \ell . \tag{5.6}$$

Symmetry arguments demonstrate that

$$y'(0) = -y'(1) ,$$

and oscillation theory implies that

$$y'(0) > 0 .$$

The first term in (5.6) is, therefore, positive. Inspection of (5.5) and (5.6) then reveals that the rod is unstable out of the plane whenever it is unstable in the plane; but that there are configurations with ℓ sufficiently small and positive that are unstable out of the plane whilst being stable in the plane. The stability results are summarised in Figures 3 and 4.

ACKNOWLEDGEMENT

This investigation was commenced at the University of Oxford, where it was supported by The Carnegie Trust for the Universities of Scotland. The research was completed with support from the National Science Foundation, the Office of Naval Research, the Office of Army Research, and the Air Force Office of Scientific Research.

REFERENCES

[1] Antman, S.S., and Rosenfeld, G.: 1978, *Global behaviour of buckled states of nonlinearly elastic rods*, SIAM Review, pp.20. Corrections and additions, Ibid., 1980, 22.

[2] Caflisch, R.E., and Maddocks, J.H.: *Dynamical theory of the elastica*, in preparation.

[3] Hestenes, M.R.: 1966, '*Calculus of variations and optimal control theory*', John Wiley & Sons, New York.

[4] Love, A.E.H.: 1927, '*A treatise on the mathematical theory of elasticity*', Dover, New York, 4th edition.

[5] Maddocks, J.H. (a): *Stability of nonlinearly elastic rods*, in manuscript.

[6] Maddocks, J.H. (b): *Restricted quadratic forms and their application to bifurcation and stability in constrained variational principles*, submitted to SIAM J. Math. Anal.

GROUP THEORETIC CLASSIFICATION OF CONSERVATION LAWS IN ELASTICITY

Peter J. Olver

School of Mathematics, University of Minnesota,
Minneapolis, MN 55455, U.S.A.

1. INTRODUCTION

Although Noether's celebrated theorem relating symmetry groups and conservation laws for systems arising from variational principles has been available for over sixty years, and despite the well-acknowledged importance of group theory in the study of elasticity, for historical reasons I do not fully understand there has never been a systematic classification of general symmetry groups and their associated conservation laws for even the simplest elastic materials. In fact, not until the work of Günther, [7], and Knowles and Sternberg, [10], was even a limited variant of Noether's general theorem applied in this subject. This is especially surprising in that the relevant techniques are completely constructive, amenable to straight-forward computational methods. In this note, the basic theory underlying Noether's theorem will be outlined, and illustrated by some recent applications to elasticity, work which is very much still in progress. It is a pleasure to thank John Ball for sparking my interest in applying these general techniques to elasticity, and to all the organizers of the NATO/LMS institute for a most enjoyable and productive conference.

2. HYPERELASTICITY, [9]

Although most techniques apply much more generally, we specialize to the case of homogeneous elastostatics in the absence of body forces. The variational problem is

$$I = \int_{\Omega} W(\nabla u)\,dx \,, \tag{2.1}$$

J. M. Ball (ed.), Systems of Nonlinear Partial Differential Equations, 323–331.

where W is the stored energy function, $x \in \mathbb{R}^p$, $u \in \mathbb{R}^q$ (usually $p = q = 2$ or 3) and $\nabla u = (u^i_\alpha) \equiv (\partial u^i/\partial x^\alpha)$ the deformation gradient. The Euler-Lagrange equations are

$$E^i = \Sigma\, W^{ij}_{\alpha\beta} u^j_{\alpha\beta} = 0, \qquad W^{ij}_{\alpha\beta} \equiv \partial^2 W/\partial u^i_\alpha \partial u^j_\beta \,. \tag{2.2}$$

Occasionally, the Legendre-Hadamard condition that the $q \times q$ matrix $Q(\xi)$ with entries

$$Q_{ij}(\xi) = W^{ij}_{\alpha\beta}\xi^\alpha \xi^\beta \,, \qquad \xi \in \mathbb{R}^p \,,$$

be positive definite for $\xi \neq 0$, will be imposed.

3. CONSERVATION LAWS, [1], [10], [13], [15].

A divergence expression

$$\mathrm{Div}\, A = \Sigma\, D_\alpha A^\alpha = 0 \tag{3.1}$$

is a conservation law if it vanishes for all solutions of (2.2). Equivalently, $\int_\Gamma A \cdot dS$ is path - (or surface) independent. Such integrals, of which Eshelby's energy momentum tensor, [5], was the first important example, are of importance in the analysis of strain concentrations at crack tips, [2], [18].

Trivial examples arise when a) $A = 0$ for all solutions and b) (3.1) holds identically; equivalently $A^\alpha = \Sigma\, D_\beta B^{\alpha\beta}$ for some skew-tensor B. Many of the results described here rely on a new characterization, [14], of the second variety of trivial laws. For simplicity, we only consider laws in which $A = A(x,u,\nabla u)$, i.e. no higher derivatives of u enter.

4. SYMMETRY GROUPS, [1], [13], [17].

A symmetry group of a system of differential equations, e.g. (2.2), is characterized by its property of transforming solutions of the system to other solutions. Geometrical or Lie symmetries are (local) diffeomorphisms $g\colon (x,u) \to (\tilde{x},\tilde{u})$ and act on solutions by point-wise transforming their graphs (think of a rotation in the x,u - plane). Leaving aside discrete symmetries (e.g. reflections), each one-parameter subgroup is characterised by its infinitesimal generator - the vector field

$$\vec{v} = \sum \eta^\alpha(x,u)\frac{\partial}{\partial x^\alpha} + \sum \phi_i(x,u)\frac{\partial}{\partial u^i}, \tag{4.1}$$

from which the group transformations are recovered as solutions of the system of o.d.e.'s

$$dx^\alpha/d\varepsilon = \eta^\alpha, \qquad du^i/d\varepsilon = \phi_i,$$

ε being the group parameter. The group transformations, and hence infinitesimal generator, "prolong" to the derivatives of u, so

$$\mathrm{pr}\,\vec{v} = \vec{v} + \sum \phi_i^\alpha \frac{\partial}{\partial u_\alpha^i} + \ldots ,$$

where

$$\phi_i^\alpha = D_\alpha \psi_i + \sum u_{\alpha\beta}^i \eta^\beta , \tag{4.2}$$

$$\psi_i = \phi_i - \sum u_\beta^i \eta^\beta . \tag{4.3}$$

Theorem

For the system $E = 0$ nondegenerate (cf. [14]) (4.1) generates a symmetry group if and only if

$$\mathrm{pr}\,\vec{v}(E) = 0 \quad \text{whenever} \quad E = 0. \tag{4.4}$$

Condition (4.4) forms a large system of elementary differential equations for the coefficients η^α, ϕ_i of $\vec{v}$, whose solution (a straight-forward, albeit tedious, task) gives the most general symmetry of the system.

A generalised symmetry has the same form for its infinitesimal generator, but now η^α, ϕ_i can depend on derivatives of u. The symmetry criterion (4.4) is unchanged, and again can be solved. The standard form of such a symmetry is

$$\vec{v}_\psi = \sum \psi_i \, \partial/\partial u^i , \tag{4.5}$$

cf. (4.3), and it easily follows that $\vec{v}$ is a symmetry if and only if $\vec{v}_\psi$ is. The corresponding group transformations are realized by solving the evolutionary system

$$\partial u^i/\partial\varepsilon = \psi_i(x,u,\nabla u,\ldots), \qquad u|_{\varepsilon=0} = u_0,$$

so that if $u_0(x)$ is a solution, so is $u(x,\varepsilon)$ for any fixed ε.

5. NOETHER'S THEOREM, [1], [3], [11], [13]

For systems arising from variational principles a (geometrical) symmetry is a variational symmetry if the integral (2.1) is unchanged by the relevant transformation for arbitrary subdomains $\Omega \subset \mathbb{R}^p$. The corresponding infinitesimal criterion is

$$\text{pr}\,\vec{v}(W) + W\,\text{Div}\,\eta = 0. \tag{5.1}$$

Bessel-Hagen, [3], replaced the right hand side of (5.1) by Div B for some unspecified B, calling the corresponding symmetries <u>divergence transformations</u> (here the precise geometrical significance is less clear). Equivalently,

$$\text{pr}\,\vec{v}_\psi(W) = \text{Div}\,\tilde{B} \tag{5.2}$$

for some $\tilde{B}$, cf. (4.5). Any generalised vector field satisfying (5.2) will be termed a variational symmetry. A nontrivial computation, [13], proves every variational symmetry is a symmetry of the corresponding Euler-Lagrange equations, but <u>not</u> conversely, scale symmetries being the most common counterexamples. Often, the easiest way to compute variational symmetries is to first find all symmetries of the Euler-Lagrange equation and then check which of these satisfy (5.2).

Noether's theorem comes from the trivial identity

$$\Sigma\,\psi_i E^i = \text{Div}\,A\,, \tag{5.3}$$

holding for any variational symmetry, where A, which yields a conservation law, can be explicitly given in terms of ψ, W and $\tilde{B}$. Thus each variational symmetry gives a conservation law and, under mild nondegeneracy hypotheses, each conservation law is equivalent (up to addition of a trivial law) to one satisfying (5.3) and hence yields a corresponding symmetry.

6. THE SYMMETRY EQUATIONS, [15]

A simplifying feature in the search for conservation laws of (2.2) is that if $A(x,u,\nabla u)$ satisfies (3.1), so does $A(x_0,u_0,\nabla u)$ for fixed x_0,u_0. Thus, the x,u - independent laws, say $A^{(1)}(\nabla u),\ldots,A^{(m)}(\nabla u)$ are calculated first, from which the most general law takes the form

$$A = \Sigma\, \gamma^{(k)}(x,u)A^{(k)} + \Sigma\, \delta^{(k)}(x,u)T^{(k)}$$

for suitable $\gamma^{(k)}$, $\delta^{(k)}$, and where $T^{(1)}, \ldots, T^{(n)}$ form a complete set of trivial x,u-independent laws [14].

For (2.2), the conditions (5.3) reduce to the system.

$$\frac{\partial A^{\alpha}}{\partial u^{i}_{\beta}} + \frac{\partial A^{\beta}}{\partial u^{i}_{\alpha}} = \Sigma\, \psi_j G^{ij}_{\alpha\beta},\quad G^{ij}_{\alpha\beta} = W^{ij}_{\alpha\beta} + W^{ji}_{\alpha\beta} \quad . \tag{6.1}$$

For $q = 1$, these are the equations for an infinitesimal conformal symmetry for metric $g_{\alpha\beta} = G^{11}_{\alpha\beta}$; this immediately leads to bounds on the number of x,u-independent conservation laws ($\leqslant \frac{1}{2}(p + 1)(p + 2)$ if $p \geqslant 3$). Presumably, the "vector conformal equations" (6.1) lead to similar results, but have not been treated before.

Proposition

A quadratic variational problem (2.2) admits nonvariational symmetries (respectively generalised symmetries) only if there is a $q \times q$ matrix $N(\xi) \neq 0$ (resp. $\neq n(\xi)I$) of linear functions of ξ such that $Q(\xi)N(\xi)$ is skew symmetric (respectively symmetric) for all ξ. (This can be applied to nonlinear materials by linearising the Euler-Lagrange equations and the symmetry at each fixed ∇u_o).

For $p = q = 3$, with the Legendre-Hadamard condition, QN skew symmetric implies that $Q(\xi) = C^T Q_0(\xi)C$ (C independent of ξ) with Q_0 either diagonal or of the form $Q_0 = p(\xi)I + \ell(\xi) \otimes \ell(\xi)$. For linear elasticity this implies the material is isotropic; the meaning for nonlinear materials is less apparent. The condition QN symmetric imposes restrictions on the form of Q, but I have been unable to discern their meaning.

7. LINEAR, ISOTROPIC ELASTICITY [8], [16]

To illustrate the theory, we classify all conservation laws with $A = A(x,u,\nabla u)$ for the case of linear, isotropic elasticity. Here

$$W = \mu\| \nabla u + \nabla u^T \|^2 + \frac{1}{2}\lambda(\nabla . u)^2 \quad , \tag{7.1}$$

where μ, λ are the Lamé moduli, so (2.2) reads

$$E = \mu\Delta u + (\mu + \lambda)\nabla(\nabla . u) = 0 \quad . \tag{7.2}$$

Strong ellipticity is implies by $\mu > 0$, $2\mu + \lambda > 0$, but we only require $\mu(\mu+\lambda)(2\mu+\lambda) \neq 0$. Conserved densities are given in tensorial form, so $A = (A_i^\alpha)$ yields the laws $\sum D_\alpha A_i^\alpha = 0$ for for each i. The identity tensor is denoted by $\underset{\sim}{1}$.

Theorem

Assume $p = q = 3$. For $7\mu + 3\lambda \neq 0$, the following densities arise from the indicated symmetry groups:

A. Geometrical symmetries ($E(3)$ & $\mathbb{R}$ - Euclidean group and scaling).

$$S = 2\mu\nabla u + (\mu+\lambda)(\nabla . u)\underset{\sim}{1} ,$$

$$P = \mu\nabla u^T \nabla u + (\mu+\lambda)\nabla u^T(\nabla . u) - \frac{1}{2}[\mu\|\nabla u\|^2 + (\mu+\lambda)(\nabla . u)^2]\underset{\sim}{1} ,$$

$$R = x \wedge P + u \wedge S ,$$

$$Y = x^T P + \frac{1}{2}u^T S ,$$

B. Generalized symmetries ($E(3)$)

$$Q = \mu(2\mu+\lambda)\nabla u(\nabla . u) + \mu^2 \nabla u(\nabla u - \nabla u^T) + \frac{1}{2}(\mu+\lambda)(2\mu+\lambda)(\nabla . u)^2\underset{\sim}{1} ,$$

$$T = (\mu+\lambda)x \wedge Q + \mu(3\mu+\lambda)u \wedge S + \frac{1}{2}\mu^2(\mu+\lambda)[(u \wedge \nabla u)^T - \mathrm{tr}(u \wedge \nabla u)\underset{\sim}{1}] ,$$

C. Addition of solutions $\varepsilon(x)$ of (7.2)

$$K_\varepsilon = \varepsilon^T S - u^T(\mu\nabla\varepsilon + (\mu+\lambda)(\nabla . \varepsilon)\underset{\sim}{1}) .$$

If $7\mu + 3\lambda = 0$, the following additional densities, corresponding to the full conformal group $O(3,1)$ of geometrical symmetries and a conformal group $O(3,1)$ of generalized symmetries, hold:

$$I = 2(x \otimes x)P - |x|^2 P + (x \otimes u - 2u \otimes x)S + 2(x.u)S - 2\mu u \otimes u - \frac{1}{2}\mu|u|^2\underset{\sim}{1} ,$$

$$Z = x^T Q + \mu u^T S + \mu^2[(\nabla . u)u - u^T \nabla u] ,$$

$$J = 2(x \otimes x)Q - |x|^2 Q + \mu(2x \otimes u - u \otimes x)S + \mu(x.u)S + \mu^2[2x \otimes (u\nabla . u - (\nabla u)u) + u^T(\nabla u)x\underset{\sim}{1} - (\nabla u^T u) \otimes x + x \wedge (\nabla u^T \wedge u) - \mathrm{tr}(u \wedge \nabla u)x \wedge \underset{\sim}{1}] .$$

Knowles and Sternberg, [10], derived the densities in A; they restricted attention just to geometrical symmetries satisfying (5.1), a point raised by Edelen, [4], in his critique of their claims of completeness of conservation laws. In particular, P is Eshelby's energy-momentum tensor, [5], [18]. The K_ε correspond to Betti reciprocity; the remaining laws are new.

The conformal case $7\mu + 3\lambda = 0$ is curious; I am unaware of any physical interpretation. Even if $7\mu + 3\lambda \neq 0$, interesting divergence identities of the form

$$\text{Div } A = \|\nabla u\|^2 \quad \text{or} \quad (\nabla . u)^2 \quad \text{or} \quad x^\alpha \|\nabla u\|^2 \quad \text{or} \quad x^\alpha (\nabla . u)^2 \qquad (7.3)$$

can be constructed by taking suitable combinations of I,Z,J, so the conformal group still plays an important role. For two dimensional elasticity, the results are even more striking.

<u>Theorem</u>

Assume $p = q = 2$. Using complex notation $z = x^1 + ix^2$, $w = u^1 + iu^2$, $\xi = 2\partial w/\partial\bar{z}$, $\eta = \mu(u_1^2 - u_2^1) + i(2\mu + \lambda)(u_1^1 + u_2^2)$. If $(3\mu + \lambda) \neq 0$, the following complex densities are conserved.

$2\mu(2\mu + \lambda)\xi\ \partial B/\partial\eta + (\mu + \lambda)i\bar{B}$, $\quad B(\bar{z},\eta)$ analytic

$C(\bar{z},\eta)$ $\quad C$ analytic

$w\eta + iz\eta^2$

$i(\tilde{w}\eta - \tilde{w}\eta)$ $\quad \tilde{w}$ an arbitrary solution with corresponding $\tilde{\eta}$.

(If $A = A^1 + iA^2$, then $D_1A^1 + D_2A^2 = 0$ is the conservation law.)

If $3\mu + \lambda = 0$, (7.2) reduces to $\partial^2w/\partial z^2 = 0$, and there are many more conservation laws. This theorem reflects the fact that any complex-analytic function gives rise to a conformal transformation in the plane.

8. FURTHER DIRECTIONS OF INVESTIGATION

1) Classify all conservation laws and symmetries for elastic materials with relatively simple constitutive relations, e.g. elastic fluid, Mooney-Rivlin materials, anisotropic linear elasticity, etc., [9].

2) Extend the results to elastodynamics. Fletcher,[6], has shown how each of Knowles and Sternberg's laws has a dynamic counterpart. Presumably, Hamiltonian techniques in [12], will be of use here.

3) Discuss applications of the laws found to crack problems, cf. [2].

4) The appearance of conformal symmetries seems especially significant. For nonlinear wave equations, Morawetz and Strauss, cf. [19], have shown the importance of conservation laws arising from conformal invariance in scattering problems. The identities (7.3) could be equally useful for scattering in elastic bodies with nonlinear body forces. The two dimensional case looks especially promising.

5) For linear problems each symmetry gives rise to an infinite family of generalized symmetries, and, often, conservation laws depending on higher order derivatives of u, [13]. Are these of use? (This question has not even been investigated for the wave equation!) For nonlinear equations, the existence of infinite families of symmetries is closely tied with complete integrability of the system, the prototypical example being the Korteweg-de Vries equation, [1]. Do any elastic materials have such a property?

ACKNOWLEDGEMENT

This research was supported in part by grants from the U.K. Science and Engineering Research Council of Great Britain and National Science Foundation NSF MCS 81-00786.

REFERENCES

[1] Anderson, R.L. and Ibragimov, N.H.: 1979, *Lie-Bäcklund Transformations in Applications*, SIAM Studies in Applied Math, Vol. 1, Philadelphia.

[2] Atkinson, C.: 1979, *Stress singularities and fracture mechanics*, Appl. Mech. Rev. 32, pp. 123-135.

[3] Bessel-Hagen, E.: 1921, *Über die Erhaltungssätze der Elektrodynamik*, Math. Ann. 84, pp. 258-276.

[4] Edelen, D.G.B.: 1981, *Aspects of variational arguments in the theory of elasticity: fact and folklore*, Int. J. Solids Structures 17, pp. 729-740.

[5] Eshelby, J.D.: 1956, *The continuum theory of lattice defects*, in Solid State Physics, (ed. F. Seitz and D. Turnbull) vol. 3, Academic Press, New York.

[6] Fletcher, D.C.: 1976, *Conservation laws in linear elastodynamics*, Arch. Rat. Mech. Anal. 60, pp. 329-353.

[7] Günther, W.: 1962, *Über einige Randintegrate der Elastomechanik*, Abh. Braunschw. wiss. Gess, 14, pp.53-72.

[8] Gurtin, M.E.: 1972, *The linear theory of elasticity*, Handbuch der Physik, (ed. C. Truesdell) Vla/2, Springer, New York, pp. 1-295.

[9] Gurtin, M.E.: 1981, *An Introduction to Continuum Mechanics*, Academic Press, New York.

[10] Knowles, J.K. and Sternberg, E.: 1972, *On a class of conservation laws in linearized and finite elastostatics*, Arch. Rat. Mech. Anal. 44, pp. 187-211.

[11] Noether, E.: 1918, *Invariante Variations Problems*, Kgl. Ger. Wiss. Nachr. Gottingen, Math-Physick, Kl 2, pp. 235-257.

[12] Olver, P.J.: 1980, *On the Hamiltonian structure of evolution equations*, Math. Proc. Camb. Phil. Soc. 88, pp. 71-88.

[13] Olver, P.J.: 1980, *Applications of Lie groups to differential equations*, Lecture Notes, University of Oxford.

[14] Olver, P.J.: 1982, *Conservation laws and null divergences*, University of Minnesota Mathematics Report, No. 81-162.

[15] Olver, P.J.: *Conservation laws in elasticity I. General results*, in preparation.

[16] Olver, P.J.: 1982, *Conservation laws in elasticity II. Linear homogeneous isotropic elastostatics*, University of Minnesota Mathematics Report, No. 81-175.

[17] Ovsiannikov, L.V.: 1982, *Group Analysis of Differential Equations*, Academic Press, New York.

[18] Rice, J.R.: 1968, *A path-independent integral and the approximate analysis of strain concentration by notches and cracks*, J. Appl. Mech. 35, pp. 379-386.

[19] Strauss, W.A.: 1978, *Nonlinear invariant wave equations*, in Invariant Wave Equations, Springer-Verlag, Lecture Notes in Physics, No. 73, New York, pp. 197-249.

PHASE TRANSITIONS VIA BIFURCATION FROM HETEROCLINIC ORBITS

Jack Carr

Department of Mathematics
Heriot-Watt University
Edinburgh, EH14 4AS
Scotland

1. INTRODUCTION

This paper discusses some recent work with M. Gurtin and M. Slemrod on the displacement boundary value problem of a one-dimensional elastic material capable of exhibiting exchanges of phase. This study was motivated by some work of Ericksen [4], who showed that by taking a non-convex stress strain law, elasticity could model phase changes (see also Antman [2], and Antman and Carbone [3]).

Consider a homogeneous bar of length $2L$ fixed at one end. We study the hard device in which the bar is pulled to length $M > 2L$ and then held fixed. The energy of the bar is assumed to be

$$E = \int_{-L}^{L} [W(V'(x)) + \varepsilon^2 (V''(x))^2] dx \tag{1.1}$$

where $\varepsilon > 0$ is small and W' has the form shown in Figure 1 with $W(u) \to \infty$ as $u \to 0$ and $u \to \infty$. The mathematical problem is then:

$$\text{minimise } E \text{ subject to } V(-L) = 0, \; V(L) = M. \tag{1.2}$$

The above problem in the case $\varepsilon = 0$ has been studied by Ericksen [4] and James [5]. Here we briefly record some of the reasons for including the $(\varepsilon V'')^2$ term.

(1) The $(\varepsilon V'')^2$ term denotes a higher order correction to the constitutive relation.

J. M. Ball (ed.), Systems of Nonlinear Partial Differential Equations, 333–341.

(2) The case $\varepsilon = 0$ has infinitely many global minimisers. We examine solutions of (1.2) as $\varepsilon \to 0$. In this way we pick out a class of admissible solutions to the problem with $\varepsilon = 0$.

(3) In [1], Andrews and Ball study the dynamic equations of a viscoelastic material for which problem (1.2) represents the equilibrium solutions of their equation. It turns out that the qualitative description of the asymptotic behaviour of solutions of their equation is much easier in the case $\varepsilon > 0$. Thus the study of (1.2) represents one step in the study of the detailed dynamics of their equation for small ε.

In order to formulate problem (1.2) mathematically we have to give other data such as the space in which we study E. We omit these details in this paper. The main difficulty in problem (1.2) is in the identification of the global minimiser. In what follows we only indicate the main ideas behind our study of (1.2); details will be presented later.

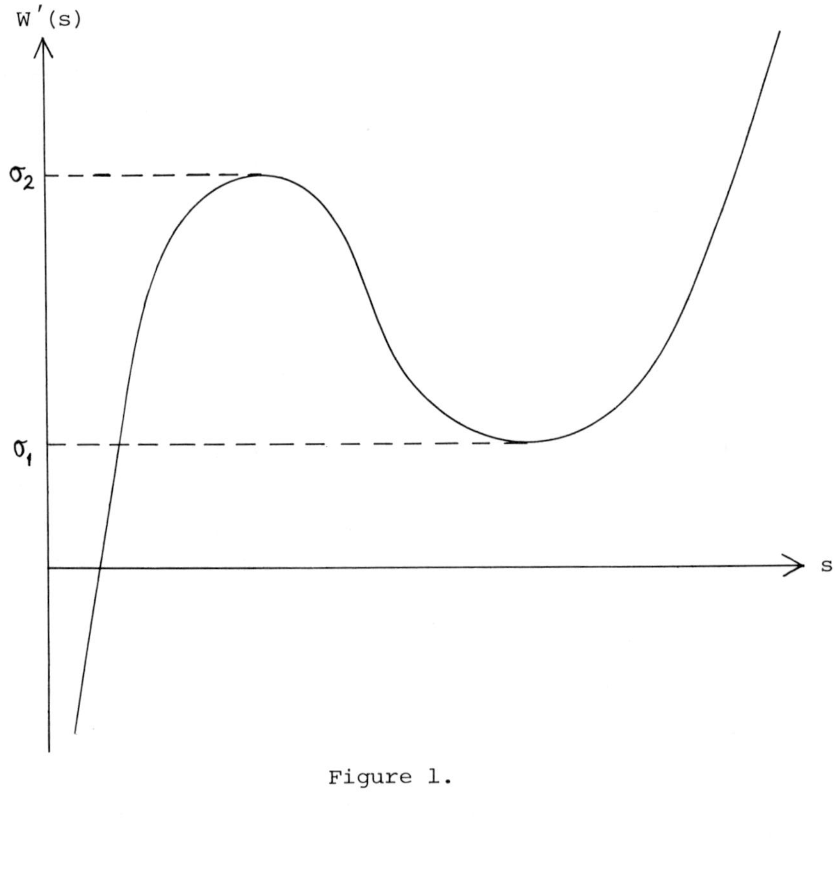

Figure 1.

2. THE CASE $\varepsilon = 0$

Problem (1.2) in the case $\varepsilon = 0$ has been studied by Ericksen [4] and James [5] and we briefly record some of their results. Under appropriate growth conditions on W, stationary values of E satisfy the Euler-Lagrange equation

$$W'(V'(x)) = \sigma , \tag{2.1}$$

$$V(-L) = 0, \quad V(L) = M \tag{2.2}$$

where σ is a constant. If $\sigma_1 < \sigma < \sigma_2$ (see Figure 1) then the equation $W'(s) = \sigma$ has three solutions α, β, γ with $\alpha < \gamma < \beta$. Then (2.1) has solutions V of the form

$$\begin{aligned} V'(x) &= \alpha \quad \text{for } x \in S_1 \\ V'(x) &= \beta \quad \text{for } x \in S_2 \end{aligned} \tag{2.3}$$

where S_1 and S_2 are disjoint measurable subsets of $[-L,L]$ with $S_1 \cup S_2 = [-L,L]$. We do not study solutions for which $V'(x) = \gamma$ since these cannot be stable. If $\alpha < (M/2L) < \beta$, it turns out that we can satisfy the boundary conditions (2.2). The boundary conditions fix the measures of S_1 and S_2 uniquely but do not fix the sets S_1 and S_2. If $S_1 = [-L,p]$, $S_2 = [p,L]$ for some p, then the corresponding solution V is called a two phase solution. There is a similar definition for a many phase solution.

Because of the arbitrary nature of σ and the almost arbitrary nature of the sets S_1 and S_2 we have two sources of non-uniqueness in the solution of (2.1-2.2). For each σ, the above solutions have a corresponding energy E which is independent of S_1 and S_2. We therefore wish to find σ so that the corresponding solution yields the global minimum of E.

It is easy to check that there is a unique σ such that $W(\beta) - W(\alpha) = \sigma(\beta - \alpha)$, i.e. the Maxwell equal area rule. We denote the unique σ by σ_0 and the corresponding α, β by α_0, β_0 so that

$$W(\beta_0) - W(\alpha_0) = \sigma_0(\beta_0 - \alpha_0) \tag{2.4}$$

$$W'(\alpha_0) = W'(\beta_0) = \sigma_0. \tag{2.5}$$

Simple computations show that if

$$\alpha_0 < (M/2L) < \beta_0 \tag{2.6}$$

then the global minimum of E occurs when $\sigma = \sigma_0$. However there are infinitely many global minimisers since S_1 and S_2 are arbitrary, modulo their measures. Finally we note that if (2.6) is not satisfied then the global minimum of E is achieved at a solution of the form

$$V'(x) = \text{constant for all } x \in [-L,L].$$

3. THE CASE $\varepsilon > 0$

For $\varepsilon > 0$ the Euler-Lagrange equation for (1.2) is

$$\begin{aligned} &2\varepsilon^2 V''' - W'(V') + \sigma = 0, \\ &V(-L) = 0,\ V(L) = M,\ V''(-L) = V''(L) = 0, \end{aligned} \tag{3.1}$$

where σ is a constant. Set $u(x) = V'(x)$, $x = \varepsilon t$, $z(t) = u(x) = u(\varepsilon t)$. Then (3.1) is equivalent to

$$\dot{z}^2 - W(z) + \sigma z = b \tag{3.2}$$

$$\dot{z}(-L\varepsilon^{-1}) = \dot{z}(L\varepsilon^{-1}) = 0, \quad \int_{-L\varepsilon^{-1}}^{L\varepsilon^{-1}} z(s)\,ds = M\varepsilon^{-1}, \tag{3.3}$$

where b is a constant and the dot denotes differentiation with respect to t.

The potential $\sigma z - W(z)$ and the corresponding phase portraits of (3.2) are sketched in Figures 2 and 3, (the phase portraits for $\sigma_2 > \sigma > \sigma_0$ are similar). We note from Figure 3 that in order to satisfy (3.3) we need only consider the periodic orbits inside either the heteroclinic orbit or the homoclinic orbit.

The complete story of all solutions of (3.2-3.3) is quite complicated. In order to get some insight into these solutions we first study solutions for which $\dot{z}(t) \geqslant 0$ and $\sigma = \sigma_0 - h$ where $h \geqslant 0$ is small. In this case, by integrating (3.2) and using the boundary conditions (3.3) the problem reduces to

$$I_0 = 2L\varepsilon^{-1}, \quad I_1 = M\varepsilon^{-1}, \tag{3.4}$$

where

$$I_r = \int_{z_1}^{z_2} \frac{z^r}{\sqrt{f(z)}}\,dz, \qquad r = 0,1, \tag{3.5}$$

and $f(z) = b + W(z) - \sigma z$, $f(z_1) = f(z_2) = 0$, $f'(z_1) > 0$, $f'(z_2) < 0$. Clearly, to satisfy (3.4) for small ε we must be close to either the heteroclinic orbit for $h = 0$ or the homoclinic orbit for $h > 0$. To simplify the terminology we will refer from now on to the homoclinic orbit with the understanding that we are referring to the heteroclinic orbit if $h = 0$.

(a) (b) (c)

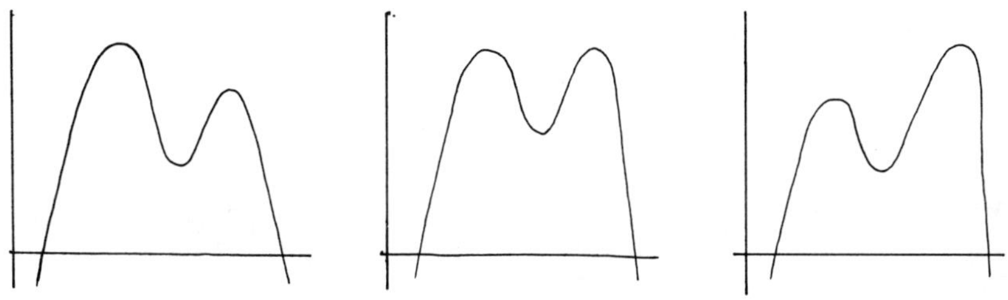

Figure 2. Graph of $\sigma z - W(z)$ when (a) $\sigma_0 > \sigma > \sigma_1$ (b) $\sigma = \sigma_0$ (c) $\sigma_2 > \sigma > \sigma_0$.

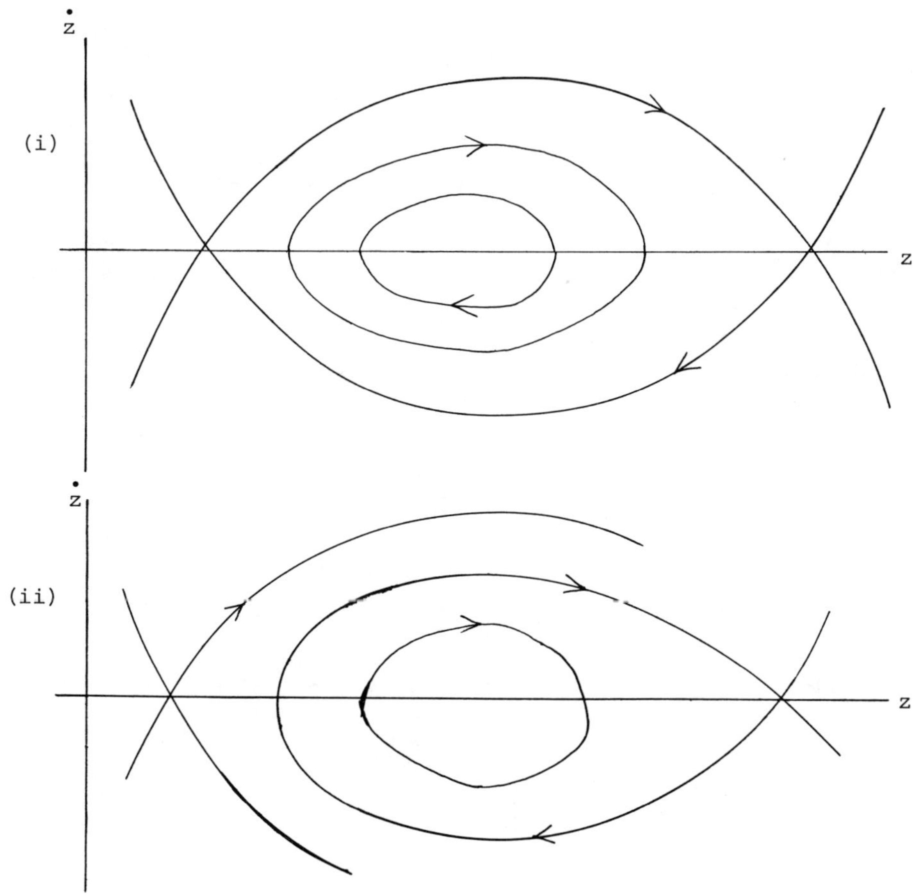

Figure 3. Phase portraits of (3.2) when (i) $\sigma = \sigma_0$ (ii) $\sigma_0 > \sigma > \sigma_1$.

The b corresponding to the homoclinic orbit is given by $b_\infty = (\sigma_0 - h)\beta - W(\beta)$. For $b = b_\infty$ the integrals I_0 and I_1 are infinite. We parametrize b by $b = b_\infty - k$, $k > 0$. Hence we have to find $h \geqslant 0$, $k > 0$ such that

$$I_0(h,k) = 2L\varepsilon^{-1}, \quad I_1(h,k) = M\varepsilon^{-1}, \tag{3.6}$$

where h is small, by hypothesis, and k is small since we must be close to the homoclinic orbit.

The main contribution to the integral I_0 comes from the singularities in the integrand at $z = z_1$ and $z = z_2$. If T_1 and T_2 denote the contributions from z_1 and z_2 then $I_0 \sim T_1 + T_2$ approximately. Since the integrand in I_1 weights the singularities by z_1 and z_2 we have that $I_1 \sim \alpha_0 T_1 + \beta_0 T_2$ where we have used $z_1 \sim \alpha_0$ and $z_2 \sim \beta_0$. Thus we have

$$(\beta_0 - \alpha_0)T_1 \sim (2L\beta_0 - M)\varepsilon^{-1}, \quad (\beta_0 - \alpha_0)T_2 \sim (M - 2L\alpha_0)\varepsilon^{-1}. \tag{3.7}$$

Clearly, for consistency, in (3.7) we must have $\alpha_0 < (2L/M) < \beta_0$, a condition that plays a crucial role in the analysis for $\varepsilon = 0$.

In order to solve (3.7) we must find out how T_1 and T_2 depend on h and k. Again we only give an approximate result,

$$T_1 \sim A\ln\lambda_1^{-1}, \quad T_2 \sim B\ln\lambda_2^{-1},$$

$$\lambda_1 \sim A[(\beta_0 - \alpha_0)h + k]^{\frac{1}{2}}, \quad \lambda_2 \sim Bk^{\frac{1}{2}},$$

$$A = \left(\frac{2}{W''(\alpha_0)}\right)^{\frac{1}{2}}, \quad B = \left(\frac{2}{W''(\beta_0)}\right)^{\frac{1}{2}}.$$

We emphasize that the above relations are only approximate and that many terms have been omitted.

Using the above relations in (3.7) we find that to satisfy (3.6) we must have

$$\begin{aligned} h &= \exp\left(-\frac{(2L\beta_0 - M + \varepsilon F)}{\varepsilon A(\beta_0 - \alpha_0)}\right) \\ k &= \exp\left(-\frac{(M - 2L\alpha_0 + \varepsilon G)}{\varepsilon B(\beta_0 - \alpha_0)}\right) \end{aligned} \tag{3.8}$$

where F and G are not determined by this simple order of magnitude calculation.

Motivated by the above crude calculations, we calculate $I_r(h,k) = I_r(F,G,\varepsilon)$ with h and k defined by (3.8). After many calculations we find that (3.6) has a solution if and only if

$$\begin{aligned} F + G &= R_1(F,G,\varepsilon) \\ \alpha_0 F + \beta_0 G &= R_2(F,G,\varepsilon) \end{aligned} \tag{3.9}$$

where $R_k(F,G,0) = c_k$, a constant independent of F and G. Moreover, the derivatives of R_k with respect to F and G are zero at $\varepsilon = 0$. Hence we can solve (3.9) by the implicit function theorem.

The above argument only proves existence for a certain range of values of $(M/2L)$. Combining this with a similar argument for small negative h, we prove that for ε small enough and $\alpha_0 < (M/2L) < \beta_0$, (3.2-3.3) has a solution $z(t)$ which is close to the homoclinic orbit. Thus $z(t)$ spends most of the time close to either α_0 or β_0.

After descaling, we obtain a solution $V_2(x)$ to (3.1). Using the above properties of $z(t)$, $V_2(x)$ is close to the two phase solution of section 2, that is, $V_2'(x) \sim \alpha_0$ for $x \in [-L,p_1]$, $V'(x) \sim \beta_0$ for $x \in [p_2,L]$ where $p_2 - p_1$ is very small. Note also that $\overline{V}_2(x) = M - V_2(-x)$ is a solution to (3.1) with the same energy E.

In the above analysis, we studied solutions for which $\dot{z}(t) \geqslant 0$ and $\sigma - \sigma_0$ is small. There are also solutions with $\sigma - \sigma_0$ not small and $\dot{z}(t) \geqslant 0$, but they have higher energy than the above solutions.

4. OTHER SOLUTIONS

We now study solutions $z(t)$ of (3.2-3.3) for which $\dot{z}(t)$ can change sign. We call $z(t)$ an N-solution if $\dot{z}(t) = 0$ for N distinct values of t in $[-L\varepsilon^{-1}, L\varepsilon^{-1}]$. The corresponding descaled functions inherit the same terminology so that $V_2(x)$ is a 2-solution.

If $N \geqslant 2$ and $\sigma - \sigma_0$ is small, there exists an N-solution if and only if

$$(N-1)I_0 = 2L\varepsilon^{-1}, \quad (N-1)I_1 = M\varepsilon^{-1}, \tag{4.1}$$

where I_0 and I_1 are defined in (3.5). If $N\varepsilon$ is sufficiently small then we can use the analysis in section 3 to prove existence of solutions of (4.1). We prove:

Theorem 1

For $\varepsilon, N\varepsilon$ small enough and $\alpha_0 < (M/2L) < \beta_0$, (3.1) has an N-solution which is close to the N phase solution for the case $\varepsilon = 0$.

We now calculate the energy of the N-solutions given by Theorem 1. We have

$$E_N = \int_{-L}^{L} [\varepsilon^2 (u_N'(x))^2 + W(u_N)]\,dx$$

$$= \varepsilon \int_{-L\varepsilon^{-1}}^{L\varepsilon^{-1}} [\dot{z}_N^2 + W(z_N)]\,dt$$

$$= \varepsilon \int_{-L\varepsilon^{-1}}^{L\varepsilon^{-1}} [2\dot{z}_N^2 + \sigma(N) z_N - b(N)]\,dt$$

$$= 2(N-1)\varepsilon \int_{z_1(N)}^{z_2(N)} \sqrt{f_N(s)}\,ds + \sigma(N)M - 2Lb(N).$$

From (3.8) we see that $\sigma(N) = \sigma(2) + O(e^{-c\varepsilon^{-1}}) = \sigma_0 + O(e^{-c\varepsilon^{-1}})$, where c is a positive constant, with similar formulae for the other variables. Hence

$$E_N = E_2 + 2(N-2)\varepsilon \int_{z_1(N)}^{z_2(N)} \sqrt{f_N(s)}\,ds + O(e^{-c\varepsilon^{-1}}). \tag{4.2}$$

From (4.2) we see that the 2-solution has the lowest energy.

In the above analysis we restricted attention to the case $N\varepsilon$ small and $\sigma - \sigma_0$ small. There are solutions when either of these conditions does not hold but they have higher energy than the 2-solution. Thus we have:

Theorem 2

For ε small enough and $\alpha_0 < (M/2L) < \beta_0$, the global minimun of E occurs at V_2 and $\overline{V}_2$.

REFERENCES

[1] Andrews, G., and Ball, J.M.: 1982, *Asymptotic behaviour and changes of phase in one-dimensional viscoelasticity*, J. Differential Equations 44, pp.306-341.

[2] Antman, S.S.: 1973, *Nonuniqueness of equilibrium states for bars in tension*, J. Math. Anal. Appl. 44, pp.333-349.

[3] Antman, S.S., and Carbone, E.R.: 1977, *Shear and necking instabilities in nonlinear elasticity*, J. Elasticity 7, pp.125-151.

[4] Ericksen, J.L.: 1975, *Equilibrium of bars*, J. Elasticity 5, pp.191-201.

[5] James, R.D.: 1979, *Co-existent phases in the one-dimensional static theory of elastic bars*, Arch. Rational Mech. Anal. 77, pp.99-140.

BIFURCATION UNDER CONTINUOUS GROUPS OF SYMMETRIES

E.N. Dancer

Department of Mathematics
University of New England
Armidale, N.S.W. 2351
Australia

The type of problem we wish to consider is the following. Assume that G is a compact Lie group and $\{T_g\}_{g\in G}$ is a continuous linear representation of G on R^n. (In other words, T_g is a linear map for each $g \in G$, $T_{gh} = T_g T_h$ for $g,h \in G$ and the map $g \longrightarrow T_g x$ is continuous for each $x \in R^n$). In addition, assume that $F : R^n \times R \longrightarrow R^n$ is continuously differentiable and G-invariant (that is, $F(T_g x,t) = T_g F(x,t)$ for $g \in G$, $x \in R^n$, $t \in R$). Finally, assume that $k \in R^n$ and $F(k,0) = 0$. Because of the symmetries, we see that $F(m,0) = 0$ for $m \in M$, where $M = \{T_g k: g \in G\}$. Note that M is a smooth manifold (cp. Bredon [1]).

One basic problem is to discuss the existence and uniqueness of solutions of $F(x,t) = 0$ near $M \times \{0\}$. Our main interest here is in obtaining results which do not depend upon a knowledge of $F(x,t)-F(x,0)$. (In many cases, $F(x,0)$ is significantly easier to calculate than $F(x,t)-F(x,0)$). In particular, we emphasise some unusual features of the problem. At times, we will weaken either the differentiability assumption or the G-invariance assumption on $F(x,t)-F(x,0)$.

It might be asked where such problems arise. Firstly the existence of the representation commuting with F often holds for autonomous ordinary differential equations (depending on boundary conditions) or for partial differential equations on

J. M. Ball (ed.), Systems of Nonlinear Partial Differential Equations, 343–350.

symmetric domains. Note that both these problems are really problems on infinite-dimensional Banach spaces rather than R^n. We will only consider the case of R^n to avoid technicalities. (The Banach space case is discussed in [5]). In addition, if one looks for small solutions, one frequently obtains a problem of the type above after a Liapounov-Schmidt reduction and a scaling transformation. We discuss this in a little more detail later. Note that our basic problem occurs when one tries to continue branches of solutions in the symmetric case.

Note that we may not change axes to obtain $k = 0$ without changing the linear action. Thus the theory in Poenaru [9] or similar articles does not apply directly.

Finally, since $F(m,0) = 0$ for $m \in M$, we see by differentiating along M that $T_k(M) \subset N(F_1'(k,0))$ where $T_k(M)$ denotes the tangent space at M at k and $N(F_1'(k,0))$ denotes the kernel of the partial derivative $F_1'(k,0)$. Thus, if M is not discrete (which will be the case of main interest to us), then $F_1'(k,0)$ has a non-trivial kernel. Thus, we cannot use the implicit function theorem directly.

1. THE BASIC CASE

Here we assume that the conditions of the introduction hold and that, in addition, $N(F_1'(k,0)) = T_k(M)$. Thus we are assuming that the kernel is as small as possible. Now the tubular neighbourhood theorem (cp. [1]) implies that every x near M can be uniquely expressed as $m + n$ where $m \in M$, $n \in T_m(M)^{\perp}$ and m is small. We split the equation $F(m+n,t) = 0$ into one normal to M and one tangential to M. Now the normal equation can be solved for n as a function of m and λ by the ordinary implicit function theorem. (There are some technicalities here which we will ignore). Thus we are left with the tangential equation. It turns out that either this tangential equation vanishes identically or it is difficult to deal with. To be more precise, we need some definitions.

Let $G_k = \{g \in G: T_g k = k\}$. G_k is called the isotropy group of k (for the action of G). We say that $\{G,G_k\}$ has Property P if $\dim N_G(G_k) = \dim G_k$, where $N_G(G_k)$ denotes the normaliser of G_k in G (that is, $N_G(G_k) = \{g \in G: g^{-1}G_k g \subset G_k\}$. We will discuss this condition a little later. We now state the main result of this section.

Theorem 1

Assume that the conditions of the introduction hold and that $T_k(M) = N(F_1'(k,0))$.

(i) *If* $\{G,G_k\}$ *has Property* P, *then, for each small* t, *there is a unique orbit of solutions of* $F(x,t) = 0$ *near* M *and this orbit depends continuously on* t. *(More precisely, there is a continuous* G*-invariant map* $n: M \times (-\varepsilon,\varepsilon) \longrightarrow R^n$ *such that* $n(m,0) = 0$ *and* $m + n(m,t)$ *are the only solutions near* M).

(ii) *If* $\{G,G_k\}$ *does not have Property* P, *then smooth* G*-invariant higher order terms* $F(x,t) - F(x,0)$ *can be found such that the equation* $F(x,t) = 0$ *has no solutions near* M *for every non-zero* t.

This was proved in [4]. (An earlier version appeared in [3]).

Thus we see that whether $\{G,G_k\}$ has Property P is of critical importance in how the system behaves under perturbation. In fact, if $\{G,G_k\}$ does not have Property P, then one can never prove the existence of solutions of $F(x,t) = 0$ by looking at only a finite number of terms in the Taylor expansion of $F(x,t) - F(x,0)$ at k. Thus, in this case, the problem is very badly behaved. Fortunately, most problems in applications seem to satisfy Property P. One interesting point on Property P is that it is independent of the map F and depends only on k through the isotropy group G_k. There is a much more geometric condition which is equivalent to $\{G,G_k\}$ having Property P. It can be shown that $\{G,G_k\}$ has Property P if and only if every G-invariant vector field v on M is trivial (that is, $v(x) = 0$ on M). When we say that v is G-invariant, we mean that $v(T_g x) = T_g v(x)$ for $x \in M$ and $g \in G$. It follows easily that $\{G,G_k\}$ has Property P if and only if

$$\{x \in T_k(M) : T_g x = x \text{ for all } g \text{ in } G_k\} = \{0\}$$

This last condition tends to be the most convenient way to check Property P in applications.

As a very simple example, one easily checks that if G is the circle group S^1 and $G_k = \{e\}$, then Property P fails while, if G is the group $O(2)$ of orthogonal matrices and $G_k = \{I, \begin{pmatrix} 1 & 0 \\ 0 & 1 \end{pmatrix}\}$, then Property P holds. This illustrates that extra finite symmetries (that is, ones not increasing the dimension of the symmetry group) can turn a bad problem into a good problem. Thus the extra finite symmetries are very important. The difference in the two examples above is reflected in differential equations. We discuss this very briefly. If one looks for periodic solutions of fixed period bifurcating from zero for an autonomous equation $x''(t) + f(x,x',\lambda) = 0$ (where λ is a parameter and $f(0,0,\lambda) = 0$), then typically the symmetry group is S^1 (due to translations). Our comments above can be used to show that this is a bad problem and in general we can only obtain a good problem by varying the period. However, if f is even in x', we now have an $O(2)$ symmetry (the extra symmetry is the map $x(t) \longrightarrow x(-t)$). We now have a good problem if we can look for solutions with fixed period (under suitable hypotheses).

Under the assumptions of Theorem 1(i), the solutions of $F(x,\lambda) = 0$ near $M \times \{0\}$ have essentially the same symmetry as k. Thus there is no symmetry breaking. If there is a scalar product on R^n such that $F(\ ,t)$ is the gradient of $f(\ ,t)$ for each t, then the conclusion of Theorem 1(i) holds without the assumption that $\{G,G_k\}$ has Property P. Thus we see that a gradient structure on F changes the problem significantly.

Finally, for this section, I briefly mention some applications. Firstly, as shown (briefly) in [3], our methods can be used (after rescaling) to study the Bénard problem. It seems that our methods need fewer calculations than the methods in [2] and [10].

Secondly, our methods can be used to determine a necessary and sufficient condition for a problem $\lambda x = C(x) + \text{h.o.t.}$ on R^n to have G-codimension zero in the sense of [7]. Here we assume that all the terms in the equation are G-invariant, that G *acts irreducibly on* R^n and that C is a homogeneous polynomial. We intend to discuss this in detail elsewhere.

2. THE CASE OF LARGER KERNEL

In this section, we retain our earlier assumptions except that we no longer assume that $T_k(M) = N(F_1'(k,0)$ and we often assume that $F(x,t) - F(x,0)$ is continuous rather than continuously differentiable. We follow the notation of §1. By analogy with the case of no symmetries, we now do not normally expect uniqueness theorems. Thus we ask purely the existence question. Once again we are looking for conditions independent of the perturbation $F(x,t) - F(x,0)$.

We say that M is C^0 G-stable for $F(x,0)$ if every continuous G-invariant map $\tilde{F}(x,t)$ with $\tilde{F}(x,0) = F(x,0)$ has a zero near M for each small t. We define C^1 G-stability analogously except that we require $\tilde{F}$ to be C^1 (rather than C^0). The existence question we were asking above is a G-stability question. It is possible to obtain a rather complicated necessary and sufficient condition for C^0 G-stability. It involves a number of degrees and conditions like Property P. In addition, it turns out that it is possible to obtain a sufficient condition for C^1 G-stability by similar methods. One interesting and unexpected (at least to me) consequence appears. It turns out that it is possible that M is C^1 G-stable but not C^0 G-stable. In fact, this can happen with $M = \{0\}$ and G the symmetric group S_3. Thus it seems that one has to think very carefully in applications what type of perturbations one permits (because it may change whether solutions persist). It seems to be an open question whether similar behaviour can occur when there is no symmetry group. (Some partial results appear in Simon and Titus [11]). The above results are proved in [6]. The proofs depend in an essential way on a corrected version of a theorem of Rubinsztein [12] on G-homotopy groups.

3. BREAKING OF SYMMETRIES

We now assume the conditions of §1 except that we weaken the invariance assumption of $F(x,t) - F(x,0)$ by only requiring that there is a closed subgroup S of G (possibly $S = \{e\}$) such that $F(x,t) - F(x,0)$ is S-invariant. Note that we still assume that $F(x,0)$ is G-invariant. Thus the perturbation at least partially breaks the symmetries. This case frequently occurs in applications. For example, $F(x,t) - F(x,0)$ might be some kind of small forcing term. As in §1, we assume that F is C^1 and that $T_k(M) = N(F_1'(k,0))$. Once again we are interested in conditions for the existence of zeros which are independent of $F(x,t) - F(x,0)$.

We use the same idea of splitting into normal and tangent equations as in §1. By similar arguments to there, our problem reduces to the zeros of an S-invariant vector field on M. (A vector field v on M is said to be S-invariant if $v(T_g x) = T_g v(x)$ for $x \in M$ and $g \in S$). More precisely, there is a continuous S-invariant map $n : M \times (-\varepsilon,\varepsilon) \longrightarrow R^n$ such that $n(m,0) = 0$ and a continuous S-invariant vector field $V(\,,t)$

defined on M for each $t \in (-\varepsilon,\varepsilon)$ such that the zeros of F near $M \times \{0\}$ are

$$\{m + n(m,t) : V(m,t) = 0\}$$

Conversely, every S-invariant vector field V occurs for a suitable perturbation $F(x,t) - F(x,0)$. Another of the oddities of these problems now appears. It may happen that there is a point k on M such that every S-invariant vector field on M vanishes at k (and thus $k + n(k,t)$ is a solution of our original problem). *We call such a point* k *a forced zero.* In fact, it seems to turn out in many applications that the physically important solutions are forced zeros. (*For example*, this happens in the Bénard problem. Note that the problem has to be arranged so that the above theory is applicable). Now, for the above idea to be useful, one needs a simple test for forced zeros. It turns out that k is a forced zero if and only if it has the following two properties.

(i) $\{S,S_k\}$ has Property P, where

$$S_k = \{g \in S: \ T_g k = k\}; \quad \text{and}$$

(ii) there are no points in $M\backslash S(k)$ near $S(k)$ which have isotropy group S_k *for the action of* S *on* M. Here $S(k)$ denotes the S-orbit $\{T_g k: g \in S\}$.

In applications, these conditions are often quite easy to check. (For the second, it is often convenient to use the orbit space). The existence of forced zeros was observed independently in [3] and [8]. It might be asked what the analogous theorem is if F has a gradient structure. In this case, $V(\ ,t)$ is the gradient of a functional on M for each t and, in our test for points k where V must vanish, we should delete (i).

Finally, we sketch rather briefly how some of the above ideas apply to the Bénard problem in the case where the bifurcation equation has no quadratic term. In this case, the bifurcation equation is of the form

$$\lambda x = C(x) + H(x,\lambda) \tag{1}$$

where C is a homogeneous cubic, H is a higher order term and all terms are invariant under a 2-dimensional group G^2 (which is a "non-commutative product" of the torus T^2 and the dihedral group D_6). However, it turns out that C is invariant under a 3-dimensional group G^3. This has the consequence that $\{x: Cx = x\}$ has a 3-dimensional component Z. If we look for solutions

$x = \lambda^{\frac{1}{2}}z$ of (1) with $\lambda > 0$, λ small and z near Z, then our equation becomes

$$z - Cz - \lambda^{-\frac{3}{2}} H(\lambda^{\frac{1}{2}}z,\lambda) = 0 \qquad .$$

We can think of this as a breaking of symmetry problem with $G = G^3$, $M = Z$ and $S = G^2$. We apply a result discussed earlier in this section. The solutions near Z are determined by the zeros of a G^2 invariant vector field on Z. This can be easily studied (since the orbit space Z/G^2 is 1-dimensional). In particular, we easily find that, for suitable higher order terms, there can be arbitrarily large numbers of G^2 orbits of solutions near Z for suitable smooth higher-order terms. (These can occur in a stable way). Thus the structure of the solutions of the general bifurcation equation in this case is very complicated. The connection with the work in Buzano and Golubitsky [2] is that they restrict the possible higher order terms by requiring a certain term has a non-zero (and not too small) coefficient. The above solutions are type V in the sense of [2] except for two orbits which are of type III and which are in fact forced zeros. Finally, the result of §1 and scaling can be easily used to study the solutions determined by the two-dimensional components of $\{x: C(x) = x\}$.

REFERENCES

[1] Bredon, G.: 1972, *'Introduction to compact transformation groups'*, Academic Press, New York.

[2] Buzano, E., and Golubitsky, M.: 1981, *Bifurcation on the hexagonal lattice and the planar Bénard problem*, preprint.

[3] Dancer, E.N.: 1980, *On the existence of bifurcating solutions in the presence of symmetries*, Proc. Royal Soc., Edinburgh, 85A, pp.321-336.

[4] Dancer, E.N.: 1980, *An implicit function theorem with symmetries and its application to nonlinear eigenvalue problems*, Bull. Austral. Math. Soc., 21, pp. 81-91.

[5] Dancer, E.N.: 1982, *On the G-invariant implicit function theorem in infinite dimensions*, Proc. Royal Soc., Edinburgh, 92A, pp.13-30.

[6] Dancer, E.N.: *Perturbation of zeros in the presence of symmetries*, submitted.

[7] Golubitsky, M., and Schaeffer, D.: 1979, *Imperfect bifurcation in the presence of symmetry*, Comm. Math. Phys., 67, pp.205-232.

[8] Michel, L.: 1980, *Symmetry defects and broken symmetry. Configurations. Hidden symmetry*, Rev. Modern Phys., 52, pp.617-651.

[9] Poenaru, V.: 1976, *'Singularities C^∞ en présence de Symétrie'*, Lecture Notes in Mathematics No. 510, Springer-Verlag, Berlin.

[10] Sattinger, D.: 1978, *Group representation theory, bifurcation theory and pattern formation*, J. Functional Anal., 28, 58-101.

[11] Simon, C.P., and Titus, C.: 1975, *Removing index zero singularities with C^1 small perturbations*, in Lecture Notes in Mathematics Vol. 468, Springer-Verlag, Berlin.

[12] Rubinsztein, R.: 1976, *On the equivariant homotopy of spheres*, Dissertationes Mathematicae, 134, pp.1-48.

MORSE DECOMPOSITIONS AND GLOBAL CONTINUATION OF PERIODIC SOLUTIONS FOR SINGULARLY PERTURBED DELAY EQUATIONS

John Mallet-Paret

Michigan State University
Department of Mathematics
East Lansing, Michigan 48824

A delay differential equation, arising in optical and biological models, is studied as a dynamical system. As a parameter $\sigma \to 0$, a family of periodic solutions $x(t, \sigma)$ can be globally continued. For small σ we have a singular perturbation problem, and at $\sigma = 0$ a non-uniform convergence to a square wave (similar to a Gibbs phenomenon) occurs. For all σ, the dynamical system possesses a Morse decomposition. This is studied with the topological theory of C. Conley.

1. INTRODUCTION

We consider the class of differential delay equations

$$\sigma\dot{x}(t) = -x(t) + f(x(t-1)) \tag{1}$$

where $x \in R$, $\sigma > 0$, and $f : R \to R$ satisfies at least

$$f(0) = 0,$$

$$a = f'(0) < -1,$$

$$xf(x) < 0 \quad \text{for all} \quad x \neq 0, \tag{2}$$

$$|f(x)| < |x| \quad \text{for large} \quad |x|, \quad \text{and} \quad f \quad \text{is} \quad C^{\infty}. \tag{3}$$

If the parameter σ is small we have a singular perturbation problem. This is a situation of special interest, although much of our analysis concerns arbitrary σ.

J. M. Ball (ed.), Systems of Nonlinear Partial Differential Equations, 351–365.

Equations of this sort arise in physiology [15], [21], in models of population growth [18], and in optics [17], [19], [20].

We study here some global dynamical properties of (1) and some related bifurcation phenomena as σ varies. Our main results concern global continuation of periodic solutions from Hopf bifurcations; the asymptotic form of such solutions as $\sigma \to 0$; and existence of a Morse decomposition for the dynamical system of (1), (that is, existence of a gradient-like structure on part of the phase space). The mathematical techniques used are both analytical (based on the theory of functional differential equations; see Hale [16]) and topological (fixed point indices for maps, indices for periodic orbits, such as the Fuller [14] and Orbit index [4], [23], and Conley's Morse index [10] for more general invariant sets.)

2. HOPF BIFURCATIONS

As σ decreases, it passes through values $s_1 > s_3 > s_5 > \cdots \to 0$ at which (1) undergoes a simple Hopf bifurcation from the origin. Indeed (see [24]) for

$$s_{2n-1}(a) > \sigma > s_{2n+1}(a)$$

the origin is a hyperbolic equilibrium with an unstable manifold of dimension $2n$. (For $\sigma > s_1(a)$ the origin is stable.) Figure 1 shows the curves $\sigma = s_{2n+1}(a)$ in the (a, σ) parameter space. (Using only odd subscripts here makes for clearer notation later on.) The periodic solutions bifurcating

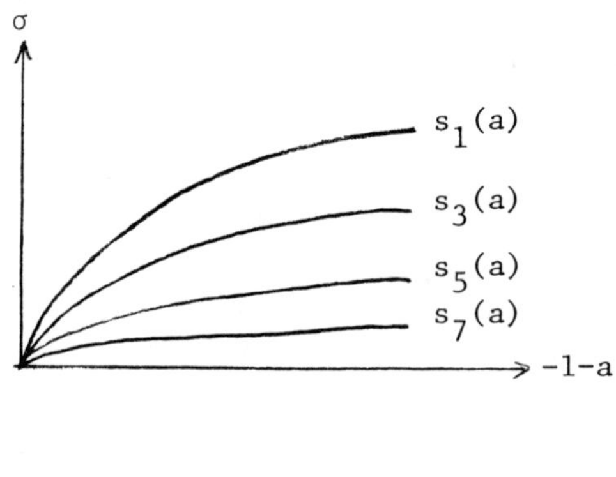

Figure 1

from $(x,\sigma) = (0,s_{2n+1})$ may be either subcritical or supercritical, depending on the higher order derivatives $f''(0)$ and $f'''(0)$. See [6], [16] for general results on Hopf bifurcation in differential delay equations.

3. GLOBAL CONTINUATION OF THE HOPF ORBITS

A global continuation from a given periodic orbit $(x^*(t),\sigma^*)$ means a connected set $P \subseteq C \times (0,\infty)$ such that each $(\phi,\sigma) \in P$ lies on a non-constant periodic orbit of (1) with parameter value σ, and such that P contains the given orbit (x_t^*,σ^*). Here $C = C[-1,0]$ denotes the phase space for the semiflow of (1), as in [16]. We write $x_t(\Theta) = x(t+\Theta)$, $\Theta \in [-1,0]$, when $x(t)$ is any solution; thus $t \to x_t \in C$ is a trajectory for the semiflow.

A number of authors have studied global continuation of periodic orbits of ordinary differential equations; see [1], [2], [3], [4], [5], and [23]. For delay differential equations Nussbaum has made extensive use of cone mappings and related topological indices to obtain global continuations; see [25] and the references therein. Chow and Mallet-Paret in [7] used the Fuller index also to find global continuations for delay equations. The following result, obtained jointly with R. Nussbaum can be proved using these techniques.

Theorem 1. There exists for each $n \geq 0$ a global continuation P_{2n+1} of the Hopf bifurcation orbits from $(x,\sigma) = (0,s_{2n+1})$ such that the minimal period $\tau : P_{2n+1} \to (0,\infty)$ is a continuous function on P_{2n+1}, and such that P_{2n+1} is maximal with respect to this property. Moreover

(a) $P_{2n+1}(\sigma) \neq \emptyset$ for each $0 < \sigma < s_{2n+1}$, where $P_{2n+1}(\sigma) = \{\phi \in C \mid (\phi,\sigma) \in P_{2n+1}\}$;

(b) $P_{2n+1} \cap P_{2m+1} = \emptyset$ if $n \neq m$;

(c) P_{2n+1} is bounded in $C \times (0,\infty)$ and precompact in $C \times [\varepsilon,\infty)$ for each ε;

(d) the closure of P_{2n+1} in $C \times (0,\infty)$ is $P_{2n+1} \cup \{(0,s_{2n+1})\}$;

(e) the minimal periods satisfy $\tau(\phi,\sigma) \in ((n+\frac{1}{2})^{-1}, n^{-1})$ for $(\phi,\sigma) \in P_{2n+1}$; each solution $x(t)$ in P_{2n+1} has exactly two zeros, both simple, in its minimal period. ///

Saying that the minimal period τ varies continuously on P_{2n+1} means that multiple period bifurcations (e.g. period doubling bifurcations) from points in P_{2n+1} are excluded from P_{2n+1}. (Of course such bifurcations may, and in fact do, occur for certain f.) Thus P_{2n+1} may not be the maximal global continuation from the Hopf orbits - it is only maximal with respect to the property that τ is continuous.

Part (c) is a reflection of the fact that there is a maximal compact invariant set $\Psi(\sigma) \subseteq C$, which is an attractor, and that for large σ (1) has no periodic solutions.

The period bound (e) reflects the fact (proved in [7]) that (1) has no solution of period $\frac{2}{m}$ for any integer m, and that $\tau(\phi,\sigma) \in ((n+\frac{1}{2})^{-1}, n^{-1})$ near the Hopf bifurcation.

4. A MORSE DECOMPOSITION

More generally, we may study the global dynamics of the semiflow of (1). A main result of ours [22] is that this semiflow possesses a Morse decomposition (see Conley [10] for the definition). This means the dynamics have a gradient-like structure after finitely many compact invariant sets S_i are identified to points.

To be precise let us first restrict out attention to the maximal compact attractor Ψ defined below. Condition (3) implies that there exists $K > 0$ (independent of σ, in fact) such that $\limsup_{t\to\infty} |x(t)| \le K$ for any solution, and such that $\sup_{(-\infty,\infty)} |x(t)| \le K$ for any solution bounded on $(-\infty,\infty)$. It follows that the set

$$\Psi = \Psi(\sigma) = \{\phi \in C \mid \text{there exists a solution } x(t) \text{ on } (-\infty,\infty), \text{ with } \sup_{(-\infty\ \infty)} |x(t)| < \infty, \text{ such that } x_0 = \phi\}$$

of initial conditions for solutions defined and bounded for all $t \in (-\infty,\infty)$, is compact, invariant, and an attractor. Thus $x_t \to \Psi$ as $t \to \infty$ for any solution of (1). In general, a set $A \subseteq C$ is invariant (positively and negatively here) for a semiflow if for each $\phi \in A$ there exists a trajectory with $x_t \in A$ for all $t \in (-\infty,\infty)$, and $x_0 = \phi$. An invariant set is called isolated if, in some neighborhood of itself, there is no larger invariant set.

The sets S_i in the Theorem below make up what is called a Morse decomposition; each S_i is called a Morse set. An integer valued Lyapunov function is used to define these sets and prove they do form a Morse decomposition. This is discussed further in Section 5.

Theorem 2. Let $\sigma \notin \{s_1, s_3, s_5, \ldots\}$ be fixed, and let $m \geq 0$ be such that $s_{2m-1} < \sigma < s_{2m+1}$ (or $m = 0$ if $\sigma > s_1$). Then there exist sets

$$S_{2n+1} = S_{2n+1}(\sigma) \subseteq \Psi - \{0\}, \quad n \geq 0$$

defined in Section 5 below, which, together with

$$S_{2m} = \{0\}$$

form a Morse decomposition. Namely,

(a) S_i is a compact isolated invariant set for each $i \in \{2m, 1, 3, 5, \ldots\}$;

(b) $S_i \cap S_j = \emptyset$ if $i \neq j$;

(c) $S_{2n+1} = \emptyset$ for large n;

(d) if $x_t \in \Psi$ for all $t \in (-\infty,\infty)$, then there exist $i \leq j$ such that $x_t \to S_i$ as $t \to \infty$ and $x_t \to S_j$ as $t \to -\infty$; and if $i = j$ then $x_t \in S_i$ for all t.

Moreover, we have

(e) $P_{2n+1}(\sigma) \subseteq S_{2n+1}(\sigma)$; and

(f) the Conley index $h(S_i)$ of S_i is

$$h(S_1) = \overline{0} \cup S^1 \quad \text{if} \quad 1 < 2m$$

$$h(S_{2n+1}) = \Sigma^{2n} \vee \Sigma^{2n+1} \quad \text{if} \quad 1 < 2n+1 < 2m$$

$$h(S_{2m}) = \Sigma^{2m}$$
$$h(S_{2n+1}) = \overline{0} \quad \text{if} \quad 2n+1 > 2m. \qquad ///$$

The Conley index $h(A)$ is a topological index defined for compact isolated invariant sets A. See [10], [11], [28] for details of the theory. This index takes values in the class of pointed topological spaces under the equivalence relation of homotopy equivalence. That is, $h(A)$ is an equivalence class of pointed spaces. Above, Σ^k denotes the pointed k-sphere, S^1 the usual 1-sphere, $\sqcup$ the disjoint union, v the wedge product of pointed spaces, and $\overline{0}$ is the one-point pointed space. The indices for S_{2n+1}, $2n+1 < 2m$, are in fact the same as those of a hyperbolic periodic orbit with a $2n+1$ dimensional orientable unstable manifold (or an attracting orbit if $n = 0$).

Conley's theory has been developed for flows; for simplicity of exposition we state our results here in terms of the semiflow of (1) and omit the description of a technical construction which embeds this semiflow in a genuine flow. Alternatively, Rybakowski [26] has developed Conley's theory directly for semiflows, and this could be used. See also [27].

The index has the property that if $h(A) \neq \overline{0}$, then $A \neq \emptyset$. Moreover, as the index is stable under perturbations of the dynamical system, this situation is maintained for nearby semiflows. Indeed, the concept of a continuation of an isolated invariant set, analogous to the case of a periodic orbit, can be discussed.

The Conley index also gives information about orbits connecting different Morse sets. Consider two adjacent sets, say S_{2n+1} and S_{2n+3} where $2n+3 < 2m$. Let

$$C_{2n+1}^{2n+3} = \{\phi \in \Psi \mid \text{there exists } x_t \to S_{2n+1} \text{ as } t \to \infty, \text{ and } x_t \to S_{2n+3} \text{ as } t \to -\infty, \text{ with } x_0 = \phi\}.$$

Then

$$S_{2n+1}^{2n+3} = S_{2n+3} \cup C_{2n+1}^{2n+3} \cup S_{2n+1}$$

is an isolated invariant set. Its index is not equal to $h(S_{2n+1}) \vee h(S_{2n+3})$, and a result of Conley implies

$C^{2n+3}_{2n+1} \neq \emptyset$.

<u>Theorem 3</u>. If $2n+3 < 2m$ then

$$h(S^{2n+3}_{2n+1}) = \Sigma^{2n} \vee \Sigma^{2n+3} \quad (\text{or } \overline{0} \sqcup S^3 \text{ if } n = 0)$$
$$\neq h(S_{2n+1}) \vee h(S_{2n+3}).$$

If $m \geq 1$ then

$$h(S^{2m}_{2m-1}) = \Sigma^{2m-2}$$
$$\neq h(S_{2m-1}) \vee h(S_{2m}).$$

In either case there is a connecting orbit between the Morse sets: $C^{2n+3}_{2n+1} \neq \emptyset$ and $C^{2m}_{2m-1} \neq \emptyset$. ///

5. NODAL PROPERTIES OF SOLUTIONS

The Morse sets S_{2n+1} are defined by nodal properties of the solution: $2n+1$ represents the number of zeros of $x(t)$ in a unit interval $(\tau-1,\tau]$ when $x(\tau) = 0$. For an arbitrary solution zeros become less frequent as t increases. The number of zeros can be thought of as a Lyapunov function, and this accounts for the gradient-like nature of the system.

For simplicity, suppose $x(t)$ is a solution of (1) on $(-\infty,\infty)$ all of whose zeros are simple. Let $t_1 < t_2$ be two consecutive zeros, with, say

$$\dot{x}(t_1) > 0 \qquad \dot{x}(t_2) < 0.$$

Assumption (2) and the form of the differential equation then imply

$$x(t_1 - 1) < 0 \qquad x(t_2 - 1) > 0.$$

It follows immediately that $x(t)$ has at least as many zeros in $(t_1 - 1, t_1]$ as in $(t_2 - 1, t_2]$ and the number in each interval is odd. See Figure 2.

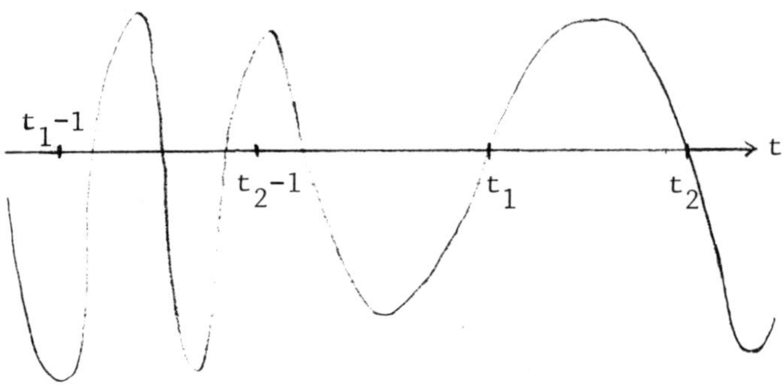

Figure 2

In general we define

$$V : \Psi - \{0\} \to \{1,3,5,\ldots\},$$

a function which is non-increasing along trajectories, as follows. Let $x(t)$ be the solution with initial condition $x_0 = \phi \in \Psi - \{0\}$, and let $t_1 \geq 0$ be the first nonnegative zero of $x(t)$. Define $V(\phi)$ to be the number of zeros (counting multiplicity) of $x(t)$ in $(t_1 - 1, t_1]$. If t_1 does not exist set $V(\phi) = 1$. It is a theorem [22] that V is bounded on $\Psi - \{0\}$, and that

$$V(x_t) \text{ is non-increasing in } t.$$

Now set

$$S_{2n+1} = \{\phi \mid \text{there exists a solution } x(t) \text{ bounded on } (-\infty,\infty), \text{ with } x_0 = \phi, \text{ and } V(x_t) = 2n+1 \text{ for all } t, \text{ such that } x_t \not\to 0 \text{ as } t \to \pm\infty\}.$$

On the set S_{2n+1}, V has the constant value $2n+1$, and trajectories starting there do not tend to zero as $t \to \pm\infty$. Because V is non-increasing, orbits can flow from S_{2n+1} to

S_{2k+1} only if $k \leq n$. (That is, the α- and ω-limit sets of these orbits lie in these Morse sets respectively.) The proof of this is actually rather subtle, due to the presence of multiple zeros of solutions $x(t)$. Semicontinuity properties of the function V are significant here.

We also have $S_{2m} = \{0\}$ for a particular m depending on the linearization of (1). A solution flowing from S_{2n+1} to S_{2m} means one with $V(x_t) \to 2n+1$ as $t \to -\infty$, and $x_t \to 0$ as $t \to \infty$. For this to happen we must have $2n+1 > 2m$. This is proved using detailed knowledge of the semiflow near the origin, in part given by the stable manifold theorem. For solutions flowing from S_{2m} to S_{2n+1} a similar situation holds.

The lowest set S_1 of the Morse decomposition, when $\sigma < s_1$, consists of solutions whose zeros are all simple and are spaced a distance greater than one apart. Such solutions are called slowly oscillating; see [29], [30].

6. A SINGULAR PERTURBATION

Lastly, we consider some aspects of (1) for small σ. Setting $\sigma = 0$ gives formally the difference equation

$$x_{n+1} = f(x_n). \tag{4}$$

The dynamics of (4) have been studied intensively by many authors (see [9], [12], [13]) and a very rich structure involving successive period doubling bifurcations, with an eventual transition to chaotic behavior, has been found when f depends on a parameter α.

The relation between the discrete system (4) and the delay system (1) is still far from clear, although many facts are emerging. It is not hard to show that with a constant initial condition $x \equiv a_0$, for small σ the solution $x(t)$ is near $a_1 = f(a_0)$, $a_2 = f(a_1), \ldots$ on successive unit intervals, with an abrupt transition between a_n and a_{n+1} near $t = n$; see Figure 3. This suggests that points of period m of (4) should be associated with square wave periodic solutions of (1), of period near m, for small σ. Indeed, Glass and Mackey [15] numerically have found a cascade of period-doubling bifurcations for (1) similar to that of (4), for fixed small σ.

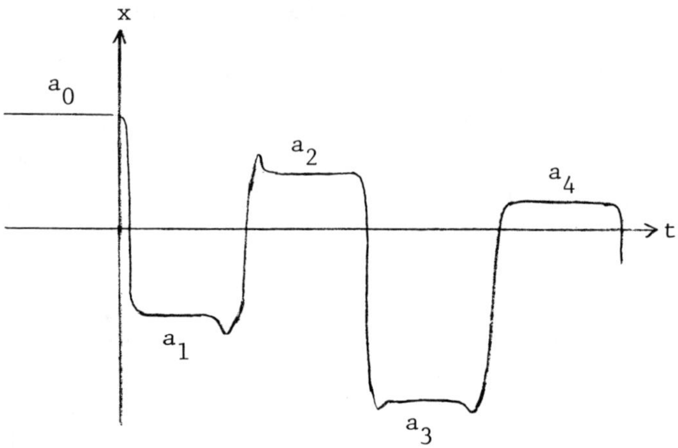

Figure 3

On the other hand, numerical studies by this author indicate the discrete system and delay equation may, for certain step functions f, have different qualitative behavior for arbitrarily small σ. For example, (4) may have a stable point of period two which attracts each point of $R - \{0\}$ upon iteration, while (1) displays no stable solutions with period near two, for any small σ. This difference in behavior seems due to the very different stability criteria for periodic solutions of the discrete and the delay systems.

To better understand the periodic behavior in these systems, a boundary layer equation ((5) below) can be introduced; see Chow and Mallet-Paret [8]. We assume $f(a_0) = a_1$ and $f(a_1) = a_0$ for some $a_0 < a_1$; then solutions of (5), with appropriate limits as $\tau \to \pm\infty$, describe the fine structure of the square wave periodic solutions of (1) which oscillate between a_0 and a_1. In particular, the fact that the solutions of (5) often are not monotone, and take on values outside the interval $[a_0, a_1]$, indicates the presence of a <u>non</u>-uniform convergence of the periodic solutions of (1) to the square wave. This sort of convergence resembles that of the classical Gibbs' phenomenon of Fourier series.

Using techniques of local bifurcation and global continuation (in the parameter α), S.-N. Chow and this author show existence of solutions to the boundary layer equation in some cases. Qualitative properties are also studied, and uniqueness proved in some cases.

One may also consider a fixed nonlinearity f and investigate the relation between the solutions in P_1 and solutions of the boundary layer equation by varying σ. Results in this direction were obtained in collaboration with R. Nussbaum.

Consider a sequence $x(t, \sigma_n)$ of periodic solutions in $P_1(\sigma_n)$ for some sequence $\sigma_n \to 0$. Recall from Theorem 1 that these solutions have minimal period greater than two, and have exactly two zeros, both simple, in each period.

Theorem 4. Assume without loss $x(0, \sigma_n) = 0$ and $\dot{x}(0, \sigma_n) > 0$. Then the next two zeros of $x(t, \sigma_n)$ are located at

$$t_{n1} = 1 + K_n \sigma_n$$

$$t_{n2} = 2 + L_n \sigma_n > 2$$

where K_n and L_n are bounded sequences. Here t_{n2} is the minimal period of $x(t, \sigma_n)$.

Moreover, there exist C_1, C_2, C_3 and C_4 independent of n such that

$$|x(t, \sigma_n)| \leq C_1, \qquad |\dot{x}(t, \sigma_n)| \leq \frac{C_2}{\sigma_n} \quad \text{for all } t;$$

$$|x(t, \sigma_n)| \leq C_3 \Rightarrow |\dot{x}(t, \sigma_n)| \geq \frac{C_4}{\sigma_n}.$$

This result says that $x(t, \sigma_n)$ suffers abrupt jumps in neighborhoods of the zeros t_{n1} and t_{n2} of size $O(\sigma_n)$. See Figure 4. To study the fine structure of these jumps we introduce the scaled time variable τ. As in [8] let

$$y^n(\tau) = x(-\sigma_n \tau, \sigma_n)$$

$$z^n(\tau) = x(\tfrac{1}{2} t_{n2} - \sigma_n \tau, \sigma_n).$$

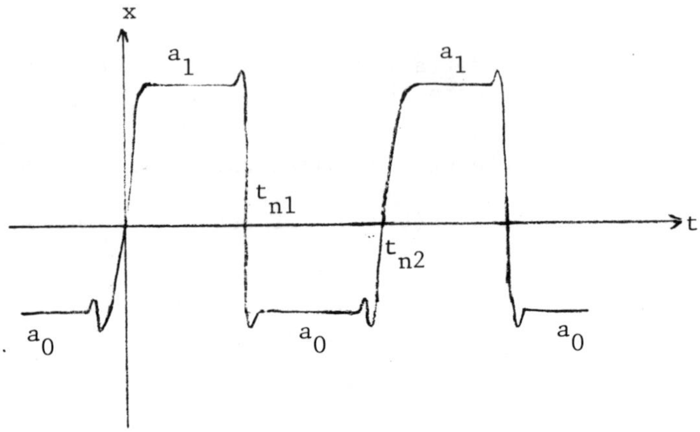

Figure 4

Theorem 4 implies that for a subsequence we have, uniformly on compact intervals

$$y^n(\tau) \to y(\tau), \quad z^n(\tau) \to z(\tau) ,$$

$$|y(\tau)|, \quad |z(\tau)| \le C_4 \quad \text{for all} \quad \tau,$$

where for $r = \lim\limits_{n\to\infty} \frac{1}{2} L_n \ge 0$

$$\begin{cases} y'(\tau) = y(\tau) - f(z(\tau - r)) , \\ z'(\tau) = z(\tau) - f(y(\tau - r)) . \end{cases} \tag{5}$$

Moreoever we have the boundary conditions and monotonicity properties

$$\begin{cases} y(\tau) \ge C_2 \quad \text{and} \quad z(\tau) \le -C_2 \quad \text{as} \quad \tau \to -\infty, \\ y(\tau) \le -C_2 \quad \text{and} \quad z(\tau) \ge C_2 \quad \text{as} \quad \tau \to \infty , \end{cases} \tag{6}$$

$$|y(\tau)| \le C_2 \Rightarrow y'(\tau) < 0,$$

$$|z(\tau)| \le C_2 \Rightarrow z'(\tau) > 0 .$$

A final assumption on the dynamics of the map (4) implies the solution $(y(\tau), z(\tau))$ has appropriate limits at $\pm\infty$; this can be used to show the solution $x(t, \sigma_n)$ is alternately near a_0 and a_1 on intervals of length $1 + O(\sigma_n)$.

Theorem 5. Assume $a_0 < 0 < a_1$ satisfies $f(a_0) = a_1$, $f(a_1) = a_0$, and that this period-two orbit is a global attractor for non-zero iterates of f: namely, $x_n \to \{a_0, a_1\}$ for x_n as in (4) and any $x_0 \neq 0$. Then any solution $(y(\tau), z(\tau))$ of (5) (with $r > 0$) which satisfies (6), in fact satisfies

$$y(\tau) \to a_1 \quad \text{and} \quad z(\tau) \to a_0 \quad \text{as} \quad \tau \to -\infty,$$

$$y(\tau) \to a_0 \quad \text{and} \quad z(\tau) \to a_1 \quad \text{as} \quad \tau \to \infty .$$

ACKNOWLEDGEMENTS

Research partially supported by NSF Grant MCS-8201768.

REFERENCES

[1] J.C. Alexander and J.A. Yorke, "Global bifurcation of periodic orbits", Am. J. Math. 100 (1978), pp. 263-292.

[2] J.C. Alexander and J.A. Yorke, "Families of periodic orbits of differential equations in R^3". To appear J. Diff. Eq.

[3] K.T. Alligood, J. Mallet-Paret and J.A. Yorke, "Families of periodic orbit: local continuability does not imply global continuability", Jour. Diff. Geom. 16 (1981), pp. 483-492.

[4] K.T. Alligood, J. Mallet-Paret and J.A. Yorke, "An index for the global continuation of relatively isolated sets of periodic orbits". Geometric Dynamics, Proc. Symp. on Dynamical Systems, IMPA, 1981.

[5] K.T. Alligood, and J.A. Yorke, "Families of periodic orbits: virtual periods and global continuability." Submitted, J. Diff. Eq.

[6] S.-N. Chow and J. Mallet-Paret, "Integral averaging and bifurcation", J. Diff. Eq. 26 (1977), pp. 112-159.

[7] S.-N. Chow and J. Mallet-Paret, "The Fuller index and global Hopf bifurcation", J. Diff. Eq. 29 (1978), pp. 66-85.

[8] S.-N. Chow and J. Mallet-Paret, "Singularly perturbed delay-differential equations". To appear, Physica D.

[9] P. Collet and J.-P. Eckmann, "Iterated Maps on the Interval as Dynamical Systems". Birkhauser, 1980.

[10] C. Conley, "Isolated invariant sets and the Morse index". N.S.F. CBMS Lecture Notes 38, Amer. Math. Soc., Providence, 1978.

[11] C. Conley and J. Smoller, "Algebraic and topological invariants for reaction-diffusion equations", these Proceedings.

[12] M. Feigenbaum, "Quantitative universality for a class of non-linear transformations", J. Statist. Phys. 19 (1978), pp. 25-52.

[13] M. Feigenbaum, "The universal metric properties of non-linear transformations", J. Statist. Phys. 21 (1979), pp. 669-706.

[14] F.B. Fuller, "An index of fixed point type for periodic orbits", Amer. Jour. Math. 89 (1967), pp. 133-148.

[15] L. Glass and M. Mackey, "Oscillation and chaos in physiological control systems", Science 197 (1977), pp. 287-289.

[16] J.K. Hale, "Theory of Functional Differential Equations", Springer-Verlag, 1977.

[17] F. Hopf, D. Kaplan, H. Gibbs and R. Shoemaker, "Bifurcation to chaos in optical bistability", Phys. Rev. 25 (1982), pp. 2172.

[18] F.C. Hoppensteadt, "Mathematical theories of population: demographics, genetics and epidemics". S.I.A.M. Publ., Philadelphia, 1976.

[19] K. Ikeda, "Multiple-valued stationary state and its instability of the transmitted light by a ring cavity system", Opt. Commun. 30 (1979), pp. 257-261.

[20] K. Ikeda, H. Daido, and O. Akimoto, "Optical turbulence: chaotic behavior of transmitted light from a ring cavity", Phys. Rev. Lett 45 (1980), pp. 709-712.

[21] A. Lasota, "Ergodic problems in biology", Astérisque 50 (1977), pp. 239-250.

[22] J. Mallet-Paret, "Morse decompositions for delay differential equations". In preparation.

[23] J. Mallet-Paret and J.A. Yorke, "Snakes: oriented families of periodic orbits, their sources, sinks, and continuation", Jour. Diff. Eq., 43 (1982), pp. 419-450.

[24] M. Martelli, K. Schmitt and H. Smith, "Periodic solutions of some nonlinear delay-differential equations", Jour. Math. Anal. Appl., 74 (1980), pp. 494-503.

[25] R.D. Nussbaum, "Periodic solutions of nonlinear autonomous functional differential equations". Springer Lecture Notes in Math., 730, pp. 283-325. Springer-Verlag, 1979.

[26] K.P. Rybakowski, "On the homotopy index for infinite-dimensional semiflows", Trans. Amer. Math. Soc. 269 (1982), pp. 351-382.

[27] K.P. Rybakowski and E. Zehnder, "A Morse equation in Conley's index theory for semiflows on metric spaces". Preprint.

[28] J. Smoller, "Shock Waves and Reaction-Diffusion Equations", Springer-Verlag, 1983.

[29] H.-O. Walther, "On instability, ω-limit sets and periodic solutions to nonlinear autonomous differential delay equations". Springer Lecture Notes in Math., 730, pp. 489-503. Springer-Verlag, 1979.

[30] H.-O. Walther, "On density of slowly oscillating solutions of $x'(t) = -f(x(t-1))$", Jour. Math. Anal. Appl. 79 (1981), pp. 127-140.

BIFURCATION AND LINEARIZATION STABILITY IN THE TRACTION PROBLEM

Jerrold E. Marsden

Department of Mathematics, University of California, Berkeley, California 94720, U.S.A.

In this lecture I will provide some insight into the recent work of Chillingworth, Marsden and Wan [1982] on the traction problem in elasto-statics. This work extends earlier results of Signorini and Stoppelli on the same problem. After briefly describing the principal features of the approach, I shall explain how they are related to linearization stability and to other bifurcation problems in mechanics by means of a series of remarks.

Let $\mathcal{B} \subset \mathbb{R}^3$ be the reference configuration and $\phi:\mathcal{B} \longrightarrow \mathbb{R}^3$ a configuration. Let $F = D\phi$ be the associated deformation gradient and $W(F)$ a materially frame indifferent stored energy function; that is, W depends on F only through the point values of the Cauchy-Green tensor $C = F^T F$, where F^T is the transpose of F. Let $\ell = (B,\tau)$ denote a (dead) load, where $B:\mathcal{B} \longrightarrow \mathbb{R}^3$ is the body force and $\tau:\partial\mathcal{B} \longrightarrow \mathbb{R}^3$ is the surface traction. If $P = \dfrac{\partial W}{\partial F}$ denotes the first Piola-Kirchhoff stress, the elasto-static equations are

$$\left.\begin{aligned} &\mathrm{DIV}\, P + B = 0 \quad \text{in} \quad \mathcal{B} \\ &P\cdot N = \tau \qquad\quad\ \ \text{on} \quad \partial\mathcal{B}, \end{aligned}\right. \qquad \text{(E)}$$

where N is the unit outward normal to $\partial\mathcal{B}$ and P is evaluated on the (unknown) configuration ϕ. Letting

$$\langle \ell,\phi \rangle = \int_{\mathcal{B}} B\cdot\phi \, dV + \int_{\partial\mathcal{B}} \tau\cdot\phi \, dA,$$

we recall that under the assumption of sufficient regularity, ϕ

J. M. Ball (ed.), Systems of Nonlinear Partial Differential Equations, 367–372.

is a solution of (E) if and only if ϕ is a critical point of the function

$$V_\ell(\phi) = \int_{\mathcal{B}} W(F)\, dV - \langle \ell, \phi \rangle$$

If (ϕ, ℓ) solves (E) then the divergence theorem shows that the total force is zero: $\int \ell = \int_{\mathcal{B}} B\, dV + \int_{\partial\mathcal{B}} \tau dA = 0.$ By using a suitable rotation we can also assume that ℓ is equilibrated relative to the reference configuration:

$$\int \ell \times X = \int_{\mathcal{B}} B(X) \times X\, dV(X) + \int_{\partial\mathcal{B}} \tau(X) \times X\, dA(X) = 0.$$

Let $SO(3) = \{Q | Q$ is a linear map of $\mathbb{R}^3$ to $\mathbb{R}^3$, $Q^T Q$ = Identity = Id and det $Q = 1\}$ denote the rotation group in $\mathbb{R}^3$. Observe that from material frame indifference,

$$V_{Q\ell}(Q\phi) = V_\ell(\phi) \quad \text{for} \quad Q \in SO(3)$$

We make three assumptions:

(H1) *the reference configuration is stress-free; that is,* $P = 0$ *when* $\phi = \mathrm{Id}$

(H2) *the reference configuration is strongly elliptic*

and in fact, for purposes of stability calculations, assume the stronger condition

(H3) *the classical elasticity tensor*

$$c = 4 \left. \frac{\partial^2 W}{\partial C \partial C} \right|_{C = \mathrm{Id}}$$

is stable:

$$c(e,e) \geqslant \eta \|e\|^2$$

for some $\eta > 0$ *and all symmetric matrices* e.

For $\ell = 0$, any $\phi = Q \in SO(3)$ is a solution of (E). These are the trivial solutions. Let ℓ_0 be given and look at loads $\lambda\ell$ for ℓ near ℓ_0 and λ small. The kernel of the equations of linear elasticity, i.e. equations (E) linearized at $\phi = Q$ and with $\lambda = 0$ is $T_Q SO(3)$ the tangent space to $SO(3)$ at $Q \in SO(3)$. In fact, it follows from (H3) that $SO(3)$ is a non-degenerate critical manifold for V_0 and so every point of $SO(3)$ is a possible bifurcation point.

By using a variant of the Liapunov-Schmidt procedure one shows that there is a function $f_{\lambda\ell}: SO(3) \longrightarrow \mathbb{R}$ whose critical points are in one to one correspondence with solutions of (E) near SO(3). By using Liusternick-Schnirelmann category one concludes that (E) has at least four solutions, one of which is a (local) minimum of the energy.

Some computation shows that $f_{\lambda\ell}$ has the form

$$f_{\lambda\ell}(Q) = -\lambda \langle Q\ell, Id \rangle + \frac{\lambda^2}{2} \langle Q\ell_0, u_{Q\ell_0} \rangle + \text{h.o.t.}$$

(h.o.t. means 'higher order terms'), where $u_{Q\ell_0}$ satisfies the equations of linear elasticity with loads $Q\ell_0$ (projected to the equilibrated loads). The term $\langle Q\ell_0, Id \rangle$ on SO(3) will have degeneracies if ℓ_0 has some 'symmetries'. Let S_{ℓ_0} denote the set of critical points of the function $Q \mapsto \langle Q\ell_0, Id \rangle$. We classify the loads according to the topology of S_{ℓ_0} as shown in the accompanying table. The critical points of $f_{\lambda\ell}$ can be studied by a further Liapunov-Schmidt reduction to S_{ℓ_0}; after removing a factor of λ, one obtains a function $\tilde{f}: S_{\ell_0} \longrightarrow \mathbb{R}$,

$$\tilde{f}(Q) = -\langle Q\ell, Id \rangle - \frac{\lambda}{2} \langle Q\ell_0, u_{Q\ell_0} \rangle + \text{h.o.t.}$$

whose critical points are in one to one correspondence with solutions of (E). A study of f leads to the results on the number of solutions shown in the accompanying table.

Type of ℓ_0	S_{ℓ_0}	Number of Solutions = n
0	4 points	4
1	2 points $\cup\ \mathbb{RP}^1$	$4 \leqslant n \leqslant 6$
2	1 point $\cup\ \mathbb{RP}^2$	$4 \leqslant n \leqslant 14$
3	$\mathbb{RP}^1 \cup \mathbb{RP}^1$	$4 \leqslant n \leqslant 8$
4	$SO(3) \cong \mathbb{R}\mathbb{R}^3$	$4 \leqslant n \leqslant 40$

The number of solutions is sharp (examples show that the maximum can be achieved) and assumes that the Betti form, defined by $\beta(Q) = \langle Q\ell_0, u_{Q\ell_0} \rangle$ has non-degenerate critical points.

The numbers in this table come from Bezout's theorem applied to an associated system of cubic polynomials on the double covering: the number of solutions branching from $\mathbb{RP}^s$ can be as many as $(3^{s+1}-1)/2$.

For loads ℓ near type 1 loads ℓ_0 and λ small, the solutions are arranged as follows. There are two unique solutions near the two points of S_{ℓ_0} and two, three or four near the set $\mathbb{RP}^1 \approx S^1$. These solutions on $\mathbb{RP}^1$ can bifurcate according to the astroid bifurcation, which is a symmetric marriage of four cusps. The stabilities can be explicitly calculated.

We now make some remarks to put the above results in perspective.

1. The role of the group SO(3) in this problem differs from that considered by several other authors (see the lectures of Schaeffer and Dancer in these proceedings) in the following two ways:

(a) the group SO(3) acts freely on the set of trivial solutions

and (b) the group SO(3) acts on the parameter space (the space of loads) as well as on the configuration space (the space of ϕ's).

Despite the differences there is the common feature that bifurcation points can be described and classified by their degree of symmetry, which is closely linked with their degree of degeneracy.

2. The techniques can be adapted to more degenerate situations. For example, for the problem of the Rivlin cube discussed in Schaeffer's lectures, one can show that for small normal loads and any isotropic material, there are homogeneous solutions in one to one correspondence with $\mathbb{RP}^2$.

3. As its name implies, linearization stability refers to the stability of solvability of nonlinear equations under the process of linearization. In elasticity, if (u_1, ℓ_1) is a solution of the linearized problem (with ℓ_1 equilibrated) one asks if there is a curve

$$\phi(\lambda) = \mathrm{Id} + \lambda u_1 + \lambda^2 u_2 + \text{h.o.t.},$$

$$\ell(\lambda) = \lambda \ell_1 + \lambda^2 \ell_2 + \text{h.o.t.}$$

satisfying the nonlinear equations. There is a well-known

obstruction to this at order λ^2, namely the Signorini compatibility conditions:

$$\int \ell_1 \times u_1 = 0 .$$

Even if the Betti form is degenerate, this can be shown to be the *only* obstruction. To obtain an accompanying perturbation scheme one needs a non-trivial refinement of the classical schemes proposed by Signorini and Truesdell and Noll.

4. There is an interesting parallel between the results quoted in the preceding remark and some recent results in classical relativistic field theory. The space of solutions of the Yang-Mills or Einstein Yang-Mills equations on a spatially compact spacetime has a quadratic singularity at each point with symmetry; that is at each solution with an isotropy group of dimension $\geqslant 1$ relative to the action of the appropriate gauge group. As one continually breaks symmetry by descending through the lattice of subgroups, one arrives at the generic solutions with at most discrete symmetries, at which point the solution set is a smooth manifold. When one attempts to write solutions in a perturbation series, there is a constraint at second order called the Taub conditions which is analogous to the Signorini conditions. Again it is somewhat remarkable that this is the only obstruction. However, despite this close analogy, the technical details apparently have little to do with one another. For details, references and related bifurcation properties of momentum maps (Noether quantities) in mechanics, see Arms, Marsden and Moncrief [1981].

5. Connections between symmetry and bifurcation are also prevalent in dynamics as well as in statics. We mention one example for illustration. The Lagrange top, i.e. the heavy top with two equal moments of inertial is an integrable system because of its symmetry. If this symmetry is broken by slightly altering the moments of inertia, there is a bifurcation from a homoclinic orbit to aperiodic solutions by the introduction of Smale horseshoes. See Holmes and Marsden [1982] for details.

All of these examples suggest that there must be a deep unified connection between symmetry and bifurcation which surfaces in different ways for different types of problems.

ACKNOWLEDGEMENTS

Research partially supported by NSF grant MCS-81-07086, the Miller Institute and DOE contract DE-AT03-82ER12097.

REFERENCES

[1] Arms, J.M., Marsden, J.E., and Moncrief, V.: 1981, *Symmetry and bifurcation of momentum mappings*, Comm. Math. Phys. 78, pp.455-478.

[2] Holmes, P.J., and Marsden, J.E.: 1982, *Horseshoes and Arnold diffusion for Hamiltonian systems on Lie groups*, Indiana Univ. Math. J. (to appear).

[3] Chillingworth, D.R.J., Marsden, J.E., and Wan, Y.H.: 1982, *Symmetry and bifurcation in three dimensional elasticity I*, Arch. Rat. Mech. An. (to appear), *II* Arch. Rat. Mech. An. (to appear), *III* (preprint).

[4] Marsden, J.E., and Wan, Y.H.: 1982, *Linearization stability and Signorini series for the traction problem in elasto-statics* (preprint).

SINGULAR ELLIPTIC EIGENVALUE PROBLEMS FOR EQUATIONS AND SYSTEMS

J.F. Toland

School of Mathematics
University of Bath

ABSTRACT This lecture reviews work by the author and others on singular elliptic eigenvalue problems. Particular emphasis is placed on equations and systems which yield solutions of solitary wave type, because in these cases it is often possible to obtain *a priori* bounds on the L_p norm of solutions in terms of the L_∞ norm. These bounds play a central role in an existence theory for such solutions, and though this theory has been worked out, it will not be considered further here. While the lecture concentrates on the theory of solutions of a special type, the solutions studied here are often the solutions which are sought in practice.

THE ESTIMATES

This lecture is concerned with a method for obtaining *a priori* bounds on the L_p norm of certain solutions u of non-linear ordinary and partial differential equations and systems, in terms of the L_∞ norm of u, when the domain of u is unbounded. The equations in question arise in various physical applications, and the solutions under consideration might be generally described as solitary waves, though the method extends to a wider class. These estimates play a central role in our approach to the existence question for these equations, and in studies of the corresponding bifurcation phenomena. In passing, we remark that because of the singular nature of the problems, standard bifurcation theory does not apply. We begin with an elementary example.

J. M. Ball (ed.), Systems of Nonlinear Partial Differential Equations, 373–384.

EXAMPLE (I) Travelling Wave solution of Boussinesq's equations.

Boussinesq's equations for water-waves are

$$\xi_t + u_x + (u\xi)_x = 0, \tag{1a}$$

$$u_t + \xi_x + uu_x - \frac{1}{3}u_{xxt} = 0, \tag{1b}$$

(see BENJAMIN(1974), Eqn.(2.14)). Here ξ denotes the amplitude and u the velocity of the wave. If we seek a travelling wave solution of (1a,b), $\xi = \xi(x-ct)$, $u = u(x-ct)$, then we find a system of ordinary differential equations to be satisfied by (ξ,u):

$$-c\xi' + u' + (u\xi)' = 0, \tag{2a}$$

$$-cu' + \xi' + uu' + \frac{c}{3}u''' = 0. \tag{2b}$$

Now if the travelling wave solution is supposed to be a solitary wave, then $\xi(\pm\infty) = u(\pm\infty) = 0$, and (2a) yields

$$\xi = u/(c-u) \tag{3}$$

and so (2b) gives an equation for u alone:

$$c^2(u - \frac{1}{3}u'') = u + \left(\frac{1}{c-u} + \frac{c}{2}\right)u^2, \quad x \in \mathbb{R}, \tag{4}$$

$$u(\pm\infty) = 0;$$

in addition we require

$$u'(x) < 0, \quad x > 0, \tag{5}$$

$$u(x) = u(-x), \quad x \in \mathbb{R} \tag{6}$$

REMARK If we change the sign of u and c, then we get another solution of (4) corresponding to the same ξ. This corresponds to the same wave propagating in the opposite direction.

A simple comparison argument shows that for solutions to exist

$$u < c$$

and

$$c^2 > 1.$$

For fixed $c > 1$, the phase portrait for the solitary wave is given by

$$c(u^2 - \frac{1}{3}u'^2) + 2u + 2c\ln(1-\frac{u}{c}) - \frac{u^3}{3} = 0. \tag{7}$$

Since $u'(0) = 0$, we find that c and $u(0)$ satisfy the equation

$$cu(0)^2 + 2u(0) + 2c\ln(1-\frac{u(0)}{c}) - \frac{u^3(0)}{3} = 0. \tag{8}$$

Hence when $u(0)$ is small and c is close to 1 (in other words close to bifurcation) we find that

$$(c-1/c) = \frac{1}{3}(u(0) + \frac{2u(0)}{3c^2}) + o(u(0)). \tag{9}$$

Now we can estimate the L_p norm of u, using (7) as follows:

$$\int_{\mathbb{R}} u(x)^p dx = 2\int_0^\infty u(x)^p dx$$

$$= 2\left(\frac{c}{3}\right)^{\frac{1}{2}} \int_0^{u(0)} \frac{u^p du}{\left\{cu^2 + 2u + 2c\ln(1-\frac{u}{c}) - \frac{u^3}{3}\right\}^{\frac{1}{2}}}$$

$$= 2\left(\frac{c}{3}\right)^{\frac{1}{2}} \int_0^{u(0)} \frac{u^{p-1} du}{\left\{c - \frac{u}{3} + \frac{2c}{u^2}\left(1u(1-\frac{u}{c}) + \frac{u}{c}\right)\right\}^{\frac{1}{2}}}$$

$$\leq 2\left(\frac{c}{3}\right)^{\frac{1}{2}} \int_0^{u(0)} \frac{u^{p-1} du}{\sqrt{Q(u(0)-u)}}$$

where

$$Q = (c-1/c)/u(0)$$

since $R(u) = c - \frac{u}{3} + \frac{2c}{u^2}(1u(1-\frac{u}{c}) + \frac{u}{c})$ is concave as a function of u (all the coefficients in its Taylor expansion in powers of u are negative).

Therefore

$$\int_{\mathbb{R}} u(x)^p dx \le \frac{2}{\sqrt{3}} \frac{cu(0)^p}{\sqrt{c^2-1}} \left(\int_0^1 \frac{v^{p-1}}{\sqrt{1-v}} dv \right) .$$

Therefore by (9),

$$\int_{\mathbb{R}} u(x)^p dx \to 0 \quad \text{as } u(0) \to 0$$

provided $p > \frac{1}{2}$.

Using (8) we find that

$$R'(u(0)) = -1 + \frac{2}{u(0)} \left(c - \frac{1}{c-u(0)} \right)$$

and

$$R'(u(0)) \to -1 \text{ as } u(0) \to 0, \text{ by } (9).$$

Hence since R is concave, we find that

$$\int_{\mathbb{R}} u(x)^p dx \ge 2(\frac{c}{3})^{\frac{1}{2}} \int_0^{u(0)} \frac{u^{p-1} du}{\sqrt{R'(u(0))[u_o - u]}}$$

$$= 2\left(\frac{c}{3R'(0)}\right)^{\frac{1}{2}} \left(\int_0^1 \frac{v^{p-1} dv}{\sqrt{1-v}} \right) u(0)^{p-\frac{1}{2}}$$

$$\to \infty \text{ as } u(0) \to 0 \text{ if } p < \tfrac{1}{2}.$$

REMARKS This simple calculation exemplifies the method. We note that the pth power integral of u does not depend continuously on u(0) when $p < \frac{1}{2}$, and it does if $p > \frac{1}{2}$.

Though we started with a system (2), the travelling wave problem for Boussinesq's equations reduces to an autonomous ordinary differential equation, where an analysis of the phase plane yields the necessary information. To proceed to more complicated examples, it is necessary to see how to treat non-autonomous ordinary differential equations.

EXAMPLE (II) Non-autonomous o.d.e.'s

We consider the problem

$$-u''(x) = \lambda u(x) + r(x)u(x)^{1+\sigma}, \tag{10}$$

$$u(x) \to 0 \quad \text{as} \quad x \to \infty, \tag{11}$$

$$u(-x) = u(x) > 0 \quad \text{on } \mathbb{R}, \tag{12}$$

$$u'(x) < 0 \text{ on } (0,\infty). \tag{13}$$

We will assume that r is non-increasing on $[0,\infty)$, and is even on $\mathbb{R}$. For such a u to exist λ must be less than 0. But because r is not necessarily constant, phase plane analysis is no longer available. An integration over (x,∞), after multiplying by u' yields the inequality

$$u'(x)^2 + \lambda u(x)^2 + 2r(x)u(x)^{2+\sigma}/(2+\sigma) \geq 0, \quad x \in (0,\infty). \tag{14}$$

Let $\underline{\overline{X}} \in [0,\infty)$ be the unique point where

$$\lambda + 2r(\underline{\overline{X}})\, u(\underline{\overline{X}})^{\sigma}/(2+\sigma) = 0. \tag{15}$$

Such a point exists because both u and r are non-increasing, and in particular (13) holds.

Hence on $(\underline{\overline{X}},\infty)$, the inequality (14) yields

$$\begin{aligned} u'(x)^2 &\geq \{|\lambda| + 2r(x)u(x)^{\sigma}/(2+\sigma)\}u(x)^2 \\ &= 2\{r(\underline{\overline{X}})u(\underline{\overline{X}})^{\sigma} - r(x)u(x)^{\sigma}\}u(x)^2/(2+\sigma) \\ &\geq 2\{u(\underline{\overline{X}})^{\sigma} - u(x)^{\sigma}\}r(\underline{\overline{X}})u(x)^2/(2+\sigma) > 0. \end{aligned}$$

Hence

$$\begin{aligned} \int_{\underline{\overline{X}}}^{\infty} u(x)^p dx &\leq \left(\frac{2+\sigma}{2r(\underline{\overline{X}})}\right)^{\frac{1}{2}} \int_0^{u(\underline{\overline{X}})} \frac{u^{p-1}dp}{\{u(\underline{\overline{X}})^{\sigma} - u^{\sigma}\}^{\frac{1}{2}}} \\ &= \left(\frac{2+\sigma}{2r(\underline{\overline{X}})}\right)^{\frac{1}{2}} \left(\int_0^1 \frac{v^{p-1}dv}{\{1-v^{\sigma}\}^{\frac{1}{2}}}\right) u(\underline{\overline{X}})^{(p-\sigma/2)} \qquad (16) \\ &= (\text{const.})\ u(\underline{\overline{X}})^p/\sqrt{|\lambda|}\ . \\ &\leq (\text{const.})\ u(0)^p/\sqrt{|\lambda|}\ . \qquad (17) \end{aligned}$$

On $(0,\underline{\overline{X}})$,

$$-u''(x) = \lambda u(x) + r(x)u(x)^{1+\sigma}$$

$$\geq \frac{-2r(x)u(x)^{1+\sigma}}{\sigma+2} + r(x)u(x)^{1+\sigma}$$

$$= \left(\frac{\sigma}{2+\sigma}\right) r(x)u(x)^{1+\sigma} \quad .$$

An integration over $(0,x)$ after multiplying by u' now yields

$$u'(x)^2 \geq \frac{2\sigma r(x)}{(\sigma+2)^2}\{u(0)^{2+\sigma} - u(x)^{2+\sigma}\}$$

$$\geq \frac{2\sigma r(\underline{\overline{X}})}{(\sigma+2)^2}\{u(0)^{2+\sigma} - u(x)^{2+\sigma}\} > 0, \quad x\in(0,\underline{\overline{X}}) \quad .$$

Hence

$$\int_0^{\overline{X}} u(x)^p dx \leq \frac{(\text{const.})}{\sqrt{r(\underline{\overline{X}})}} \int_{u(\underline{\overline{X}})}^{u(0)} \frac{u^p du}{\{u(0)^{2+\sigma} - u^{2+\sigma}\}^{\frac{1}{2}}} \tag{18}$$

$$= \frac{(\text{const.})u(\underline{\overline{X}})^{\sigma/2}}{\sqrt{|\lambda|}} \left(\int_{u(\underline{\overline{X}})/u(0)}^{1} \frac{v^p dv}{\{1-v^{2+\sigma}\}^{\frac{1}{2}}}\right) u(0)^{p-\sigma/2} \tag{19}$$

$$\leq (\text{const.})u(0)^p/\sqrt{|\lambda|} \quad .$$

Hence

$$\int_{\mathbb{R}} u(x)^p dx \leq (\text{const.})u(0)^p/\sqrt{|\lambda|} \ .$$

On the other hand, (10) gives

$$-u'' \geq \lambda u(x),$$

whence

$$u'(x)^2 \leq |\lambda| u(x)^2 \ , \quad x\in\mathbb{R} \ ,$$

Therefore

$$\int_0^{\infty} u(x)^p dx \geq \int_0^{u(0)} \frac{u^{p-1} du}{\sqrt{|\lambda|}}$$

$$= (\text{const.}) u(0)^p / \sqrt{|\lambda|} \;.$$

We have established in an extremely elementary way the following inequality

$$(\text{const.}) u(0)^p / \sqrt{|\lambda|} \leq \int_{\mathbb{R}} u(x)^p dx \leq (\text{const.}) u(0)^p / \sqrt{|\lambda|} \;. \tag{20}$$

It is remarkable that the constants involved here are independent not only of u and λ but also of r. Returning to (14) we find that

$$\lambda + 2r(0) u(0)^{\sigma} / 2+\sigma \geq 0,$$

whence

$$\frac{|\lambda|}{u(0)^{\sigma}} \leq (\text{const.}) \;,$$

and so by (20)

$$\int_{\mathbb{R}} u(x)^p dx \geq (\text{const.}) u(0)^{(p-\sigma/2)} \;.$$

Hence the pth power integral of u is not a continuous functional of u(0) if $p < \sigma/2$. To study the case $p > \sigma/2$ in the space available, we must restrict attention to the case when $r \geq \alpha > 0$ on $\mathbb{R}$. For the general case, see TOLAND(1982). In this case it is easy to see from (16) and (19) that

$$\int_0^{\infty} u(x)^p dx \leq (\text{const.})(u(\underline{\overline{X}})^{p-\sigma/2} + u(0)^{p-\sigma/2})$$

$$\leq (\text{const.})\; u(0)^{p-\sigma/2} \;, \quad p \geq \sigma/2 \;.$$

Hence

$$\int_{\mathbb{R}} u(x)^p dx \le (\text{const.})u(0)^{p-\sigma/2}$$

$$\to 0 \text{ as } u(0) \to \infty \text{ if } p > \sigma/2.$$

This ties in with our observation in the discussion of the Boussinesq equation. There the nonlinearity is essentially quadratic, and $p = \frac{1}{2}$ is the critical value: in general when the nonlinearity is a $(1+\sigma)$-power, then $p = \sigma/2$ is critical.

REMARK Example (II) is the equation for solitary wave solutions of the Korteweg-de Vries equation with $r = \frac{1}{2}$, $\sigma = 1$.

The above two examples concern ordinary differential equations or problems which reduce to ordinary differential equations. The next three examples cannot be reduced in such a simple way.

EXAMPLE (III) A system of o.d.e.s.

The system of equations of BONA and SMITH(1976) for the two way propagation of water-waves resembles that of example (I), but is superior to it from the point of view of the simplicity and well-posedness of the initial value problem.

It is

$$\eta_t + u_x + (u\eta)_x - \frac{1}{3}\eta_{xxt} = 0$$

$$u_t + \eta_x + uu_x - \frac{1}{3}(u_t+\eta_x)_{xx} = 0.$$

The travelling solitary wave problem reduces to a system of ordinary differential equations:

$$c(\eta-\frac{1}{3}\eta'') = u + u\eta , \tag{21}$$

$$c(u-\frac{1}{3}u'') = \eta-\frac{1}{3}\eta'' + u^2/2 , \tag{22}$$

$$u(x) = u(-x) > 0, > \eta(x) = \eta(-x) > 0, \quad x\in\mathbb{R} , \tag{23}$$

$$u(x) \to 0, \quad \eta(x) \to 0 \text{ as } x \to \infty , \tag{24}$$

$$u'(x) < 0, \quad \eta'(x) < 0, \quad x > 0.$$

In order to implement the procedures of examples (I) and (II), it is first necessary to obtain *a priori* bounds on the functions u and η. It is easy to see that $c > 1$ if there are to be solutions of this system. Clearly (22) and the Maximum Principle give

$$cu(x) > \eta(x), \quad x \in \mathbb{R} .$$

Let $\alpha = u(0)/(2+cu(0))$. Then (21) and (22) give

$$(1-\tfrac{1}{3}\partial_x^2)(\alpha c\eta + \eta - cu) = u(\alpha(1+\eta)-\tfrac{1}{2}u)$$

$$= u\left\{\frac{2u(0) - 2u(x) + u(0)(2\eta(x) - cu(x))}{2(2 + cu(0))}\right\}$$

$$\geq u\left\{\frac{2u(0) - 2u(x) + u(0)[\alpha c\eta(x) + \eta(x) - cu(x)]}{2(2 + cu(0))}\right\}$$

since $\alpha c \leq 1$. If $\alpha c\eta + \eta - cu = w$, then

$$\frac{1}{3}w''(x) - \left(1-\frac{u(0)u(x)}{2(2 + cu(0))}\right) w(x) \leq 0.$$

If $u(0) \leq 2$, then the coefficient of w is negative. The maximum principle then says that w has no non-positive local minimum in $\mathbb{R}$. Since $w(x) \to 0$ as $|x| \to \infty$ we conclude that

$$w(x) = \left(1 + \left(\frac{cu(0)}{2+cu(0)}\right)\right)\eta(x) - cu(x) > 0 \text{ on } \mathbb{R}$$

provided $u(0) < 2$. (A more general result than this is ture, see TOLAND(1981), THM.3.5). Here we have proved that if $u(0) \leq 2$, then

$$\eta(x) < cu(x) < \left(1 + \left(\frac{cu(0)}{2+cu(0)}\right)\right)\eta(x), \quad x \in \mathbb{R}.$$

A substitution of these inequalities into the equations (21), (22), and a manipulation along the lines of example (II) gives the conclusion that

$$\int_{\mathbb{R}} u(x)^p dx \begin{cases} \to 0 \text{ as } u(0) \to 0, & p > \tfrac{1}{2} \\ \to \infty \text{ as } u(0) \to 0, & p < \tfrac{1}{2} \end{cases}$$

and $c \to 1$ if and only if $u(0) \to 0$.

REMARK A theory of solitary wave solutions of the BONA-SMITH equations is given in TOLAND(1981). The potential for generalisation is clear.

EXAMPLE (IV) Semilinear problems on a two-dimensional strip.

Consider the problem

$$-\Delta u(x,y) = \lambda u(x,y) + r(x,y)u(x,y)^{1+\sigma}, \quad (x,y)\in S \tag{25}$$

$$u(x,y) = 0, \quad (x,y)\in\partial S \tag{26}$$

$$u(x,y) = u(x,-y) = u(-x,y) > 0, \quad (x,y)\in S \tag{27}$$

$$u(x,y) \to 0 \text{ as } |x| \to \infty, \tag{28}$$

where

$$S = \mathbb{R}\times(-\pi/2,\pi/2).$$

The argument for obtaining *a priori* bounds for solutions of this problem with

$$u_x < 0, \quad x > 0; \quad u_y < 0, \; y > 0, \; (x,y)\in S, \tag{29}$$

is presented in detail in AMICK and TOLAND(1983), and depends on two observations. The first is simply that the first Fourier coefficient of u in the y-direction

$$U(x) = \frac{2}{\pi}\int_{-\pi/2}^{\pi/2} u(x,y)\cos y\, dy$$

satisfies the integro-differential identity:

$$-U''(x) = (\lambda-1)U(x) + \int_{-\pi/2}^{\pi/2} u(x,y)^{1+\sigma} dy, \tag{30}$$

and that

$$U(0) \to 0 \quad \text{as} \quad \lambda \to 1.$$

The second is that

$$|U(x)\cos y - u(x,y)|/U(x) \to 0 \text{ as } \lambda \to 1, \tag{31}$$

uniformly for $(x,y)\in S$. Space does not permit a presentation of the argument for (31) (AMICK and TOLAND(1983)). However, using it in (30) we find that U satisfies a differential inequality

$$(1-\varepsilon)U(x)^{1+\sigma} \le -U''(x)-(\lambda-1)U(x) \le (1+\varepsilon)U(x)^{1+\sigma} ,$$

where $\varepsilon \to 0$, as $\lambda \to 1$ or equivalently as $U(0) \to 0$. By a sequence of arguments precisely analogous to those in Example (II) we conclude that

$$\int_{\mathbb{R}} U(x)^p dx \begin{cases} \to 0 \text{ as } & U(0) \to 0, \quad p > \sigma/2 \\ \to \infty \text{ as } & U(0) \to 0, \quad p < \sigma/2 \end{cases} .$$

However, this statement about U, the first Fourier coefficient in the y-direction of u, has implications for u, because of (31), namely

$$\int_S u(x,y)^p dxdy \begin{cases} \to 0 \text{ as } \lambda \to 1, & p > \sigma/2 \\ \to \infty \text{ as } \lambda \to 1, & p < \sigma/2 \end{cases} .$$

EXAMPLE (V) Radially symmetric problems on $\mathbb{R}^N$.

Because of recent work of GIDAS, NI and NIRENBERG(1979) there has recently been a great deal of interest in radially symmetric solutions of semi-linear problems in $\mathbb{R}^N$. Here we will restrict attention to the problem

$$-\Delta u(x) = \lambda u(x) + ru(x)^{1+\sigma} \quad \text{on } \mathbb{R}^N , \tag{32}$$

$$u(x) \to 0 \text{ as } |x| \to \infty, \tag{33}$$

$$u > 0 \text{ and radially symmetric.} \tag{34}$$

Supposing that r is radially symmetric and radially non-increasing, and that u is a classical solution of (32)-(34) we can re-write (32) as an ordinary differential equation

$$-(x^{N-1}u'(x))' = x^{N-1}(\lambda u(x)+r(x)u(x)^{1+\sigma}).$$

Now manipulations in the spirit of example (II) yield the result that if $r(x) \ge \alpha > 0$, then

$$\int_{\mathbb{R}^N} u^p\, dx \to 0 \text{ as } u(0) \to 0, \quad p > \frac{\sigma N}{2} .$$

These and similar results are presented in TOLAND(1983).

REFERENCES

AMICK, C.J. and TOLAND, J.F. (1983), Nonlinear elliptic eigenvalue problems on an infinite strip - global theory of bifurcation and asymptotic bifurcation. (To appear in Math. Annalen.)

BENJAMIN, T.B. (1974), Lectures on nonlinear waves. In Lectures in Applied Mathematics, Volume 15., A.M.S. Providence, R.I.

GIDAS, B., NI, W-M., and NIRENBERG, L. (1979), Symmetry and related properties via the maximum principle. Comm. Math. Phys. 68, pp.209-243.

TOLAND, J.F. (1981), Solitary wave solutions for a model of the two-way propagation of water-waves in a channel. Math. Proc. Camb. Phil. Soc. 90, pp.343-360.

TOLAND, J.F. (1982), Global bifurcation for Neumann problems without eigenvalues. J. Diff. Eqns. 44, pp. 82-110.

TOLAND, J.F. (1983), Positive solutions of nonlinear elliptic equations - existence and non-existence of solutions with radial symmetry in $L_p(\mathbb{R}^N)$. (Submitted for publication.)

REGULARIZATION OF NON ELLIPTIC VARIATIONAL PROBLEMS

B. Dacorogna

Département de mathématiques
École Polytechnique Fédérale de Lausanne
1015 Lausanne, Switzerland

INTRODUCTION

In this article we discuss the existence of solutions for

(P) $\inf\{\int_\Omega f(x,u(x),\nabla u(x))\,dx : u \in u_0 + W_0^{1,\alpha}(\Omega;\mathbb{R}^m)\}$

where

(i) Ω is a bounded open set of $\mathbb{R}^n$; $u: \Omega \subset \mathbb{R}^n \to \mathbb{R}^m$ and thus $\nabla u \in \mathbb{R}^{nm}$.

(ii) $f: \Omega \times \mathbb{R}^m \times \mathbb{R}^{nm} \to \mathbb{R}$ is continuous and satisfies for every $(x,u,F) \in \Omega \times \mathbb{R}^m \times \mathbb{R}^{nm}$

$$a_1(x) + c_1|F|^\alpha \leqslant f(x,u,F) \leqslant a_2(x) + b|u|^\alpha + c_2|F|^\alpha \tag{0.1}$$

where $a_1, a_2 \in L^1(\Omega)$, $b \geqslant 0$, $c_2 \geqslant c_1 > 0$ and $\alpha > 1$.

(iii) $W^{1,\alpha}(\Omega;\mathbb{R}^m) = \{u = (u_1,\ldots,u_m) : u_j \in L^\alpha(\Omega)$ and $\text{grad}\, u_j \in L_n^\alpha(\Omega)$, $j = 1,\ldots,m\}$. By $u \in u_0 + W_0^{1,\alpha}(\Omega;\mathbb{R}^m)$ we just mean that $u, u_0 \in W^{1,\alpha}(\Omega;\mathbb{R}^m)$ and $u = u_0$ on $\partial\Omega$ in the sense of traces (see Adams [Ad1], for details).

In order to show that the problem (P) has a solution one usually uses the so called "direct methods of the calculus of variations", i.e., if

$$I(u,\Omega) = \int_\Omega f(x,u(x),\nabla u(x))\,dx \tag{0.2}$$

one tries to show that I is (sequentially) weakly lower semicontinuous i.e.,

J. M. Ball (ed.), Systems of Nonlinear Partial Differential Equations, 385–400.

$$I(u,\Omega) \leqslant \liminf_{u_\nu \xrightarrow{W^{1,\alpha}} u} I(u_\nu,\Omega). \tag{0.3}$$

(By $\rightharpoonup$, we denote weak convergence.)

Roughly speaking, the property (0.1) ensures (with $\alpha > 1$) that minimizing sequences are bounded in a reflexive space, and thus are (up to a subsequence) weakly convergent; property (0.3) then implies that the weak limit is the actual minimum of I.

A functional I having the property (0.3) is said to be regular. Morrey ([Mo1],[Mo2]) has isolated a necessary and sufficient condition on f, called quasiconvexity, in order that (0.3) holds. A function f is said to be quasiconvex if

$$\int_D f(x_0,u_0,F_0 + \nabla\phi(x))dx \geqslant f(x_0,u_0,F_0)\,\text{meas}\,D, \tag{0.4}$$

for every bounded domain $D \subset \mathbb{R}^n$, for every $(x_0,u_0,F_0) \in \Omega \times \mathbb{R}^m \times \mathbb{R}^{nm}$ and for every $\phi \in W_0^{1,\infty}(D;\mathbb{R}^m)$.

For more details about the exact conditions imposed on f in order that quasiconvexity and weak lower semicontinuity be equivalent, as well as for precise references concerning the next remarks see [Mo1],[Mo2],[Me1],[AF1] or [Da4].

Remarks:

(i) If $n = 1$ or $m = 1$, then the convexity of f (with respect to the last argument) is equivalent to the quasiconvexity of f, while if $n,m > 1$, then one only has that: convexity $\Rightarrow$ quasiconvexity.

(ii) If $m = n$ and there exists $g\colon \mathbb{R} \to \mathbb{R}$ continuous so that $f(\nabla u) = g(\det\nabla u)$, then the quasiconvexity of f is equivalent to the convexity of g.

(iii) If f is quasiconvex then f is rank one convex, i.e.,

$$f(\lambda F + (1-\lambda)G) \leqslant \lambda f(F) + (1-\lambda)f(G), \tag{0.5}$$

for every $\lambda \in [0,1]$, $F,G \in \mathbb{R}^{nm}$ with rank $(F-G) \leqslant 1$. If $f \in C^2(\mathbb{R}^{nm})$ then (0.5) is equivalent to the Legendre-Hadamard (or ellipticity) condition

$$\sum_{i,j,\alpha,\beta} \frac{\partial^2 f(F)}{\partial F_{i\alpha}\partial F_{j\beta}} \lambda_i\lambda_j\mu_\alpha\mu_\beta \geqslant 0 \quad \text{for all} \quad \lambda \in \mathbb{R}^n,\ \mu \in \mathbb{R}^m.$$

Without the quasiconvexity hypothesis on f, then, in general (P) does not have a solution in $W^{1,\alpha}$ and one is thus led to look for "generalized solutions" of (P). One possible approach for definitions of such solutions is the so called relaxation theorem that we describe now. Let Qf be the lower quasiconvex envelope of f, i.e.,

$$Qf(x,u,\cdot) = \sup\{\Phi(x,u,\cdot) \leqslant f(x,u,\cdot) \text{ and } \Phi(x,u,\cdot) \text{ quasi-convex}\}. \tag{0.6}$$

Let

$$\bar{I}(u,\Omega) = \int_{\Omega} Qf(x,u(x),\nabla u(x))\,dx \tag{0.7}$$

and define

$$(QP) \quad \inf\{\bar{I}(u,\Omega)\,;\ u \in u_0 + W_0^{1,\alpha}(\Omega;\mathbb{R}^m)\}$$

We will then prove that

Theorem:

(i) (QP) is a regular problem

(ii) $\inf(P) = \inf(QP)$

(iii) For every $\bar{u} \in W^{1,\alpha}(\Omega;\mathbb{R}^m)$, there exists a sequence $\{u^\nu\}_{\nu=1}^\infty$, $u^\nu \in W^{1,\alpha}(\Omega;\mathbb{R}^m)$ such that

$$\begin{cases} u^\nu = \bar{u} \text{ on } \partial\Omega, \quad \nu \geqslant 1, & (0.8)\\ u^\nu \rightharpoonup \bar{u} \text{ in } W^{1,\alpha}, \text{ as } \nu \to \infty, & (0.9)\\ I(u^\nu,\Omega) \to \bar{I}(\bar{u},\Omega), \text{ as } \nu \to \infty\,. & (0.10)\end{cases}$$

The theorem therefore shows that (P) and (QP) are "equivalent" with respect to weak convergence and then one may define the solutions of (QP) as "generalized solutions" of (P).

The above theorem was proved first by L.C. Young ([Yo1],[Yo2]) in the case $m = n = 1$, using his notions of generalized curves. Following the same approach it was then established by Ekeland ([Ek1]) and by Ekeland and Témam ([ET1]) in the case $m = 1$ or $n = 1$ (see also [BL1],[MS1]). The general case $(m,n > 1)$ was established in [Da3]. It has since then also been proved, using different methods and somehow weaker hypotheses on f, by Acerbi and Fusco [AF1].

In the last section of this article, we will give an example of applications of the above result to phase transitions for gases satisfying the Van der Waal's equation of state.

To end this introduction we give an elementary example illustrating the theorem.

Example:

Let $m = n = 1$, $\Omega = (0,1)$, $u_0(0) = u_0(1) = 0$ and

$$f(u,u') = (1 + u^2)(1 + (u'^2 - 1)^2). \tag{0.11}$$

So let

$$I(u) = \int_0^1 (1 + (u(x))^2)(1 + ((u'(x))^2 - 1)^2)\,dx, \tag{0.12}$$

and

(P) $\inf\{I(u) : u \in W_0^{1,\infty}(0,1)\}$.

As mentioned earlier, in the case $m = 1$ or $n = 1$ the notions of convexity and quasiconvexity are equivalent. It is easy to see that $f(u,\cdot)$ is not convex, therefore

$$Qf(u,u') = \begin{cases} f(u,u') & \text{if } |u'| \geq 1 \\ (1 + u^2) & \text{if } |u'| \leq 1. \end{cases} \tag{0.13}$$

Let

$$\bar{I}(u) = \int_0^1 Qf(u(x),u'(x))\,dx \tag{0.14}$$

and

(QP) $\operatorname{Inf}\{\bar{I}(u) : u \in W_0^{1,\infty}(0,1)\}$.

So define for $k = 0,1,2,\ldots,(\nu-1)$

$$u^\nu(x) = \begin{cases} x - \frac{k}{\nu} & \text{if } x \in [\frac{2k}{2\nu}, \frac{2k+1}{2\nu}] \\ -x + \frac{k+1}{\nu} & \text{if } x \in [\frac{2k+1}{2\nu}, \frac{2k+2}{2\nu}]. \end{cases} \tag{0.15}$$

We thus have $u^\nu \in W^{1,\infty}(0,1)$ and

$$\begin{cases} 0 \leq u^\nu(x) \leq \frac{1}{2\nu}, & x \in [0,1] \\ u^\nu(0) = u^\nu(1) = 0 & \\ |(u^\nu(x))'| = 1 & \text{a.e. in } (0,1). \end{cases} \tag{0.16}$$

It is then easy to see that u^ν is a minimising sequence of (P), since

$$1 = \bar{I}(0) = \inf(QP) \leq \inf(P) \leq I(u^\nu) \leq 1 + \frac{1}{4\nu^2} \tag{0.17}$$

We then get the theorem, i.e.,

$$\begin{cases} \inf(P) = \inf(QP) = 1 \\ u^\nu(0) = u^\nu(1) = 0 \\ u^\nu \overset{*}{\rightharpoonup} 0 \quad \text{in} \quad W^{1,\infty}(0,1), \quad \text{as} \quad \nu \to \infty, \\ I(u^\nu) \to \bar{I}(0), \quad \text{as} \quad \nu \to \infty. \end{cases} \tag{0.18}$$

(By $\overset{*}{\rightharpoonup}$, we denote weak * convergence.)

I. PRELIMINARY RESULTS

We have seen in the introduction the definition of quasi-convexity; we now turn our attention to a special type of functions Φ, called quasiaffine or null-Lagrangians, which are such that Φ and $-\Phi$ are quasiconvex. The next theorem characterizes such functions (for a proof see [Ba1],[Da4] ...).

Theorem 1:

The following properties are equivalent:

(i) Φ is quasiaffine (or null Lagrangian).

(ii) For every bounded domain $\Omega \subset \mathbb{R}^n$, for every $F \in \mathbb{R}^{nm}$ and for every $\phi \in W_0^{1,\infty}(\Omega;\mathbb{R}^m)$

$$\int_\Omega \Phi(F + \nabla\phi(x))\,dx = \Phi(F)\,\text{meas}\,\Omega. \tag{1.1}$$

(iii) Let $\text{adj}_s F$ denote the matrix of all $s \times s$ $(1 \leqslant s \leqslant \inf\{n,m\})$ subdeterminants of the matrix $F \in \mathbb{R}^{nm}$; then

$$\Phi(F) = A + \sum_{s=1}^{\inf\{n,m\}} \langle B_s; \text{adj}_s F\rangle_{\sigma(s)}, \tag{1.2}$$

where

$$\sigma(s) = \binom{m}{s}\binom{n}{s} = \frac{m!n!}{s!(m-s)!s!(n-s)!}$$

$\langle .,.\rangle_{\sigma(s)}$ denotes the scalar product in $\mathbb{R}^{\sigma(s)}$, $A \in \mathbb{R}$ and $B_s \in \mathbb{R}^{\sigma(s)}$ are constants.

Remarks:

(i) If $m = n = 2$ then (1.2) can be rewritten as

$$\Phi(F) = A + \sum_{i,j=1}^{2} B_1^{ij} F_{ij} + B_2 \det F.$$

(ii) The quasiaffine functions are the only weakly continuous functions.

(iii) The Euler Lagrange equations for the functional $\int_\Omega \Phi(\nabla u(x))dx$ reduce to an identity if Φ is a null Lagrangian.

We now turn our attention to lower quasiconvex envelopes (for a proof of the next theorem, see [Da3], [Da4]).

Theorem 2:

Let $f: \Omega \times \mathbb{R}^m \times \mathbb{R}^{nm} \to \mathbb{R}$ be continuous and such that

$$|f(x,u,F)| \leq a(x) + b|u|^\alpha + c|F|^\alpha,$$

for some $\alpha \geq 1$, $a \in L^1(\Omega)$, $b \geq 0$ and $c \geq 0$. Then

$$Qf(x_0,u_0,F_0) = \sup\{\Phi(x_0,u_0,F_0) \leq f(x_0,u_0,F_0) : \Phi(x_0,u_0,\cdot) \text{ quasiconvex}\}$$

$$= \inf_\phi\{\frac{1}{\text{meas} D}\int_D f(x_0,u_0,F_0 + \nabla\phi(y))dy : \phi \in W_0^{1,\infty}(D;\mathbb{R}^m)\}$$

where D is any bounded domain of $\mathbb{R}^n$. Furthermore Qf is continuous.

II. RELAXATION THEOREMS

In this section we will assume that $\Omega \subset \mathbb{R}^n$ is a bounded open set with Lipschitz boundary. We then have the following result established in [Da3].

Theorem 3:

Let $f: \mathbb{R}^{nm} \to \mathbb{R}$ be continuous and such that

$$\text{(H1)} \quad a + \sum_{\nu=1}^N b_\nu|\Phi_\nu(F)|^{\beta_\nu} \leq f(F) \leq c + \sum_{\nu=1}^N d_\nu|\Phi_\nu(F)|^{\beta_\nu}$$

for every $F \in \mathbb{R}^{nm}$, for some $a,c \in \mathbb{R}$, $N \geq 1$ (an integer), $\beta_\nu > 1$, $d_\nu \geq b_\nu > 0$ and where $\Phi_\nu: \mathbb{R}^{nm} \to \mathbb{R}$, $\nu = 1,\ldots,N$ are quasiaffine functions.

Then for every $\bar{u} \in W^{1,\infty}(\Omega;\mathbb{R}^m)$, there exists $\{u^s\}_{s=1}^\infty$, $u^s \in W^{1,\infty}(\Omega;\mathbb{R}^m)$ such that

$$\begin{cases} u^s = \bar{u} \text{ on } \partial\Omega & (2.1)\\ \Phi_\nu(\nabla u^s) \rightharpoonup \Phi_\nu(\nabla\bar{u}) \text{ in } L^{\beta_\nu}(\Omega),\ \nu = 1,..,N,\ s \to \infty & (2.2)\\ \int_\Omega f(\nabla u^s(x))dx \to \int_\Omega Qf(\nabla\bar{u}(x))dx. & (2.3)\end{cases}$$

Remarks:

(i) The case where f satisfies a condition

$$a + b|F|^{\beta} \leqslant f(F) \leqslant c + d|F|^{\beta}$$

is a particular case of (H1); one needs to choose $N = nm$, $\beta_{\nu} = \beta > 1$, $d_{\nu} = d \geqslant b_{\nu} = b > 0$ for $\nu = 1,\dots,N$ and for $F = (F_{ij})_{1 \leqslant i \leqslant n, 1 \leqslant j \leqslant m}$

$$\begin{cases} \Phi_1(F) = F_{11}, \dots\dots \Phi_n(F) = F_{n1}, \\ \Phi_{n+1}(F) = F_{12}, \dots \Phi_{2n}(F) = F_{n2}, \\ \dots\dots\dots\dots\dots\dots\dots\dots \\ \dots\dots\dots\dots\dots \Phi_{nm}(F) = F_{nm} . \end{cases}$$

Furthermore in this case condition (2.2) is equivalent to

$u^s \rightharpoonup \bar{u}$ in $W^{1,\beta}(\Omega;\mathbb{R}^m)$.

(ii) Similarly if $n = m$ and f satisfies

$$a + b|\det F|^{\beta} \leqslant f(F) \leqslant c + d|\det F|^{\beta}$$

then by choosing $N = 1$ and $\Phi_{\nu}(F) = \det F$, we get (H1).

(iii) One may note however that the problem of minimal hypersurface in parametric form is not contained in Hypothesis (H1), since in this case $\beta_{\nu} = 1$. However one can still prove a relaxation result for this problem (see [Da2]).

We now show that as corollaries of Theorem 3, we may allow f to depend not only on ∇u but also on x and on u.

Corollary 1:

Let $f\colon \Omega \times \mathbb{R}^{nm} \to \mathbb{R}$ be continuous and such that

$$\text{(H2)}\quad a(x) + \sum_{\nu=1}^{N} b_{\nu}|\Phi_{\nu}(F)|^{\beta_{\nu}} \leqslant f(x,F) \leqslant c(x) + \sum_{\nu=1}^{N} d_{\nu}|\Phi_{\nu}(F)|^{\beta_{\nu}}$$

for every $F \in \mathbb{R}^{nm}$, $a, c \in L^1(\Omega)$, b_{ν}, d_{ν} and Φ_{ν} as in (H1). Then for every $\bar{u} \in W^{1,\infty}(\Omega;\mathbb{R}^m)$, there exists $\{u^s\}_{s=1}^{\infty}$, $u^s \in W^{1,\infty}(\Omega;\mathbb{R}^m)$ such that

$$\begin{cases} u^s = \bar{u} \quad \text{on} \quad \partial\Omega, & (2.4) \\ \Phi_\nu(\nabla u^s) \rightharpoonup \Phi_\nu(\nabla\bar{u}) \quad \text{in} \quad L^{\beta_\nu}(\Omega),\ \nu = 1,\dots,N, & (2.5) \\ \int_\Omega f(x,\nabla u^s(x))\,dx \to \int_\Omega Qf(x,\nabla\bar{u}(x))\,dx. & (2.6) \end{cases}$$

Corollary 5:

Let $f: \Omega \times \mathbb{R}^m \times \mathbb{R}^{nm} \to \mathbb{R}$ be continuous and such that

(H3) $a_1(x) + c_1|F|^\beta \leqslant f(x,u,F) \leqslant a_2(x) + b|u|^\beta + c_2|F|^\beta$

for every $F \in \mathbb{R}^{nm}$, $a_1,a_2 \in L^1(\Omega)$, $c_2 \geqslant c_1 > 0$, $b \geqslant 0$ and $\beta > 1$. Then for every $\bar{u} \in W^{1,\beta}(\Omega;\mathbb{R}^m)$, there exists $\{u^s\}_{s=1}^\infty$, $u^s \in W^{1,\beta}(\Omega;\mathbb{R}^m)$ such that

$$\begin{cases} u^s = \bar{u} \quad \text{on} \quad \partial\Omega, & (2.7) \\ u^s \rightharpoonup \bar{u} \quad \text{in} \quad W^{1,\beta}(\Omega;\mathbb{R}^m), & (2.8) \\ \int_\Omega f(x,u^s(x),\nabla u^s(x))\,dx \to \int_\Omega Qf(x,\bar{u}(x),\nabla\bar{u}(x))\,dx. & (2.9) \end{cases}$$

We now sketch the proofs of Corollaries 4 and 5.

Proof of Corollary 4:

Fix $\varepsilon > 0$ and let

$$K(\bar{u}) = 1 + \int_\Omega Qf(x,\nabla\bar{u}(x))\,dx + \int_\Omega |a_1(x)|\,dx, \tag{2.10}$$

$$A = \{\phi \in W^{1,\infty}: \sum_{\nu=1}^{N} b_\nu \|\Phi_\nu(\nabla\phi)\|_{L^{\beta_\nu}}^{\beta_\nu} \leqslant K(\bar{u})\}. \tag{2.11}$$

Let Ω be approximated by a union of hypercubes Ω_p, $1 \leqslant p \leqslant P$, of edge length $\frac{1}{p}$ and define

$$x_p = \frac{1}{\text{meas}\Omega_p} \int_{\Omega_p} x\,dx \tag{2.12}$$

Choose p sufficiently large so that

$$\Big| \int_{\Omega - \bigcup_{p=1}^{P} \Omega_p} f(x,\nabla\bar{u}(x))\,dx \Big| \leqslant \frac{\varepsilon}{5}, \tag{2.13}$$

$$\left| \int_{\Omega - \bigcup_{p=1}^{P} \Omega_p} Qf(x,\nabla\bar{u}(x))\,dx \right| \leqslant \frac{\varepsilon}{5}\,, \tag{2.14}$$

$$\left| \int_{\bigcup_{p=1}^{P} \Omega_p} Qf(x,\nabla\bar{u}(x))\,dx - \sum_{p=1}^{P} \int_{\Omega_p} Qf(x_p,\nabla\bar{u}(x))\,dx \right| \leqslant \frac{\varepsilon}{5}\,, \tag{2.15}$$

$$\left| \int_{\bigcup_{p=1}^{P} \Omega_p} f(x,\nabla\phi(x))\,dx - \sum_{p=1}^{P} \int_{\Omega_p} f(x_p,\nabla\phi(x))\,dx \right| \leqslant \frac{\varepsilon}{5}, \text{ for all } \phi \in A. \tag{2.16}$$

Having fixed p in this way, we may use Theorem 3 on each Ω_p, $1 \leqslant p \leqslant P$, and deduce that there exist $u_p^s \in W^{1,\infty}(\Omega_p;\mathbb{R}^m)$ such that for s sufficiently large

$$u_p^s = \bar{u} \quad \text{on} \quad \partial\Omega_p \tag{2.17}$$

$$\Phi_\nu(\nabla u_p^s) \rightharpoonup \Phi_\nu(\nabla\bar{u}) \quad \text{in} \quad L^{\beta_\nu}(\Omega_p),\ 1 \leqslant \nu \leqslant N \tag{2.18}$$

$$\left|\int_{\Omega_p} f(x_p,\nabla u_p^s(x))\,dx - \int_{\Omega_p} Qf(x_p,\nabla\bar{u}(x))\,dx\right| \leqslant \frac{\varepsilon}{5P}\,. \tag{2.19}$$

Defining $u^s = u_p^s$ in Ω_p and $u^s = \bar{u}$ in $\Omega - \cup\Omega_p$ we immediately get (2.4) and (2.5). Therefore it remains to show (2.6). Observe first that, by (H2), $u^s \in A$. Combining (2.13) - (2.16) with (2.19) we immediately get the result.

Proof of Corollary 5:

Fix $\varepsilon > 0$. Apply Corollary 4 to $f(x,\bar{u}(x),\cdot)$ and $Qf(x,\bar{u}(x),\cdot)$ to get $u^s \in W^{1,\infty}(\Omega;\mathbb{R}^m)$ such that

$$\begin{cases} u^s = \bar{u} \quad \text{on} \quad \partial\Omega, & (2.20) \\ u^s \rightharpoonup \bar{u} \quad \text{in} \quad W^{1,\beta}(\Omega;\mathbb{R}^m), & (2.21) \\ \int_\Omega f(x,\bar{u}(x),\nabla u^s(x))\,dx \to \int_\Omega Qf(x,\bar{u}(x),\nabla\bar{u}(x))\,dx. & (2.22) \end{cases}$$

Using Rellich's theorem we have (up to a subsequence)

$$u^s \to \bar{u} \quad \text{a.e. and in} \quad L^\beta(\Omega). \tag{2.23}$$

Hypothesis (H3), combined with (2.22) and (2.23), gives the result.

III. APPLICATIONS

Before giving an example of applications of Theorem 3, we will give a representation of the lower quasiconvex envelope Qf for a certain particular function.

Proposition 6:

Let $n = m$, $f: \mathbb{R}^{n^2} \to \mathbb{R}$ and $g: \mathbb{R} \to \mathbb{R}$ be continuous and such that

$$f(F) = g(\det F), \tag{3.1}$$

for every $F \in \mathbb{R}^{n^2}$. Then

$$Qf(F) = g^{**}(\det F), \tag{3.2}$$

where g^{**} denotes the lower convex envelope of g, i.e.

$$g^{**} = \sup\{\Phi \leqslant f;\ \Phi \ \text{convex}\}.$$

Remark:

We have mentioned in the introduction that if f has the form (3.1), then it has been shown by Morrey ([Mo1], see also [Da4]) that the quasiconvexity of f and the convexity of g are equivalent. The above result has also been proved in [AF1] using the Relaxation Theorem.

In order to prove the above proposition, we need two elementary lemmas (the proof of the first one being obvious).

Lemma 7:

Let $F \in \mathbb{R}^{n^2}$ with rank $F \leqslant n-1$. Let $\Theta > 0$ be fixed. Then there exists $G \in \mathbb{R}^{n^2}$, with rank $G = n-1$ and $\|F-G\| = \sum_{i,j} |F_{ij} - G_{ij}| \leqslant \Theta$.

<u>Lemma 8</u>:

Let $\lambda \in (0,1)$, $d,\delta \in \mathbb{R}$ be fixed. Let $F \in \mathbb{R}^{n^2}$, with rank $F \geqslant n-1$, be such that

$$\det F = \lambda d + (1 - \lambda)\delta. \tag{3.3}$$

Then there exist $a,b \in \mathbb{R}^n$ such that

$$\begin{cases} \det(F + \lambda a \otimes b) = \delta, & (3.4) \\ \det(F - (1 - \lambda) a \otimes b) = d, & (3.5) \end{cases}$$

where $(a \otimes b)_{ij} = a_i b_j$, $1 \leqslant i, j \leqslant n$.

<u>Proof of Lemma 8</u>:

Since rank $F \geqslant n-1$, we may suppose (without loss of generality) that

$$\det \hat{F}_{11} = \begin{vmatrix} F_{22} & \cdots & F_{2n} \\ \vdots & & \vdots \\ F_{n2} & \cdots & F_{nn} \end{vmatrix} \neq 0. \tag{3.6}$$

We may then choose a and b in such a way that

$$\begin{cases} a_j = b_j = 0, \quad 2 \leqslant j \leqslant n \\ b_1 = 1 \\ a_1 = \dfrac{\delta - \det F}{\lambda \det \hat{F}_{11}} \end{cases}$$

It is then easy to show that (3.4) and (3.5) hold.

<u>Proof of Proposition 6</u>:

<u>Step 1</u>: Observe that since $g^{**} \leqslant f$ and g^{**} is convex (and thus quasiconvex) we trivially have

$$g^{**} \leqslant Qf \leqslant f = g. \tag{3.7}$$

<u>Step 2</u>: We need to show the converse of (3.7). First define

$$h(\delta) = \sup\{Qf(F) : \det F = \delta\}. \tag{3.8}$$

We immediately deduce that

$$
\begin{aligned}
Qf(F) &\leqslant h(\det F) = \sup\{Qf(G) : \det G = \det F\} \\
&\leqslant \sup\{f(G) : \det G = \det F\} \\
&\leqslant \sup\{g(\det G) : \det G = \det F\} \\
&\leqslant g(\det F) = f(F).
\end{aligned} \tag{3.9}
$$

Therefore if we show that h is convex, we will obtain that

$$
Qf(F) \leqslant h(\det F) \leqslant g^{**}(\det F) \leqslant f(F) \tag{3.10}
$$

and thus combining (3.7) and (3.10), we will obtain the result.

Step 3: It therefore remains to show that for $\lambda \in (0,1)$, $d, \delta \in \mathbb{R}$, we have

$$
h(\lambda d + (1 - \lambda)\delta) \leqslant \lambda h(d) + (1 - \lambda) h(\delta). \tag{3.11}
$$

Fix $\varepsilon > 0$. Then, by definition (3.8), there exists $F_1 \in \mathbb{R}^{n^2}$ such that

$$
\det F_1 = \lambda d + (1 - \lambda)\delta, \tag{3.12}
$$

$$
h(\lambda d + (1 - \lambda)\delta) \leqslant \frac{\varepsilon}{2} + Qf(F_1). \tag{3.13}
$$

Observe that there is no loss of generality if we suppose $\operatorname{rank} F_1 \geqslant n - 1$, otherwise use Lemma 7 and the continuity of Qf (c.f. Theorem 2) to get $F \in \mathbb{R}^{n^2}$ with

$$
\det F = \lambda d + (1 - \lambda)\delta, \quad \operatorname{rank} F \geqslant n - 1, \tag{3.14}
$$

$$
h(\lambda d + (1 - \lambda)\delta) \leqslant \varepsilon + Qf(F). \tag{3.15}
$$

Note that we trivially have

$$
\begin{aligned}
h(\lambda d + (1 - \lambda)\delta) \leqslant \varepsilon + Qf(&\lambda[F - (1 - \lambda) a \otimes b] \\
&+ (1 - \lambda)[F + \lambda a \otimes b]),
\end{aligned} \tag{3.16}
$$

for every $a, b \in \mathbb{R}^n$. Using the rank one convexity of Qf (c.f. (0.5) in the introduction: Qf quasiconvex $\Rightarrow$ Qf rank one convex), we have

$$
\begin{aligned}
h(\lambda d + (1 - \lambda)\delta) \leqslant \varepsilon &+ \lambda Qf(F - (1 - \lambda) a \otimes b) \\
&+ (1 - \lambda) Qf(F + \lambda a \otimes b),
\end{aligned} \tag{3.17}
$$

for every $a, b \in \mathbb{R}^n$. We now make a particular choice of a and b as in Lemma 8, and we get

$$
\begin{aligned}
h(\lambda d + (1 - \lambda)\delta) \leqslant \varepsilon &+ \lambda \sup\{Qf(G) : \det G = d\} \\
&+ (1 - \lambda) \sup\{Qf(H) : \det H = \delta\},
\end{aligned} \tag{3.18}
$$

i.e.,

$$h(\lambda d + (1 - \lambda)\delta) \leqslant \varepsilon + \lambda h(d) + (1 - \lambda)h(\delta) \;;$$

ε being arbitrary, we have indeed established the proposition.

Remark:

One can establish in the same way a similar result for functions $f(F) = g(\Phi(F))$ where Φ is any quasiaffine (null Lagrangian) function.

We conclude this article with an example of application of the relaxation theorem. We consider a gas an an elastic material which occupies in a reference configuration a volume $\Omega \subset \mathbb{R}^3$. We use Lagrangian coordinates. In a deformed configuration the particle $x \in \Omega$ occupies the position $u(x) \in \mathbb{R}^3$. We assume furthermore that the displacement is prescribed on the boundary of Ω, i.e., $u = u_0$ on $\partial\Omega$. The energy of the deformation is then, in absence of external forces,

$$I(u) = \int_\Omega \Big(\int_1^{\det \nabla u(x)} P(v)\,dv \Big) dx \tag{3.19}$$

where P is the pressure and v the specific volume. We will assume that P satisfies the Van der Waal's equation of state, i.e.,

$$P(v) = \frac{nRT}{v - nb} - \frac{an^2}{v^2}, \tag{3.20}$$

where T denotes the temperatures, a,b,R are constants and n is the number of moles (see [Cal] for reference). We let

$$f(\nabla u) = g(\det \nabla u) = \int_1^{\det \nabla u(x)} P(v)\,dv. \tag{3.21}$$

It is well known that for some values of the temperature T, the pressure P, in (3.20), is not a monotone function of v and thus g is not convex (and hence f is not quasiconvex). Therefore there is a range of values of v where the Euler-Lagrange equations associated to (3.19) are not elliptic. We may then consider the quasiconvex envelope of f, $Qf = g^{**}$ (Proposition 6). So let

$$\bar{I}(u) = \int_\Omega Qf(\nabla u(x))\,dx = \int_\Omega g^{**}(\det \nabla u(x))\,dx. \tag{3.22}$$

If one considers the corresponding pressure associated to Qf, one gets the usual Maxwell line (see for more details [Da4] p.98). We then have the following

Theorem 9:

Let Ω be diffeomorphic to a sphere. Let $u_0 \in C^2(\bar{\Omega};\mathbb{R}^3)$ be a homeomorphism and let

(P) $\inf\{I(u): u = u_0$ on $\partial\Omega$, $u \in C^1(\bar{\Omega};\mathbb{R}^3)\}$

(QP) $\inf\{\bar{I}(u): u = u_0$ on $\partial\Omega$, $u \in C^1(\bar{\Omega};\mathbb{R}^3)\}$.

Then

(i) (QP) has a solution $\bar{u} \in C^1$ with $\bar{u} = u_0$ on $\partial\Omega$ and $\det \nabla\bar{u}(x) > 0$ in $\bar{\Omega}$.

(ii) $\inf(P) = \inf(QP)$.

(iii) Let $\bar{u}$ be a solution of (QP) (i.e., $\bar{u} = u_0$ on $\partial\Omega$, $\bar{u} \in C^1$ and $\det \nabla\bar{u} > 0$ in $\bar{\Omega}$), then there exists a minimizing sequence of (P) with $\det \nabla u_s > 0$ in $\bar{\Omega}$, $u_s = u_0$ on $\partial\Omega$ such that

$$\begin{cases} \det \nabla u_s \rightharpoonup \det \nabla\bar{u} \quad \text{in } L^1(\Omega), & (3.23) \\ I(u_s) \to \bar{I}(\bar{u}) & (3.24) \end{cases}$$

Proof:

(i) The existence of a solution is established in [Da1].
(ii),(iii) In order to apply Theorem 3, one needs to modify the integrand f, which does not satisfy the continuity and coercivity conditions assumed in Theorem 3. One changes the integrand in the following manner

$$\int_\Omega g(\det \nabla u(x))\,dx = \int_{u_0(\Omega)} g\left(\frac{1}{\det \nabla x(u)}\right) \det \nabla x(u)\,du \tag{3.25}$$

and observe that if

$$h(t) = g(t^{-1})t \quad \text{for} \quad t > 0, \tag{3.26}$$

then

$$h^{**}(t) = g^{**}(t^{-1})t. \tag{3.27}$$

We then apply Theorem 3 to h and h**, and by inverting the minimizing sequences, we obtain (3.23) and (3.24). The fact that $\det \nabla u_s > 0$ in $\bar{\Omega}$ (in (3.23)) requires a more detailed analysis which is carried out in [Da1].

REFERENCES:

[AF1] Acerbi, E. and Fusco, N.: *Semicontinuity problems in the calculus of variations*, to appear in Arch. Rat. Mech. Anal.

[Ad1] Adams, R.A.: 1975, *Sobolev Spaces*, Academic Press, New York.

[Ba1] Ball, J.M.: 1977, *Convexity conditions and existence theorems in nonlinear elasticity*, Arch. Rat. Mech. Anal. 63, pp. 337-403.

[BL1] Berliocchi, H. and Lasry, J.M.: 1973, *Intégrandes normales et mesures paramétrées en calcul des variations*, Bull. SMF 101, pp. 129-184.

[Ca1] Callen, H.B.: 1968, *Thermodynamics*, Wiley, New York and London.

[Da1] Dacorogna, B.: 1981, *A relaxation theorem and its application to the equilibrium of gases*, Arch. Rat. Mech. Anal. 77, pp.359-386.

[Da2] Dacorogna, B.: 1982, *Minimal hypersurfaces problems in parametric form with non convex integrands*, Indiana Univ. Math. Journal 31, pp. 531-552.

[Da3] Dacorogna, B.: 1982, *Quasiconvexity and relaxation of non convex problems in the calculus of variations*, J. Funct. Anal. 46, pp. 102-118.

[Da4] Dacorogna, B.: 1982, *Weak continuity and weak lower semicontinuity of nonlinear functionals*, Lecture Notes in Math., Springer-Verlag, Vol. 922.

[Ek1] Ekeland, I: 1972, *Sur le contrôle optimal de systèmes gouvernés par des équations elliptiques*, J. Funct. Anal. 3, pp. 1-62.

[ET1] Ekeland, I. and Témam, R.: 1976, *Analyse convexe et problèmes variationnels*, Dunod, Gauthier-Villars, Paris.

[MS1] Marcellini, P. and Sbordone, C.: 1980, *Semicontinuity problems in the calculus of variations*, Nonlin. Anal. Theory, Meth. and Appl. 4, pp. 241-257.

[Me1] Meyers, N.G.: 1965, *Quasi convexity and lower semicontinuity of multiple variational integrals of any order*, Trans. Amer. Math. Soc. 119, pp. 125-149.

[Mo1] Morrey, C.B.: 1952, *Quasiconvexity and the lower semicontinuity of multiple integrals*, Pacific J. Math. 2, pp. 25-53.

[Mo2] Morrey, C.B.: 1966, *Multiple integrals in the calculus of variations*, Springer-Verlag, Berlin.

[Yo1] Young, L.C.: 1937, *Generalised curves and the existence of an attained absolute minimum in the calculus of variations*, C.R. Soc. Sci. Lett. Varsovie, Classe III 30, pp. 212-234.

[Yo2] Young, L.C.: 1969, *Lectures on the calculus of variations and optimal control theory*, W.B. Saunders, Philadelphia.

REMARKS ON THE RELAXATION OF INTEGRALS OF THE CALCULUS OF VARIATIONS

Nicola Fusco

Istituto di Matematica
v. Mezzocannone, 8
Napoli

If we want to minimize a functional of the type:

$$F(u,\Omega) = \int_\Omega f(x,u(x),Du(x))dx \tag{1}$$

where Ω is an open set in $\mathbb{R}^n$, $u:\Omega \longrightarrow \mathbb{R}^m$, $u \in W^{1,p}(\Omega;\mathbb{R}^m)$, it is useful to know whether the integral $F(u,\Omega)$ is sequentially lower semicontinuous (slsc) with respect to a topology τ on $W^{1,p}$ such that there exists a τ-compact minimizing sequence.

When F is not lsc, we can define its lsc envelope with respect to the topology τ, i.e. the greatest τ-slsc functional $\bar{F}$ less than or equal to F. Indeed, under suitable hypotheses, $\bar{F}$ has a minimum, $\operatorname{Min} \bar{F} = \operatorname{Inf} F$, and the minimum points of $\bar{F}$ can be obtained as limits of the minimizing sequences of F (see [10]).

This procedure (which is called "relaxation" of the minimum problem) has been exhaustively studied in [10] (in the scalar case) and in some particular cases in [6], [7], [8] (if $m > 1$), when $\bar{F}$ is the lsc envelope of F with respect to the weak topology of $W^{1,p}$ (weak * if $p = \infty$).

Here we want to refer to some papers in which a characterisation of $\bar{F}$ is given, following a technique introduced in [9] of splitting the problem into two steps:

i) to show that $\bar{F}$ is an integral of the same type of F; i.e. to prove that there exists an integrand g such that

J. M. Ball (ed.), Systems of Nonlinear Partial Differential Equations, 401–408.

$$\bar{F}(u,\Omega) = \int_\Omega g(x,u(x),Du(x))\,dx \tag{2}$$

ii) to give a characterization of g, possibly producing a representation formula.

To prove that $\bar{F}$ is an integral of the Calculus of Variations, first we can show that for any u the set function $\Omega \longrightarrow \bar{F}(u,\Omega)$ is the trace on the open sets of a regular Borel measure, so that there exists a locally integrable function h_u such that for any open set Ω in $\mathbb{R}^n$

$$\bar{F}(u,\Omega) = \int_\Omega h_u(x)\,dx$$

Then we can get formula (2) if we prove that, at least for smooth functions, if $u(x_o) = v(x_o)$ and $Du(x_o) = Dv(x_o)$ in some point x_o, then $h_u(x_o) = h_v(x_o)$. When we know that $\bar{F}$ is an integral of the same type as F, we can characterize the integrand g, if we know which conditions on the integrand f are equivalent to the slsc of a functional of the type (1).

With this technique the characterization is given in [13] of the slsc envelope of F with respect to the weak topology of $W^{1,p}$ (in the scalar case). Still in the scalar case, but with respect to the topology of L^p, problem i) has been studied in [5], and a characterization of g has been given, at least for quadratic functions, in [12] and [11].

Here we are interested in the case $m > 1$. We begin with some results on the swlsc envelope in $W^{1,p}$, and then we shall make some remarks about the lsc envelope of F with respect to the L^p topologies.

<u>Theorem 1</u> (see [1])

Let $f(x,s,\xi): \mathbb{R}^n \times \mathbb{R}^m \times \mathbb{R}^{nm} \longrightarrow \mathbb{R}$ *be a Carathéodory function such that for any* $x \in \mathbb{R}^n$ *a.e.,* $s \in \mathbb{R}^m$, $\xi \in \mathbb{R}^{nm}$

$$0 \leqslant f(x,s,\xi) \leqslant a(x) + b(x,\xi) \tag{3}$$

with $a(x)$ *non-negative and locally summable in* $\mathbb{R}^n$, $b \geqslant 0$ *and locally bounded in* $\mathbb{R}^m \times \mathbb{R}^{nm}$. *Let us assume that*

$$\left| f(x,s_1,\xi) - f(x,s_2,\xi) \right| \leqslant \omega(x,|s_1-s_2|)\beta(|\xi|) \tag{4}$$

where $\omega : \mathbb{R}^n \times \mathbb{R}^+$ *is a Carathéodory function such that* $\omega(x,0) = 0$ *and* $\beta \geqslant 0$ *is non-decreasing. Then there exists a Carathéodory function* $g(x,s,\xi) : \mathbb{R}^n \times \mathbb{R}^m \times \mathbb{R}^{nm} \longrightarrow \mathbb{R}$ *such that for any open set* Ω *in* $\mathbb{R}^n$ *the functional*

$$u \longrightarrow \int_\Omega g(x,u(x),Du(x))\,dx$$

*is the sw*lsc envelope on the space* $W^{1,\infty}(\Omega,\mathbb{R}^m)$ *of the functional*

$$u \longrightarrow \int_\Omega f(x,u(x),Du(x))\,dx$$

We remark that condition (4) above is satisfied if $f = f(x,\xi)$ (in this case also g does not depend on s) and that, if

$$0 \leqslant f(x,s,\xi) \leqslant a(x) + C(|s|^p + |\xi|^p) \qquad p \geqslant 1 \tag{5}$$

with $a \geqslant 0$, $a \in L^1_{loc}(\mathbb{R}^n)$, $C > 0$, Theorem 1 is still valid if we consider the swlsc envelope of F on $W^{1,p}(\Omega;\mathbb{R}^m)$.

Before showing how to characterize the integrand g we give the following:

Definition 2

A continuous function $f : \mathbb{R}^{nm} \longrightarrow \mathbb{R}$ *is quasi-convex if for any* $\xi \in \mathbb{R}^{nm}$, *for every open subset* Ω *of* $\mathbb{R}^n$ *and every function* $z \in C^1_0(\Omega;\mathbb{R}^m)$

$$\text{meas}(\Omega)\; f(\xi) \leqslant \int_\Omega f(\xi+Dz(x))\,dx \tag{6}$$

(For properties of quasi-convex functions see [16], [17], [15], [3], [4]). The following result shows that swlsc on $W^{1,p}$ of the functional (1) is equivalent to the quasi-convexity of f

in ξ.

Theorem 3 (see [1])

Let $f : \mathbb{R}^n \times \mathbb{R}^m \times \mathbb{R}^{nm} \to \mathbb{R}$ *be a Carathéodory function satisfying* (5) *(or* (3) *if* $p = \infty$). *Then the functional* (1) *is swlsc on* $W^{1,p}$ *(sw*lsc if* $p = \infty$) *if and only if* f *is quasi-convex in* ξ.

From this result follows the characterization of g:

Corollary 4

Under the same hypotheses for f *as in* Theorem 1, *for any* $x \in \mathbb{R}^n$ *a.e.,* $s \in \mathbb{R}^m$, *the function*

$$\xi \longrightarrow g(x,s,\xi)$$

is the greatest quasi-convex function less than or equal to

$$\xi \longrightarrow f(x,s,\xi)$$

In the scalar case this characterization becomes (see also [10], [13])

$$g(x,s,\xi) = f^{**}(x,s,\xi)$$

where $f^{**}(x,s,\cdot)$ is the greatest convex function less than or equal to $f(x,s,\cdot)$. Indeed, if $m = 1$, quasi-convexity is equivalent to convexity. A sharper characterization of g can be obtained in particular cases. If ξ is regarded as a $n \times m$ matrix, let us denote by $X^o\xi$ the vector whose components are the subdeterminants of ξ of highest order. With $N(n,m)$ we denote the dimension of $X^o\xi$. For example, if $n = m$, $X^o\xi = \det\xi$ and $N(n,n) = 1$. Then (see [2])

Theorem 5

Let f *satisfy the assumptions of* Theorem 1. *If there exists a Carathéodory function* $\phi : \mathbb{R}^n \times \mathbb{R}^m \times \mathbb{R}^{N(n,m)} \to \mathbb{R}$ *such that for any* $x \in \mathbb{R}^n$ *a.e.,* $s \in \mathbb{R}^m$, $\xi \in \mathbb{R}^{nm}$, $f(x,s,\xi) = \phi(x,s,X^o\xi)$, *then there exists another Carathéodory function* $\psi : \mathbb{R}^n \times \mathbb{R}^m \times \mathbb{R}^{N(n,m)} \to \mathbb{R}$ *such that* $g(x,s,\xi) = \psi(x,s,X^o\xi)$. *Moreover if* $m = n$ *or* $n + 1$, $\phi^{**}(x,s,\cdot) = \psi(x,s,\cdot)$.

Finally we make some remarks on the lsc of the functional F with respect to the topology of $L^p(\Omega;\mathbb{R}^m)$ $(1 \leqslant p \leqslant \infty)$. If $m > 1$ the situation is completely different from the scalar case. Indeed, even the functional $\int_\Omega \det Du(x)\,dx$ is not lsc (see [2]) with respect to the topology of L^p, if $p<\infty$. Only the topology of L^∞ seems to be interesting, and many semicontinuity results are known in this case ([17], [14], [2]). But the problem of characterizing the lsc envelope of F with respect to the topology of L^∞ is still open. In fact we don't know in general whether $\bar{F}$ is still an integral of the Calculus of Variations. Nevertheless we know in some cases which conditions on the integrand are equivalent to lsc with respect to the topology of L^∞ of a functional of the type (1). As an example, we conclude by stating the following:

Theorem 6

Let $f:\Omega\times\mathbb{R}^n\times\mathbb{R}\longrightarrow[0,+\infty]$ *be a continuous function,* Ω *an open subset of* $\mathbb{R}^n$, $n>1$. *Let us assume that for any* $\Sigma\subset\subset\Omega\times\mathbb{R}^n$ *there exists a continuous function* $\omega_\Sigma:\mathbb{R}^+\times\mathbb{R}^+\longrightarrow\mathbb{R}^+$, *with* $\omega_\Sigma(0,0)=0$, *such that for any* $(x,s),\ (y,t)\in\Sigma$

$$\left|f(x,s,\lambda)-f(y,t,\lambda)\right|\leqslant\omega_\Sigma(|x-y|,|s-t|)\,[1+f(x,s,\lambda)] \tag{7}$$

Then the functional

$$F(u,\Omega)=\int_\Omega f(x,u(x),\det Du(x))\,dx$$

is lsc on the space $W^{1,n}_{loc}(\Omega;\mathbb{R}^n)\cap C(\Omega;\mathbb{R}^n)$ *endowed with the topology of* $L^\infty_{loc}(\Omega;\mathbb{R}^n)$ *if and only if for any* $x\in\Omega$, $s\in\mathbb{R}^n$

$$\lambda\longrightarrow f(x,s,\lambda)\ \textit{is a convex function such that}\quad f(x,s,0)=\min_\lambda f(x,s,\lambda). \tag{8}$$

Proof: The sufficiency of condition (8) is proved [2] (Theorem III.7). If F is lsc with respect to the topology of L^∞_{loc} it is also slsc with respect to the weak topology of $W^{1,p}$ for any $p>n$. Hence, by Theorem 3, for any $(x,s)\in\Omega\times\mathbb{R}^n$ the function $\xi\longrightarrow f(x,s,\det\xi)$ is quasi-convex. This implies (see [17], Theorem 4.4.10) that f is convex in λ. Let us suppose that there exists $(x_o,s_o)\in\Omega\times\mathbb{R}^n$ such that $f(x_o,s_o,0)>\inf_\lambda f(x_o,s_o,\lambda)$. Then, since f is convex in λ, there exists $\alpha\neq 0$ such that

$$\int_0^\pi f(x_o,s_o,\alpha\sin^2 x)\,dx<f(x_o,s_o,0) \tag{9}$$

Let us choose $\varepsilon > 0$ such that $\Omega_\varepsilon = x_o +]0,2\varepsilon\pi[^n \subset \Omega$ and put

$$u_{\varepsilon,h}(x) = \frac{\varepsilon}{\sqrt[n]{h}} (\phi_{\varepsilon,h}(x^1-x_o^1), \alpha\psi_\varepsilon(x^2-x_o^2) \sin h (\frac{x^1-x_o^1}{\varepsilon}),$$
$$\psi_\varepsilon(x^3-x_o^3), \ldots, \psi_\varepsilon(x^n-x_o^n)) + s_o,$$

where $\phi_{\varepsilon,h}$, $\psi_\varepsilon \in \mathrm{Lip}(\mathbb{R})$ and $\phi_{\varepsilon,h}(t) = -\cos\frac{ht}{\varepsilon}$ if $0 \leqslant t \leqslant 2\varepsilon\pi$ and otherwise is constant, $\psi_\varepsilon(t) = \frac{t}{\varepsilon}$ if $0 \leqslant t \leqslant 2\varepsilon\pi$ and assumes constant values outside. Then $u_{\varepsilon,h} \longrightarrow s_o$ as $h \longrightarrow +\infty$. By the assumption on F:

$$F(s_o,\Omega) \leqslant \liminf_h F(u_{\varepsilon,h},\Omega)$$

But, since $\det Du_{\varepsilon,h} = 0$ in $\Omega - \Omega_\varepsilon$, we have

$$F(s_o,\Omega_\varepsilon) \leqslant \liminf_h F(u_{\varepsilon,h},\Omega_\varepsilon)$$

Then, applying (7) with a suitable Σ:

$$\int_{\Omega_\varepsilon} f(x,s_o,0) \leqslant |\Omega_\varepsilon|[1 + \max_{|\tau|\leqslant\alpha} f(x_o,s_o,t)] \times$$

$$\times \max\{\omega_\Sigma(t_1,t_2) : (t_1,t_2) \in [0,2\varepsilon\pi] \times [0,\|u_{\varepsilon,h} - s_o\|_\infty]\} +$$

$$+ |\Omega_\varepsilon| \int_0^{2\pi} f(x_o,s_o,\alpha\sin^2 x)\,dx .$$

Then, dividing by $|\Omega_\varepsilon|$ and letting $\varepsilon \longrightarrow 0^+$, the result follows by contradiction.

REFERENCES

[1] Acerbi, E., and Fusco, N.: *Semicontinuity problems in the calculus of variations*, Arch. Rational Mech. Anal., to appear.

[2] Acerbi, E., Buttazzo, G., and Fusco, N.: *Semicontinuity and relaxation for integrals depending on vector-valued functions*, to appear.

[3] Ball, J.M.: 1977, *Constitutive inequalities and existence theorems in nonlinear elastostatics*, Nonlinear analysis and mechanics: Heriot-Watt Symposium (Edinburgh, 1976), Vol. I, pp.187-241. Res. Notes in Math., No.17, Pitman, London.

[4] Ball, J.M.: 1977, *Convexity conditions and existence theorems in nonlinear elasticity*, Arch. Rational Mech. Anal., 63, pp.337-403.

[5] Carbone, L., and Sbordone, C.: 1979, *Some properties of Γ-limits of integral functionals*, Ann. Mat. Pura Appl., (4), 122, pp.1-60.

[6] Dacorogna, B.: 1981, *A relaxation theorem and its application to the equilibrium of gases*, Arch. Rational Mech. Anal., 77, pp.359-386.

[7] Dacorogna, B.: 1982, *A generic result for non convex problems in the calculus of variations*, J. Funct. Anal., 46, pp.102-118.

[8] Dacorogna, B.: *Minimal hypersurfaces problems in parametric form with non convex integrands*, India Univ. Math. J., to appear.

[9] de Giorgi, E.: 1975, *Sulla convergenza di alcune successioni di integrali del tipo dell'area*, Rend. Mat., 8, pp.277-294.

[10] Ekeland, I., and Temam, R.: 1974, *Analyse convèxe et problèmes variationnels*, Dunod, Gauthier-Villars.

[11] Fusco, N., and Moscariello, G.: 1981, *L^2-lower semicontinuity of functionals of quadratic type*, Ann. Mat. Pura Appl., (4), 129, pp.305-326.

[12] Marcellini, P.: 1978, *Some problems of semicontinuity and of Γ-convergence for integrals of the calculus of variations*, Proc. of the Int. Coll., Metodi recenti di Analisi non lineare e Applicazioni, Roma.

[13] Marcellini, P., and Sbordone, C.: 1980, *Semicontinuity problems in the calculus of variations*, Nonlinear Anal., 4, pp.241-257.

[14] Marcus, M., and Mizel, V.: 1975, *Lower semi-continuity in parametric variational problems, the area formula and related results*, Amer. J. of Math., 99, pp.579-600.

[15] Meyers, N.G.: 1965, *Quasi-convexity and lower semi-continuity of multiple variational integrals of any order*, Trans. Amer. Math. Soc., 119, pp.125-149.

[16] Morrey, C.B.: 1952, *Quasi-convexity and the lower semi-continuity of multiple integrals*, Pacific J. Math., 2, pp.25-53.

[17] Morrey, C.B.: 1966, *'Multiple integrals in the calculus of variations'*, Springer, Berlin.

A DIFFUSION EQUATION WITH A NONMONOTONE CONSTITUTIVE FUNCTION

Klaus Höllig[1,2] and John A. Nohel[1]

Mathematics Research Center
University of Wisconsin-Madison

ABSTRACT

We discuss the well-posedness of the model problem

$$u_t = \phi(u_x)_x \quad \text{on} \quad [0,1] \times [0,T],\ T > 0\ , \tag{P}$$

subject to given Neumann or Dirichlet boundary conditions at $x = 0$ and $x = 1$, and to the initial condition $u(x,0) = f(x)$; the given functions $f : [0,1] \to \mathbf{R}$, $\phi : \mathbf{R} \to \mathbf{R}$ are assumed to be smooth, $\phi(0) = 0$, ϕ satisfies the coercivity assumption $\xi\phi(\xi) \geq c\xi^2$, for some constant $c > 0$ and for $\xi \in \mathbf{R}$, and ϕ is assumed to be decreasing on an interval (a,b) with $a > 0$. We present a recent nonuniqueness result in the special case when ϕ is piecewise linear and study a related convexified problem.

1. INTRODUCTION

We discuss the initial boundary value problem

$$\begin{aligned} &u_t = \phi(u_x)_x \quad \text{on} \quad [0,1] \times [0,T] \\ &u_x(0,t) = u_x(1,t) = 0, \quad u(x,0) = f(x)\ , \end{aligned} \tag{P}$$

where subscripts denote partial derivatives, under the principal assumption:

J. M. Ball (ed.), Systems of Nonlinear Partial Differential Equations, 409–422.

$\phi : \mathbf{R} \to \mathbf{R}$ smooth, $\phi(0) = 0$, and there exists a constant $c > 0$ such that $\xi\phi(\xi) \geq c\xi^2$, $\xi \in \mathbf{R}$, and there exists an interval (a,b) with $a > 0$ such that $\phi'(\xi) < 0$, $\xi \in (a,b)$. (A)

A model case is sketched in Figure 1.

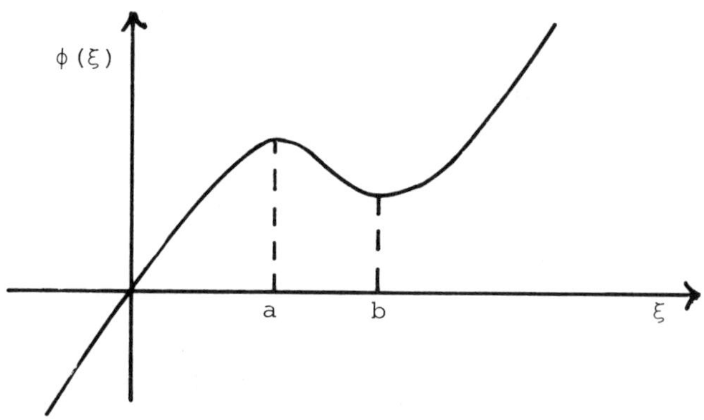

Figure 1

Concerning f we assume throughout that $f : [0,1] \to \mathbf{R}$ is as smooth as needed and satisfies the boundary conditions. Dirichlet and inhomogeneous boundary conditions can also be discussed, including the theorems in Sections 3 and 4.

One motivation for studying (P) is the theory of nonlinear diffusion. Specifically the Clausius-Duhem inequality [3, p. 79] in one space dimension can be shown to imply that the flux $\phi(u_x)$ need not be a monotone function of the gradient of temperature u_x; this is consistent with our assumption (A). If ϕ is strictly monotone increasing (as is usually assumed) standard theory guarantees that (P) has a unique solution which is, roughly speaking, as smooth as the function ϕ. In particular (see Figure 1), this is true in the model case if the data f satisfy $f'(x) \leq a$. If, however, $\phi'(f'(x_0)) < 0$ for some $x_0 \in (0,1)$, then in a neighborhood of $(x_0,0)$, (P) behaves like a backward parabolic equation which is not well posed. In particular, if in Figure 1 $\phi(\xi) = \frac{1}{3}\xi^3 - \frac{3}{2}\xi^2 + 2\xi$, then (P) cannot have a solution u such that u_x is piecewise continuous on $[0,1] \times [0,T]$ for any $T > 0$, unless f is analytic. The basic problem to be discussed here is whether (P) is well-posed in some precise sense, whenever the data $f'(x)$ fall in the critical range ((a,b) in the model case).

The a priori estimates given in Section 2 suggest that (P) may have solutions u with $u_x \in L^\infty([0,1] \times [0,T])$ for some $T > 0$. Such solutions are constructed in Section 3 for <u>piecewise linear</u> ϕ satisfying assumption (A); however, the solutions are not unique. Motivated by known results for the steady-state, non-elliptic problem $(\phi(u_x)_x = 0)$, we are lead to discussing in Section 4 the "convexified" problem associated with (P), with the objective (not yet established) of distinguishing a solution of (P) with special properties. Other open questions include the existence of solutions of (P) when ϕ is not piecewise linear, and a justification of phase changes suggested by numerical experiments.

2. REMARKS ON A PRIORI ESTIMATES

Consider (P) with either homogeneous Neumann or Dirichlet boundary conditions at $x = 0,1$, and with f smooth. Let $u \in W^{1,2}$ be a solution of (P) on $[0,1] \times [0,T]$ for some $T > 0$. Standard energy methods yield the following a priori information.

<u>a</u>. Multiplying by u and integrating the p.d.e. over $[0,1] \times [0,T]$ gives

(i) $$\operatorname*{ess\,sup}_{t\in[0,T]} \int_0^1 u^2(x,t)dx \leq \frac{1}{2}\int_0^1 f^2(x)dx, \quad \text{and}$$

(ii) $$\int_0^T\int_0^1 u_x^2(x,t)dxdt \leq \frac{1}{2c}\int_0^1 f^2(x)dx,$$

where c is the coercivity constant in (A); if u satisfies Dirichlet b.c., the Poincaré inequality applied to (ii) also yields

$$\int_0^T\int_0^1 u^2 dxdt \leq \text{const.}$$

<u>b</u>. Multiplying the p.d.e. by u_t and integrating over $[0,1] \times [0,T]$ gives

(i) $$\int_0^T\int_0^1 u_t^2 dxdt \leq \int_0^1 \Phi(f'(x))dx,$$

where $\Phi'(\cdot) = \phi(\cdot)$ and

$$\text{(ii)} \quad \operatorname*{ess\,sup}_{t\in[0,T]} \int_0^1 \Phi(u_x(x,t))dx \leq \int_0^1 \Phi(f'(x))dx$$

which, in view of the coercivity assumption, implies

$$\text{(iii)} \quad \operatorname*{ess\,sup}_{t\in[0,T]} \int_0^1 u_x^2(x,t)dx \leq \frac{2}{c}\int_0^1 \Phi(f'(x))dx .$$

It should be observed that if ϕ is strictly monotone these estimates imply that $u \in H^2$, and standard techniques yield existence, without applying e.g. the theory of maximum monotone operators (in particular, this is true if $f'(x) < a$, see Figure 1). However, one cannot do this if ϕ is not monotone everywhere ($\phi'(f'(x)) < 0$ for some $x \in \mathbf{R}$).

A more subtle estimate, proved by J. Bona, L. Wahlbin and the second author, is

c. For either the Dirichlet or Neumann problem

$$\|u_x\|_\infty \leq \frac{1}{c} \sup_{x\in[0,1]} |\phi(f'(x))| ;$$

the optimal choice of the coercivity constant $c = \inf\limits_{y\in\mathbf{R}} \frac{\phi(y)}{y} > 0$.

Proof: Multiply the p.d.e. by $\phi^{2k}(u_x)u_t$, where in the exponent k is a positive integer, and integrate with respect to x and t. Defining

$$\Psi_k(y) = \int_0^y \phi^{2k+1}(\xi)d\xi \quad (\geq 0 \text{ by hypothesis on } \phi \text{ for } y \in \mathbf{R})$$

one obtains, after using integration by parts and the boundary conditions:

$$\int_0^t\int_0^1 \phi^{2k}(u_x)u_t^2 dxdt + \frac{1}{2k+1}\int_0^1 \Psi_k(u_x)dx$$

$$= \frac{1}{2k+1}\int_0^1 \Psi_k(f'(x))dx . \qquad (1)$$

From the coercivity assumption

$$\Psi_k(y) \geq \frac{c^{2k+1}}{2k+2}\, y^{2k+2} \quad (y \in \mathbf{R}), \quad \text{and}$$

$$\int_0^1 \Psi_k(u_x)\,dx \geq \frac{c^{2k+1}}{2k+2} \int_0^1 u_x^{2k+2}\,dx \,.$$

Next let $M = \|f'\|_\infty$, $N = \sup_{x\in[0,1]} |\phi(f'(x)|$; then

$$\int_0^1 \Psi_k(f'(x))\,dx \leq MN^{2k+1} \,.$$

Dropping the positive first term in (1), and then raising both sides to the $(2k+2)^{-1}$ yields

$$\frac{c^{\frac{2k+1}{2k+2}}}{(2k+2)^{\frac{1}{2k+2}}} \|u_x(\cdot,t)\|_{(2k+2)} \leq M^{\frac{1}{2k+2}} N^{\frac{2k+1}{2k+2}}, \quad 0 \leq t \leq T \,.$$

Finally, letting $k \to \infty$ we obtain

$$\|u_x\|_\infty \leq \frac{N}{c}, \qquad 0 \leq t \leq T \,.$$

d. Maximum-minimum principle. Consider (P) in the model case of Figure 1 with homogeneous Neumann boundary conditions. Let $u \in W^{2,2}$ be a solution on $[0,1] \times [0,T]$ for some $T > 0$. Then for $0 \leq t \leq T$:

$$\min(a, \min_{0\leq x\leq 1} f'(x)) \leq u_x(x,t) \leq \max(b, \max_{0\leq x\leq 1} f'(x)) \,.$$

For classical solutions the result can be proved by a standard comparison method. We sketch a more general approach. Let $v = u_x$. Then v is a solution of the Dirichlet problem

$$\begin{aligned} &v_t = \phi(v)_{xx} \quad \text{on} \quad [0,1] \times [0,T] \\ &v(x,0) = f'(x); \quad v(0,t) = v(1,t) = 0 \,. \end{aligned} \tag{2}$$

Let $0 < a' < a$, $b' > b > 0$ (see Figure 1). Define $g : \mathbf{R} \to \mathbf{R}$ such that $g'(\cdot) = 0$ on (a',b') and $g'(\cdot) > 0$ otherwise. Let $G(s) = \int_0^s g(\xi)d\xi$. Multiply (1) by $g(v)$ and integrate over $[0,1] \times [0,t]$, $0 < t \leq T$. Integration by parts and the boundary conditions yield

$$\int_0^1 G(v(x,t))dx + \int_0^t\int_0^1 \phi'(v)g'(v)v_x^2 dxdt = \int_0^1 G(f'(x))dx .$$

By definition of g and assumptions on ϕ the double integral is positive, so that

$$\int_0^1 G(v(x,t))dx \leq \int_0^1 G(f'(x))dx, \quad 0 \leq t \leq T .$$

The successive specific choices

$$g(\cdot) = (\cdot - k)_+, \quad k = \max(b', \max_{0 \leq x \leq 1} f'(x)) ,$$

and

$$g(\cdot) = -(\tilde{k} - \cdot)_+, \quad \tilde{k} = \min(a', \min_{0 \leq x \leq 1} f'(x))$$

yield $G(v(x,0)) = G(f'(x)) = 0$. Therefore, respectively,

$$\frac{1}{2}\int_0^1 (v(x,t) - k)_+^2 dx \leq 0, \quad 0 \leq t \leq T ,$$

and

$$\frac{1}{2}\int_0^1 (\tilde{k} - v(x,t))_+^2 dx \leq 0, \quad 0 \leq t \leq T .$$

Thus

$$\tilde{k} \leq v(x,t) \leq k, \quad 0 \leq t \leq T .$$

Letting $a' \to a$ and $b' \to b$ yields the result. It should be noted that this stronger result, proved under the stronger assumption $u \in W^{2,2}$ (rather than $W^{1,2}$), is not implied by the L^∞ estimate in 2c.

3. EXISTENCE AND NONUNIQUENESS IN A SPECIAL CASE

For a piecewise linear constitutive function ϕ it was shown in [5] that the problem (P) has infinitely many solutions if

$$\{x = \phi'(f'(x)) < 0\} \neq \emptyset : \tag{3}$$

Theorem. Let ϕ be piecewise linear with $\phi'((-\infty,a) \cup (b,\infty)) > 0$ and $\phi'((a,b)) < 0$ (see Figure 1) and assume that $f'(x)$ satisfies (3). There exists $T > 0$ such that (P) has infinitely many solutions u with $\phi(u_x)$ Hölder continuous, and $u_x \in L^\infty$, $u_t \in L^p$ $(p < \infty)$ on the domain $[0,1] \times [0,T]$. Moreover, $u_x(x,t) \notin [a,b]$ for $(x,t) \in [0,1] \times (0,T]$.

We outline the proof of the Theorem in the case when the increasing parts of ϕ are parallel, i.e. ϕ is of the form:

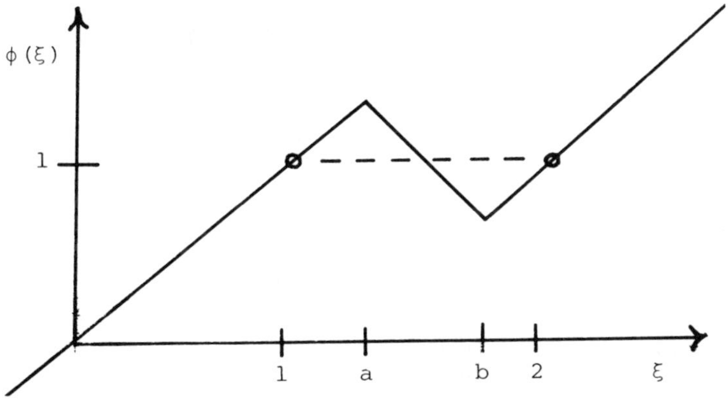

Figure 2

with $\phi'(\mathbf{R}\backslash[a,b]) = 1$. For simplicity we also assume that $f'(x) \leq 2$, $x \in [0,1]$. The key to the proof is the construction of a function w which represents the oscillating part of a solution u and reduces the problem (1) to an inhomogeneous heat equation for the smooth part v of the solution $u = v + w$.

We describe an idea of G. Strang [6] for constructing w which simplifies the original argument in [5]. Consider the following (infinite) triangulation of $[0,1]^2$

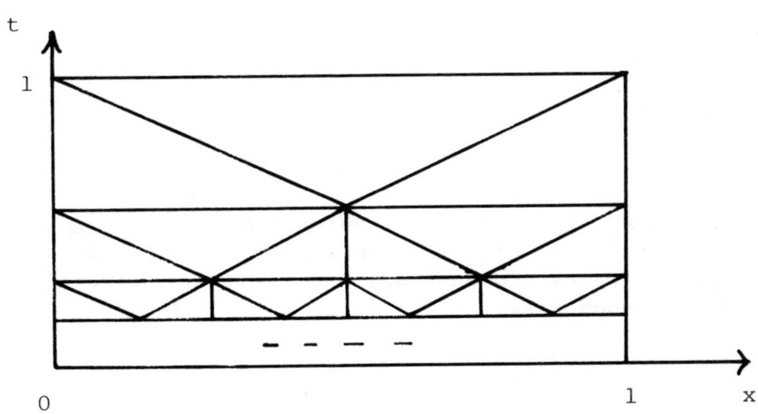

Figure 3

which is determined by the points

$$z_{jk} := (k2^{-j}, 2^{-j}), \quad k = 0,\dots,2^j, \quad j \in \mathbf{N} .$$

Set $h(0) = 0$, $h'(x) = \max(0, f'(x) - 1)$. Define w as the piecewise linear function with respect to this partition which interpolates the data

$$w_{jk} := \max\{\nu 2^{-j} : \nu 2^{-j} \leq h(k2^{-j}),\ \nu \in \mathbf{Z}\}$$

at the points z_{jk}.

By the definition of w_{jk},

$$w_{jk} \leq h(k2^{-j}) < w_{jk} + 2^{-j} ,$$

and since $0 \leq h' \leq 1$ we have

$$w_{j(k+1)} - w_{jk} \in \{0, 2^{-j}\} .$$

One then easily verifies that

$$\begin{aligned}
&\text{(i)} \quad w_x : [0,1] \times [0,T] \to \{0,1\} \\
&\text{(ii)} \quad w_t \in L^\infty \\
&\text{(iii)} \quad \lim_{t \to 0} \|w(\cdot,t) - h\|_\infty = 0 \qquad\qquad (4) \\
&\text{(iv)} \quad f'(x) \leq 1 - 2^{-j} \Longrightarrow w_x(x,t) = 0 \quad \text{for} \quad t \leq 2^{-j} .
\end{aligned}$$

Property (iv) means that for small t $w_x(\cdot,t)$ is supported in a neighborhood of the set $\{x : f'(x) > 1\}$. Let v be the solution of the problem

$$\begin{aligned}
&v_t + w_t = v_{xx} \\
&v_x(0,t) = v_x(1,t) = 0 \qquad\qquad (5) \\
&v(x,0) = f(x) - h(x) .
\end{aligned}$$

We claim that $u = v + w$ is a solution of problem (P).

Since $w_t \in L^\infty$, v_x is Hölder continuous and $v_t \in L^p$ for any $p < \infty$ by standard estimates for the heat equation. Put $\kappa := \min(a - 1, 2 - b)$. By the continuity of v_x and (4(iv)) there exists $T > 0$ such that

$$v_x(x,t) \leq 1 + \kappa, \qquad (x,t) \in [0,1] \times [0,T] ,$$

and

$$(\text{supp } w_x) \cap [0,1] \times [0,T] \subseteq$$
$$\{(x,t) : t \in [0,T], v_x(x,t) > 1 - \kappa\} .$$

In view of the relation

$$\phi(\xi + 1) = \phi(\xi), \ \xi \in [1 - \kappa, 1 + \kappa] , \tag{6}$$

it follows that

$$v_t + w_t = v_{xx} = \phi(v_x)_x = \phi(v_x + w_x)_x .$$

Clearly, the function w is not uniquely determined. For any $\lambda > 0$, $w_\lambda(x,t) := w(x,\lambda t)$ satisfies (4). Since v_x is continuous the solutions $u_\lambda = v_\lambda + w_\lambda$ are distinguished by their different discontinuity patterns. Thus there is a continuum of solutions to the problem (P).

If the monotone increasing parts of ϕ have different slopes the proof of the Theorem is considerably more complicated (see [5]). The reason is that instead of (6) we have the relation

$$\phi(\xi + (A\xi + B)) = \phi(\xi) , \tag{7}$$

where the shift is no longer independent of ξ. Nevertheless, the proof is based on the construction of an auxiliary function w which, however, depends on v and the equation for the smooth part v becomes quasilinear. For a general smooth nonmonotone ϕ the existence question is not settled.

In spite of nonuniqueness, numerical approximations to the problem (P) are stable; the x-derivatives of the numerical solution have the oscillatory behavior asserted by the Theorem.

We used a Crank-Nicholson scheme

$$u_j^{n+1} - u_j^n = \frac{\Delta t}{\Delta x} D\phi\left(\frac{\Delta x}{2} (Du_{j-1}^{n+1} + Du_{j-1}^n)\right) \tag{8}$$

where u_j^n approximates $u(j\Delta x, n\Delta t)$ and $Du_j = u_{j+1} - u_j$ denotes the forward difference. For ϕ in Figure 2 with $a = 1.25$, $b = 1.75$ and the model initial data $f(x) = 3x^2 - 2x^3$ the numerical solutions show the following typical behavior. For small t, $u\Delta t < 0.1$, Du_j^n oscillates over the interval where $\phi'(f')$ is negative as shown in figure (4) below

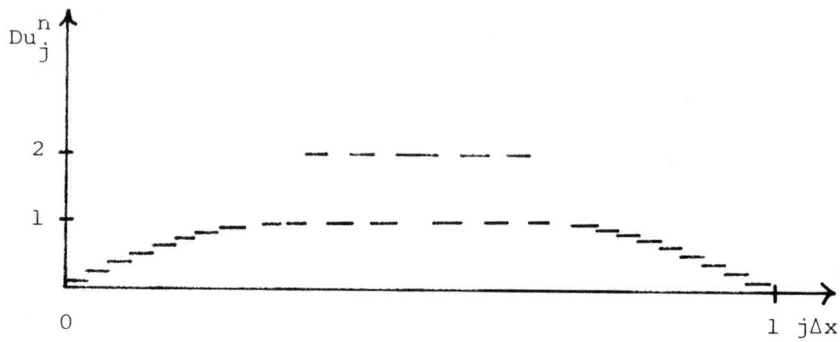

Figure 4

where $\Delta x = 2^{-5}$, $\Delta t = (\Delta x)^2$, $n\Delta t = 2^{-7}$. The oscillations are very regular. Du_j^n alternates between the values 1 and 2 which, in agreement with the Theorem, differ by 1 and do not lie in the interval [1.25, 1.75] where ϕ' is negative. Moreover, $\phi(Du_j^n)$ is (numerically) continuous.

As $t = n\Delta t$ increases the oscillations gradually disappear and u_j^n approaches the limiting state $u_x = 0$. Similar numerical approximations have been independently studied by G. Strang and A. Naby [6]. It has not been established that the numerical approximations converge and thereby possibly single out a distinguished solution of the problem (P).

4. A RELATED CONVEXIFIED PROBLEM

It is known [1] that in one space dimension the solution of the problem

$$0 = \phi(u_x)_x, \qquad x \in [0,1]$$
$$u(0) = 0, \quad u(1) = d, \quad d \text{ a constant}, \tag{9}$$

can be obtained by minimizing the functional

$$J(u) = \int_0^1 \Phi(u_x(x))dx, \qquad \Phi'(\cdot) = \phi(\cdot),$$

over all functions $u \in W^{1,2}([0,1])$ satisfying the given boundary conditions, under suitable assumptions on Φ (these do not require Φ convex). If ϕ satisfies assumptions (A), (9) is the steady state problem associated with (P). By a result of Ekeland and Témam [4] such a solution u of the variational problem also minimizes the functional

$$\bar{J}(u) = \int_0^1 \bar{\Phi}(u_x(x))dx ,$$

where $\bar{\Phi}$ is the convexification of Φ, and u solves the convexified problem:

$$\begin{aligned} &0 = \bar{\phi}(u_x)_x, \qquad x \in [0,1] \\ &u(0) = 0, \quad u(1) = d, \quad \bar{\phi}(\cdot) = \bar{\Phi}'(\cdot) ; \end{aligned} \tag{10}$$

the function $\bar{\phi}(\cdot)$ is necessarily nondecreasing on **R** (if ϕ has the graph in Figure 2, Figure 5 below is the graph of $\bar{\phi}$). An analogous variational principle has been established for problem (P), but only for ϕ monotone (see [2]).

Motivated by these results we consider the convexified problem associated with (P):

$$\begin{aligned} &u_t = \bar{\phi}(u_x)_x, \qquad (x,t) \in [0,1] \times [0,T] , \\ &u_x(0,t) = u_x(1,t) = 0, \quad u(x,0) = f(x) . \end{aligned} \tag{$\bar{P}$}$$

By the Theorem below, solutions of (P) and ($\bar{P}$) show a qualitatively different behavior and, at first sight, there is no obvious connection between them when the data f' is smooth and falls in the critical range (a,b). Nevertheless, it appears plausible that as $t \to \infty$ problems (P) and ($\bar{P}$) approach the same steady state.

For simplicity we discuss ($\bar{P}$) only for the piecewise linear constitutive function considered in Section 3. For ϕ in Figure 2, $\bar{\phi}$ is of the form

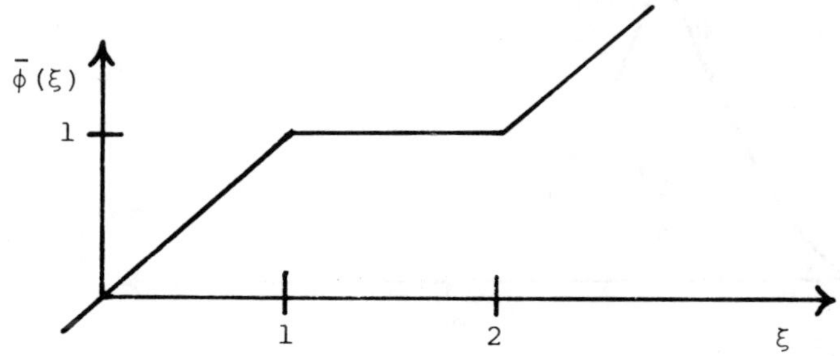

Figure 5

The graph of $\bar{\phi}$ is familiar from the Stefan problem. In fact, if the initial data satisfy

$$f'(x) \notin (1,2), \quad x \in [0,1] ,$$

$$\bar{\phi}(f') \quad \text{continuous} ,$$

$(\bar{P})$ describes a Stefan problem. If, however, f is smooth which we assume throughout this section, the evolution of the free boundaries resulting from $(\bar{P})$ is different. For simplicity we assume that

$$f'(x) \leq 2, \quad x \in [0,1] ,$$

$$\{x : f'(x) > 1\} = (r_0,s_0), \quad 0 < r_0 < s_0 < 1 , \tag{11}$$

$$f''(r_0) > 0, \quad f''(s_0) < 0 .$$

Theorem. Problem $(\bar{P})$ with $\bar{\phi}$ given in Figure 5 has a unique solution u on $[0,1] \times \mathbf{R}_+$ with

$$\|u_x\|_\infty \leq \|f'\|_\infty$$

$$\|u_t\|_\infty \leq \|\phi(f')'\|_\infty .$$

If f satisfies assumptions (11), the region $\Omega := \{(x,t) : u_x(x,t) > 1\}$ where $u_t = 0$ is bounded by two curves $r,s \in H^{1/2}$ (Hölder class) connecting the points $(r_0,0),(s_0,0)$ with the point (x_*,T_*) (Figure 6). Moreover,

$$T_* \leq \|f'\|_\infty . \tag{12}$$

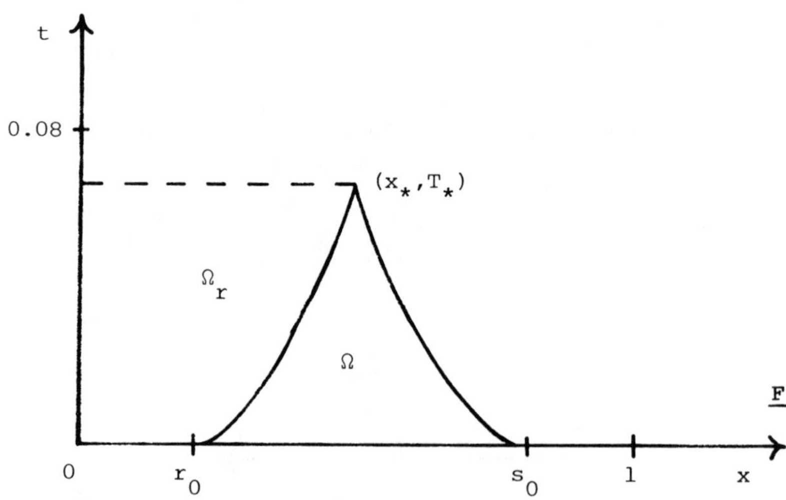

Figure 6

Note that u satisfies the heat equation on $[0,1] \times \mathbf{R}_+ \setminus \Omega$ and the regularity assertions in particular describe the behavior of u in a neighborhood of the points $(r_0,0),(s_0,0)$.

The regularity for u,r,s is proved by using the implicit semidiscrete approximation

$$u(x,(n+1)\Delta t) - u(x,n\Delta t) = \Delta t\,\phi(u_x(x,(n+1)\Delta t))_x, \quad x \in [0,1] ,$$

of the equation $u_t = \bar{\phi}(u_x)_x$. We prove the inequality (12). Since u is continuous and $u(x,t) = u(x,0)$ on Ω we have

$$u(r(t),t) = f(r(t)) . \tag{13}$$

Also, note that

$$u_x(r(t)^-,t) = 1 . \tag{14}$$

Integrating the equation $u_t = \bar{\phi}(u_x)_x$ over the domain $\Omega_r := \{(x,t) : 0 < x < r(t),\ t \in (0,T_*)\}$ we obtain

$$0 = \iint_{\Omega_r} (u_t - u_{xx})dxdt =$$

$$- \int_0^{r_0} f(x)dx - \int_0^{T_*} f(r(t))\dot{r}(t)dt + \int_0^{r(T_*)} u(x,T_*)dx$$

$$- \int_0^{T_*} 1\, dt .$$

Using (13) and $\|u_x\|_\infty \leq \|f'\|_\infty$ for the estimate, it follows that

$$T_* = \int_0^{r(T_*)} (u(x,T_*) - f(x))dx \leq \int_0^1 \int_x^{r(T_*)} 2\|f'\|_\infty dydx \leq \|f'\|_\infty .$$

Numerical computations show (Figure 6 is a particular example with $f(x) = 3x^2 - 2x^3$) that the curves r,s touch the x-axis. Formally, this can be explained as follows. Differentiating equation (13) with respect to t and using (14) we have

$$\dot{r}(t)(f'(r(t)) - 1) = u_t(r(t),t) .$$

Since $f'(r(0)) = f'(r_0) = 1$ we can rewrite this in the form

$$\dot{r}(t)(r(t) - r_0) = u_t(r(t),t)/g(r(t) - r_0)$$

where $g(\xi - r_0) := \frac{f'(\xi) - f'(r_0)}{\xi - r_0}$. If we assume that u_t, u_{xx} are continuous on $\overline{\Omega_r}$ it follows that

$$r(t) = r_0 + \sqrt{2t} + o(\sqrt{t}) \quad (t \to 0), \tag{15}$$

which is consistent with regularity of r established in the theorem above.

We have not yet succeeded in proving (15). The difficulty is that the free boundary $x = r(t)$ is characteristic for $(\overline{P})$ at the point $(r_0, 0)$.

FOOTNOTES

[1]Sponsored by the United States Army under Contract No. DAAG29-80-C-0041.

[2]This material is based upon work partially supported by the National Science Foundation under Grant No. MCS-7927062, Mod. 1.

REFERENCES

1. Aubert, C. and Tahraoui, R., Théorèmes d'existence pour des problèmes du calcul des variations du type: $\text{Inf} \int_0^L f(x,u'(x))dx$ et $\text{Inf} \int_0^L f(x,u(x),u'(x))dx$, J. Differential Equations 33 (1979), pp. 1-15.

2. Brézis, H. and Ekeland, I., Un principe variationnel associé à certaines équations paraboliques, Comptes Rendus Acad. Sc. Paris 282 (1976), pp. 971-974.

3. Day, W., The thermodynamics of simple materials with fading memory, Springer tracts in natural philosophy, vol. 22, New York, Heidelberg, Berlin (1970).

4. Ekeland, I. and Temam, R., Convex Analysis and Variational Problems, Amsterdam: North-Holland (1976).

5. Höllig, K.: Existence of infinitely many solutions for a forward backward heat equation, Trans. Amer. Math. Soc., to appear.

6. Strang, G. and Abdel-Naby, M.: Notes on softening and local instability. Lecture in Engineering, Springer Verlag, to appear.

AN ADMISSIBILITY CRITERION FOR FLUIDS EXHIBITING PHASE TRANSITIONS

Marshall Slemrod

Department of Mathematical Sciences
Rensselaer Polytechnic Institute
Troy, New York 12181 USA

ABSTRACT

This paper shows how Korteweg's theory of capillarity can be used as an admissibility criterion for weak solutions of the equations of compressible fluid flow. Included in the theory are fluids capable of exhibiting liquid-vapor changes of phase.

0. INTRODUCTION

In a recent series of papers [1], [2], [3], [4] R. Hagan and I have tried to understand shock "structure" for fluids undergoing phase transitions. The beginnings of a theory of inter-phase shocks were first put forward by H. Bethe [5] in his classic article. Since then numerous theoretical articles have appeared (see [6] for a good discussion) and experiments performed [7]; however the shock "structure" problems lay untouched. In this article I will try to highlight the essence of our ideas on this issue and show how Korteweg's theory of capillarity plays the key role in the resolution of the problem.

1. MATHEMATICAL FORMULATION

Consider the one dimensional motion of a compressible fluid along the x-axis. If $\rho(x,t)$, $u(x,t)$, $\tau(x,t)$ denote the density, velocity, and Cauchy stress at a point x and time t the equations of balance of mass and momentum (in the Eulerian coordinates) are

$$\rho_t + (\rho u)_x = 0 \tag{1}$$

J. M. Ball (ed.), Systems of Nonlinear Partial Differential Equations, 423–432.

$$(\rho u)_t + (\rho u^2)_x = \tau_x, \quad -\infty < x < \infty, \tag{2}$$

$$\rho(x,0) = \rho_0(x), \; u(x,0) = u_0(x).$$

We also assume (i) there are no body forces and (ii) that the fluid is imbedded in a "heat bath" so that the motion is isothermal i.e. the balance of energy equation is automatically satisfied with constant temperature. We denote $w(x,t) = \rho(x,t)^{-1}$ as the specific volume.

In an <u>elastic fluid</u> the constitutive relation

$$\boxed{\tau = -\, p(w)} \tag{3}$$

is assumed where p is the pressure. One example of an elastic fluid is the <u>ideal gas</u>

$$\boxed{p(w) = \frac{R\theta_0}{w}} \tag{4}$$

as depicted in Figure 1. Here θ_0 is the (assumed) constant temperature.

Figure 1. Ideal gas p-w diagram

We notice that for the ideal gas $p'(w) < 0$ and systems (1), (2) becomes a quasi-linear hyperbolic system.

While the ideal gas law is certainly important, it is not the only type of elastic fluid. Before moving on to a second kind consider the following experiment described in [8]. Put water vapor in a large container at 15° C. This is sketched in Figure 2.

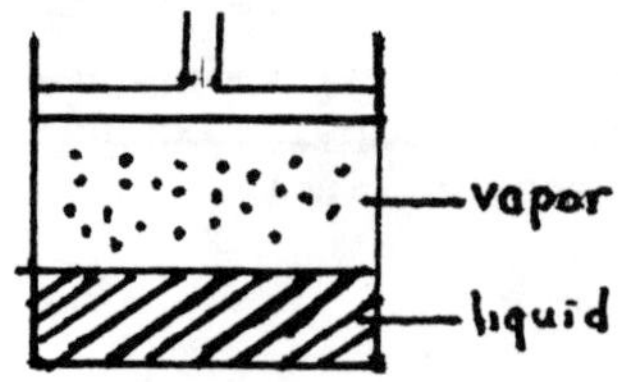

Figure 2. Water vapor in large container

Now let us start to slowly move the piston into the container. As we move the piston the density of the vapor increases and so the specific volume decreases. For a while the pressure inside the container increases. This is depicted on the $E_1 - E_2$ curve in Figure 3. When we reach E_2 liquid drops of water start to form and as we bring the piston down the pressure stays constant until we reach F_1. Here all the vapor has condensed to liquid at which point the pressure again rises with decreasing w. This is shown on the $F_1 - F_2$ curve. Hence we notice the departure from the ideal gas law.

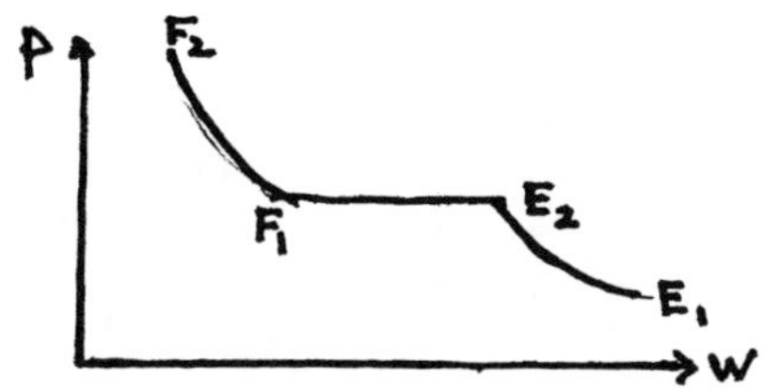

Figure 3. Water 15° C p-w diagram

To account for the possible co-existence of liquid and vapor phases found in many fluids van der Waals in 1873 introduced the constitutive relation that bears his name

$$p(w) = \frac{R\theta_0}{w-b} - \frac{a}{w^2}, \qquad \begin{array}{l} 0 < b < w < \infty \\ R, a > 0. \end{array} \tag{5}$$

In (5) the first term represents a correction to the ideal gas law based on the idea that a fluid cannot be compressed beyond some critical molecular size. The second term accounts for inter-particle attractiveness. Notice that when w is small the second term becomes dominant, i.e. precisely when the fluid is in the liquid phase the previously neglected inter-particle attractions are now accounted for. A graph of the van der Waals isotherm for θ_0 sufficiently small is given in Figure 4.

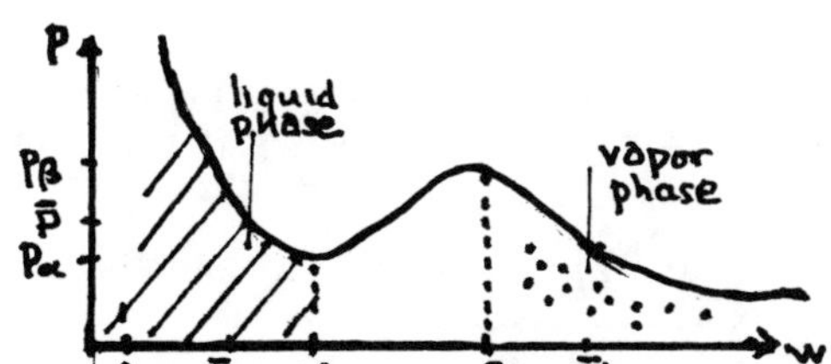

Figure 4. van der Waals isotherm

From Figure 4 we see that for a certain range of pressures, $p_\alpha < p < p_\beta$, the van der Waals fluid can simultaneously exist in liquid and vapor phases. However this is not what we observed in our hypothetical experiment for water. Specifically at equilibrium $u \equiv 0$ and (2) implies

$$p(w) = \bar{p} \tag{6}$$

where $\bar{p}$ is an arbitrary constant. Hence

$$\begin{aligned} w &= \bar{w}_\ell \quad x \in S, \\ &= \bar{w}_v \quad x \in \mathbb{R} - S, \end{aligned} \tag{7}$$

S any set in $\mathbb{R}$, will be a weak solution of (1), (2) since $p(\bar{w}_\ell) = p(\bar{w}_v) = \bar{p}$. On the other hand in our experiment there was a preferred pressure p_{eq} for the co-existence of liquid and vapor phases. So in a sense we are looking for a preferred solution of (1), (2) at equilibrium. One way to obtain such a solution was given by Maxwell using thermodynamics and this is the presentation typically found in standard texts [9], [10]. However in 1895 [11] van der Waals himself presented a resolution of this problem in a different manner which is particularly enlightening from the point of view of the dynamics of (1), (2). Van der Waals' idea has come down to us via subsequent refinements [12] and statistical derivations [13]. However in this paper I'll follow the approach of Serrin [14] who has based his arguments on Korteweg's theory of capillarity [15], [16].

In Korteweg's theory of capillarity the stress in an isotropic fluid takes on the constitutive form

$$\boxed{\tau = -p(w,\theta_0) + \mu(w,\theta_0)u_x + D(w,\theta_0)w_x^2 - A(w,\theta_0)w_{xx}} \tag{8}$$

where θ_0 is still the constant temperature. Here $\mu(w,\theta_0) > 0$ is the viscosity and the w_x^2, w_{xx} terms denote higher order corrections to the elastic fluid constitutive relation. The importance of these terms for fluid distributions like (7) where $|w_x| \approx \infty$ at at certain points is self evident.

Let us now use (8) to find the equilibrium configuration of the van der Waals fluid. At equilibrium $u \equiv 0$ and (2) implies

$$-p(w,\theta_0) + D(w,\theta_0)w_x^2 - A(w,\theta_0)w_{xx} = -\bar{p}, \quad -\infty < x < \infty, \tag{9}$$

$\bar{p}$ a constant

Now (9) may be integrated explicitly to find

$$w_x^2(x) = -2 \int_{w_-}^{w(x)} \exp\left(\int_k^{w(x)} \frac{2D(\zeta,\theta_0)}{A(\zeta,\theta_0)} d\zeta\right) A(k,\theta_0)^{-1}(p(k,\theta_0)-\bar{p})\,dk \tag{10}$$

where $w(-\infty) = w_-$. Thus for $\lim_{x\to+\infty} w(x) = w_+$ to exist we need

$$0 = \int_{w_-}^{w_+} \exp\left(\int_k^{w_+} \frac{2D(\zeta,\theta_0)}{A(\zeta,\theta_0)}\, d\zeta\right) A(k,\theta_0)^{-1}(p(k,\theta_0)-\bar{p})\, dk,$$
$$p(w_-,\theta_0) = p(w_+,\theta_0) = \bar{p}. \tag{11}$$

Thus (11) provides a "rule" on the end states for phase equilibrium. For example when $D(\zeta,\theta_0) = -\frac{1}{2}\frac{\partial A}{\partial \zeta}(\zeta,\theta_0)$ we are led to

$$\int_{w_-}^{w_+} (p(k,\theta_0)-\bar{p})\, dk = 0,$$

i.e. the Maxwell equal area rule shown in Figure 5.

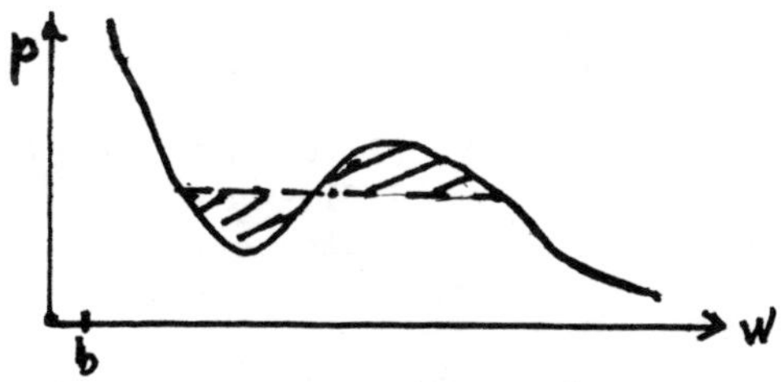

Figure 5.

On the other hand, as noted by Serrin [14], other choices of A,D when applied to (11) will lead to other "rules" for phase equilibrium.

Now let us reconsider a solution of type (7), for example,

$$w = \begin{cases} \bar{w}_\ell & x < x_0 \\ \bar{w}_v & x > x_0 \end{cases} \tag{12}$$

We may ask when (12) is a limit of solutions of equilibrium solutions to (1), (2) with τ given by Korteweg's theory as A, D → 0? Since the Korteweg theory is hopefully a better model of reality than the elastic fluid law it should provide us with the preferred or admissible solution of $p(w)_x = 0$. This motivates the following definition

Definition. A weak solution of $p(w)_x = 0$, $-\infty < x < \infty$, is (A,D) admissible if it is the limit boundedly a.e. of solutions of the equation

$$0 = (p(w^\varepsilon) + D\varepsilon^2 w_x^{\varepsilon 2} - A\varepsilon^2 w_{xx}^\varepsilon)_x \tag{13}$$

as $\varepsilon \to 0+$.

Remark 1. The condition $(*) D(\zeta,\theta_0) = -\frac{\partial A}{\partial \zeta}(\zeta,\theta_0)$ will imply that bounded a.e. limits of (13) are actually distributional solutions of the $\varepsilon = 0$ problem. Note the lack of the factor 1/2 which occurred in the criterion for the Maxwell rule. If A = constant both (*) and Maxwell's rule hold.

If we set $w^{\varepsilon}(x) = \hat{w}(\xi)$, $\xi = \frac{x-x_0}{\varepsilon}$, then (13) becomes

$$0 = (-p(\hat{w})+D\hat{w}'^2 - A\hat{w}'')' \tag{14}$$

where $' = \frac{d}{d\xi}$. Furthermore if we desire $w^{\varepsilon} \to \underline{w_{\ell}}$, $x < x_0$; $w^{\varepsilon} \to \overline{w}_v$, $x > x_0$, we need $\hat{w}(-\infty) = \overline{w}_{\ell}$, $\hat{w}(+\infty) = \underline{w}_v$. So the same computation that led to (1) shows (12) is (A,D) admissible if $w_- = w_{\ell}$, $w_+ = w_v$ satisfy (11). In particular (12) is (A,D) admissible, A constant, if w_{ℓ}, w_v satisfy Maxwell's equal area rule.

We can extend this idea to non-equilibrium cases as well.

Definition. A weak solution of (1), (2) with $\tau = -p(w)$, is (A,D) is admissible if it is the limit boundedly a.e. as $\varepsilon \to 0+$ for (x,t) $\mathbb{R} \times \mathbb{R}^+$ of solutions of the problem

$$\rho^{\varepsilon}_t + (\rho^{\varepsilon}u^{\varepsilon})_x = 0,$$

$$(\rho^{\varepsilon}u^{\varepsilon})_t + (\rho^{\varepsilon}u^{\varepsilon^2})_x = (-p(w^{\varepsilon},\theta_0) + \varepsilon u^{\varepsilon}_x + \varepsilon^2 D(w^{\varepsilon},\theta_0)w^{\varepsilon^2}_x$$

$$- \varepsilon^2 A(w^{\varepsilon}\theta_0)w^{\varepsilon}_{xx})_{,x}, \tag{15}$$

where $w^{\varepsilon} = \frac{1}{\rho^{\varepsilon}}$.

Remark 2. We note that Remark 1 also applies to the above non-equilibrium case.

As an example consider the piece-wise constant, time-independent solution of (1), (2) with $\tau = p(w)$, given by

$$\left.\begin{array}{l} u = u_- \\ w = w_- \end{array}\right\} x < 0, \qquad \left.\begin{array}{l} u = u_+ \\ w = w_+ \end{array}\right\} x > 0, \tag{16}$$

where u_-, u_+, w_-, w_+ satisfy the Rankine-Hugoniot jump conditions

$$\begin{aligned} \rho_+ u_+ &= \rho_- u_- \ , \\ \rho_+ u_+^2 + p(w_+) &= \rho_- u_-^2 + p(w_-) , \end{aligned} \tag{17}$$

where

$$w_+ = \frac{1}{\rho_+} \ , \ w_- = \frac{1}{\rho_-} \ .$$

We note that if $m = \rho_- u_-$ (17) implies

$$m^2 = \frac{-p(w_+)+p(w_-)}{w_+-w_-} \tag{18}$$

We now seek time independent solutions of (15) of the form $u^\varepsilon(x) = \hat{u}(\xi)$, $w^\varepsilon(x) = \hat{w}(\xi)$, $\xi = \frac{x}{\varepsilon}$. In order that $u^\varepsilon, w^\varepsilon$ have the desired limits as $\varepsilon \to 0_+$ we require

$$\hat{u}(-\infty) = u_-, \ u(+\infty) = u_+, \ \hat{w}(-\infty) = w_-, \ \hat{w}(+\infty) = w_+ . \tag{19}$$

A direct substitution into (15) shows

$$\begin{aligned} &A\hat{w}''(\xi) - m\hat{w}'(\xi) - D\hat{w}'(\xi)^2 + m^2(\hat{w}(\xi)-w_-) + p(\hat{w}(\xi),\theta_0) \\ &\qquad - p(w_-,\theta_0) = 0, \end{aligned} \tag{20}$$

where again $' = \frac{d}{d\xi}$. So if (19), (20) possesses a solution we see

$$w^\varepsilon(x) = \hat{w}(\tfrac{x}{\varepsilon}) \to \begin{cases} w_- & x < 0 \\ w_+ & x > 0 \end{cases} \quad \text{as } \varepsilon \to 0_+ \text{ and}$$

$$u^\varepsilon(x) = \hat{u}(\tfrac{x}{\varepsilon}) = m\hat{w}(\tfrac{x}{\varepsilon}) \to \begin{cases} u_- & x < 0 \\ u_+ & x > 0 \end{cases} \quad \text{as } \varepsilon \to 0_+$$

and (16) will be (A,D) admissible.

As an example of an (A,D) admissible solution with $A > 0$ consider the case when w_- is in the liquid phase as shown in Figure 6. We will try to find $m < 0$ and a w_+ in the vapor phase so that (16) is (A,D) admissible. First for fixed w_-, m let $w_+(m)$ denote the solution (if it exists) of (18). Again this is illustrated in Figure 6. Now let $\eta(\hat{w};m) = \hat{w}'(\xi)$ where $\hat{w}$ is a solution of (20), $\hat{w}(-\infty) = w_-$. A direct integration of (20) shows

$$\eta^2(\hat{w};m) = \int_{w_-}^{\hat{w}} G(k,\hat{w};\theta_0,m) \Big\{ 2m\eta(k;m) - 2(p(k,\theta_0) - p(w_-,\theta_0)) - 2m^2(k-w_-) \Big\} dk \tag{21}$$

where $G(k,\hat{w};\theta_0,m) = \exp(\int_k^{\hat{w}} \frac{2D(\zeta,\theta_0)}{A(\zeta,\theta_0)} d\zeta)A(k,\theta_0)^{-1}$.

$G > 0$ since we have assumed $A > 0$.

Consider the case $m = 0$. Then we see

$$\eta^2(w_+(0),0) = -2 \int_{w_-}^{w_+(0)} G(k;w_+(0);\theta_0,0)(p(k,\theta_0)-p(w_-,\theta_0))dk > 0$$

if

$$\int_{w_-}^{w_+(0)} \exp(\int_k^{w_+(0)} \frac{2D(\zeta,\theta_0)}{A(\zeta,\theta_0)} d\zeta)A(k,\theta_0)^{-1}(p(k,\theta_0)-p(w_-,\theta_0)dk < 0. \quad (22)$$

So far for $m = 0$ $w' > 0$ at $w_+(0)$. In fact the phase flow for $m = 0$ is shown in Figure 7.

Now let us decrease m from zero. Let $\bar{m} < 0$ be such that

$$\int_{w_-}^{w_+(\bar{m})} G(k,w_+(\bar{m});\theta_0\bar{m}) \left\{p(k,\theta_0)-p(w_-,\theta_0 + \bar{m}^2(k-w_-)\right\} dk = 0. \quad (23)$$

(For example in the case $A > 0$ constant, $D \equiv 0$, this means area A = area B as shown in Figure 6.) In this case $\eta^2(w_+(\bar{m}),\bar{m})$ cannot exist. For if it did we see from (21) that

$$2\bar{m} \int_{w_-}^{w_+(m)} G(k,w_+(\bar{m}),\theta_0,\bar{m})\eta(k;\bar{m})dk > 0$$

and since $\bar{m}$ $G < 0$ $\eta(k;\bar{m})$ must become negative before $w_+(m)$ and turn back towards the w' axis. Hence the phase flow has the picture shown in Figure 7.

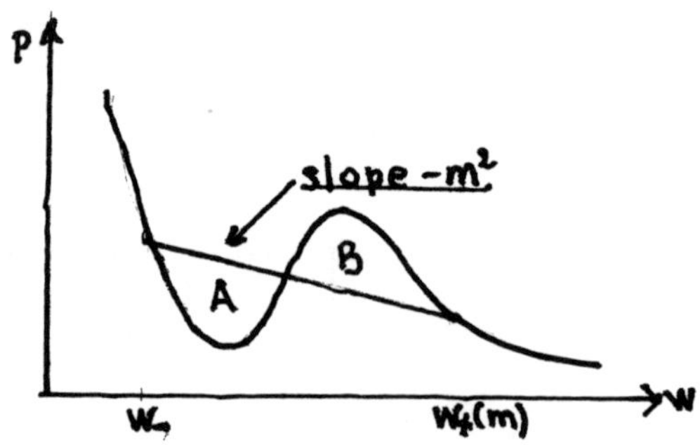

Figure 6

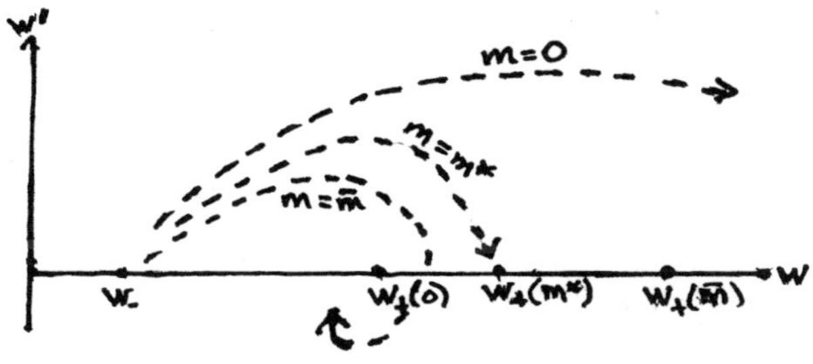

Figure 7

Examination of the two extreme cases $m = 0$, $m = \bar{m}$, shows that $w_+(\underline{m})$ has "moved through" the trajectory as m decreased from zero to $\bar{m}$. Hence for some value of m, say m^*, $\bar{m} \leq m^* < 0$, we have $\eta(w_+(m^*),m^*) = 0$. For this value of $m^*, w_+(m^*)$, (19), (20) has a solution and hence for $\rho_+ = \frac{1}{w_+(m^*)}$ and $u_+ = m^* w_+(m^*)$ (16) is (A,D) admissible. Notice in this case the state on the "left" in the liquid phase determines a state on the "right" in order to make (16) (A,D) admissible. This differs from the usual shock wave theory for materials not undergoing phase transitions where one state can be connected by a shock to any other state as long as the Rankine-Hugoniot conditions are satisfied and the shock is compressive.

ACKNOWLEDGEMENT

This research was sponsored in part by the Air Force Office of Scientific Research, Air Force Systems Command, USAF, under Contract/ Grant No. AFOSR-81-0172. The United States Government is authorized to reproduce and distribute reprints for government purposes not withstanding any copyright herein.

REFERENCES

1. M. Slemrod, Admissibility criteria for propagating phase boundaries in a van der Waals fluid, Univ. of Wisconsin MRC Technical Summary Report #2280, to appear Archive for Rational Mechanics and Analysis.

2. M. Slemrod, Dynamic phase transitions in a van der Waals fluid, Univ. of Wisconsin MRC Technical Summary Report #2298, to appear J. of Differential Equations.

3. R. Hagan and M. Slemrod, The viscosity-capillarity admissibility criterion for shocks and phase transitions, to appear as Univ. of Wisconsin MRC Technical Report and in Archive for Rational Mechanics and Analysis.

4. R. Hagan, Dynamic phase transitions, Ph.D. thesis, Dept. of Mathematical Sciences, Rensselaer Polytechnic Institute, Troy, N.Y. 12181, August, 1982.

5. H. Bethe, On the theory of shock waves for an arbitrary equation of state, Office of Scientific Research and Development Report 545, May, 1942.

6. Ya. B. Zel'dovich and Yu. P. Raizer, Physics of shock waves and high temperature hydrodynamic phenomena, Vol. II. Academic Press: New York and London, (1967).

7. G. Dettleff, P. A. Thompson, G. E. A. Meier, and H. -D. Speckman, An experimental study of liquefaction shock waves, J. Fluid Mechanics, 95, (1979), 279-304.

8. H. N. V. Temperley and D. H. Trevena, Liquids and their properties. Ellis Horwood Publishers: Chichester, England, (1978).

9. A. B. Pippard, Elements of classical thermodynamics, Cambridge University Press: Cambridge (1964).

10. A. Sommerfeld, Thermodynamics and statistical mechanics, Academic Press: New York and London (1964).

11. J. D. van der Waals, Theorie thermodynamique de la capillarité, dans l'hypothese d'une variation continue de densité, Archives Neerlandaises des Sciences exactes et naturelles, 28, (1895), 121-209.

12. J. W. Cahn & J. E. Hilliard, Free energy of a nonuniform system, Journal of Chemical Physics, 28, (1958), 258-267.

13. H. T. Davis & L. E. Scriven, Stress and structure in fluid interfaces, Advances in Chemical Physics, (1982), 357-454.

14. J. Serrin, Phase transitions and interfacial layers for van der Waals fluids, Recent Methods in Nonlinear Analysis and Applications, Ed. A. Confora, S. Rionero, C. Sborbone, and G. Trombetti, Instituto Mathematico, Universita di Napoli, Naples, (1980), 169-175.

15. D. J. Korteweg, Sur la forme que prennent les equations des mouvement des fluids si l'on tient comple des forces capillaires par des variations de densité, Archives Neerlandaises des Sciences exactes et naturelles, Series II, 6, (1901), 1-24.

16. C. A. Truesdell & W. Noll, The non-linear field theories of mechanics, Vol. III/3 of the Encyclopedia of Physics, S. Flugge, editor, Springer-Verlag: Berlin (1965).

STABILIZATION OF SOLUTIONS FOR A SYSTEM WITH A CONTINUUM OF EQUILIBRIA AND DISTINCT DIFFUSION COEFFICIENTS

Nicholas Alikakos

Purdue University
West Lafayette, IN 47905

ABSTRACT

A reaction-diffusion system is considered and the behavior of the solutions as $t\to\infty$ is studied. The mathematically interesting features are the following two. First the lack of appropriate invariant sets necessary for localizing the ω-limit sets, and this is a manifestation of the general diffusion matrix, and second the existence of connected sets of equilibria.

The objective of this note is to establish under natural hypotheses the stabilization as $t\to\infty$ of the solutions of the system

$$u_t = D\Delta u + (c-G(u))u \qquad \text{on} \quad \Omega\times(0,\infty) \tag{1}$$

$$\left.\frac{\partial u}{\partial n}\right|_{\partial\Omega} = 0 \qquad \Omega \subseteq \mathbb{R}^n, \ |\Omega| = 1$$

$u = \mathrm{col}(u_1,\ldots,u_N)$, $G(u) = \langle Au;u\rangle$, $\langle\ ,\ \rangle$ inner product in $\mathbb{R}^N$, D, A constant positive definite symmetric matrices. System (1) is of relevance in various superconductivity theories ([1]). Our results extend those in [2] where the case D=A=I, N=2 is treated. This extension derives its mathematical interest from two facts. First the introduction of distinct diffusion coefficients destroys the invariance of certain nonrectangular sets that are crucial in the analysis in [2]. Second, for N>2 there is a higher dimensional connected manifold of equilibria.

J. M. Ball (ed.), Systems of Nonlinear Partial Differential Equations, 443–441.

We introduce two hypotheses each of which has a physical explanation.

(H1) DA=AD

The commutativity guarantees that the system has a gradient structure[1] and so the ω-limit sets consist of equilibria. This assumption may be related heuristically to nonequilibrium thermodynamics. The connection goes as follows. Under appropriate circumstances the reaction-diffusion system

$$\frac{\partial u_i}{\partial t} = \sum_k d_{ik} \Delta u_k + \ldots \tag{2}$$

may be written alternatively in the form

$$\frac{\partial u_i}{\partial t} = \operatorname{div} \sum_j L_{ij} \nabla(\mu_j(u)) + \ldots \tag{3}$$

where the L_{ij} are the so called phenomenological coefficients and μ_j the chemical potentials ([3]). Taking L_{ij} as constants we obtain the system

$$\frac{\partial u_i}{\partial t} = \sum_{j,k} L_{ij} \frac{\partial \mu_j}{\partial u_k} \Delta u_k + \ldots \tag{4}$$

Onsager's requirement is that the matrix $L = (L_{ij})$ be symmetric. In our case the natural entropy functional is $J(u(\cdot)) = -\int_\Omega G(u(x))dx$ and so the chemical potentials are given by $\mu_i(u) = \frac{\partial G(u)}{\partial u_i}$. Upon comparing (2) and (4) we obtain that D=LA and so the symmetry of L is equivalent to (H1).

Let $\lambda_1, \ldots, \lambda_N$ denote the eigenvalues of A. For convenience we take $\lambda_N = 1$ and $c=1$. Let O be an orthogonal matrix that diagonalizes simultaneously D and A. Define

$$\bar{D} = S^{-1}(O^T D O)S, \ \bar{D} = (\bar{d}_{ij})$$

where

$$S = \operatorname{diag}\left(\frac{1}{\sqrt{\lambda_1}}, \frac{1}{\sqrt{\lambda_2}}, \ldots, \frac{1}{\sqrt{\lambda_{N-1}}}, 1\right).$$

We assume that for some fixed i_0 $d_{i_0 j} = 0$ for all $j \neq i_0$. Now we are ready for the second hypothesis.

(H2) $\bar{d}_{i_0 i_0} > \mu_2$, μ_2 = reciprocal of the 2nd eigenvalue of $-\Delta_N$.

(H2) becomes particularly clear in the special case $D=\text{diag}(\nu_1,\ldots,\nu_N)$, $A = I$. Then $O=S=I$ and $\bar{d}_{i_0 i_0} = \nu_{i_0}$. So we assume that the i_0th diffusion coefficient is sufficiently large with respect to the given domain. Before stating the result we describe very briefly the setting for (1). We adopt the formulation in Henry [4] and write (1) abstractly as an equation in the Banach space $X=L^p$ $p>1$. In this setting the fractional spaces X^α defined for each α as the Banach space that results if the domain of A^α $(A=-D\Delta)$ is equipped with the graph norm are employed for measuring smoothness. We choose our initial data in $X^{\bar{\alpha}}$, $2\bar{\alpha} - \frac{n}{p} \geq \nu > 0$ (some $\nu>0$). The imbedding properties of this space guarantee a unique local solution. For details we refer the reader to [4]. That the solution can be globalized has already been discussed in [5].

Theorem

Under the hypothesis (H1), (H2) for any solution $u(x,t)$ of (1) corresponding to initial data in $X^{\bar{\alpha}}$ the $v_{i_0}(x,t) = (S^{-1}O^T u(x,t))_{i_0}$ converges uniformly on Ω as $t \to \infty$ to a constant. Moreover if this constant is different from zero the vector $v(x,t)$ converges uniformly as $t \to \infty$ to a constant.

Remark

The essence of this result is captured by the special case $D=\text{diag}(\nu_1,\ldots,\nu_N)$, $A = I$. Ignoring for the moment the diffusion it is easy to see the phase portrait in $\mathbb{R}^N$. The rays emanating from the origin are invariant, the origin is unstable and all solutions with the exception of the zero one approach a point on the sphere (See Fig.).

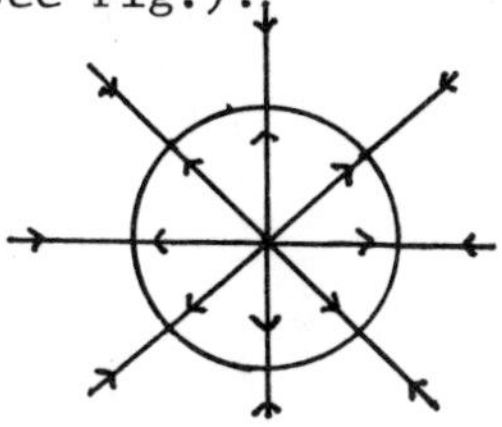

Fig.

The diffusion has the effect of destroying the invariance of the spheres $\{|z| \leq k,\ k \geq 1\}$ and of the rays. It should be noted that by the maximum principle the rectangles $\{|z_i| \leq k,\ k \geq 1\}$ are still invariant. However it turns out that this is very weak information incapable of identifying the limit states. Instead we make use of the observation, valid in general, that the sets

$$M^k = \{\xi(\cdot) \in X^{\overline{\alpha}} \ / \int_\Omega |\xi(x)|^2 dx \leq k,\ k \geq 1\}$$

are invariant for the system

$$v_t = \overline{D}\Delta v + (1-|v|^2)v \tag{5}$$

$$\left.\frac{\partial v}{\partial n}\right|_{\partial\Omega} = 0.$$

and that M^1 is a global attractor ([5]).

Proof of Theorem

We present the proof in several steps.

Step I

First note that $v(x,t) = S^{-1}{}_0{}^T u(x,t)$ satisfies (5) which in turn can be written for some scalar function $E\colon \mathbb{R}^N \to \mathbb{R}$ in the form

$$v_t = \overline{D}\Delta v - \nabla E(v). \tag{6}$$

Here the gradient is taken with respect to the argument of E. Equation (6) has the Liapunov Functional

$$V(v) = \frac{1}{2}\sum_{i,j} \overline{d}_{ij} \int_\Omega \nabla v_i \nabla v_j dx + \int_\Omega E(v)dx \tag{7}$$

which happens to be continuous on $X^{\overline{\alpha}}$. Since the orbits are relatively compact ([5]) we conclude by the LaSalle invariance principle that the ω-limit sets consist of equilibria. It follows then by the imbedding properties of $X^{\overline{\alpha}}$ that $V(\cdot,t)$ approaches uniformly in Ω the <u>set</u> of the solutions of (5).

Step II

Let $v(x)$ be in the ω-limit set of $v(x,t)$. Then $v_1(x)$ satisfies

$$\overline{d}_{11}\Delta v_1 + [1 - (v_1^2 + \ldots + v_N^2)]v_1 = 0 \quad \left.\frac{\partial v_1}{\partial n}\right|_{\partial\Omega} = 0. \tag{8}$$

Define

$$q(x) = 1 - (v_1^2 + \ldots + v_N^2).$$

By the remarks just before the proof

$$\int q(x)\,dx \geq 0. \tag{9}$$

Given k, a constant, we have

$$-\bar{d}_{11}\Delta(v_1 - k) - q(v_1 - k) = qk. \tag{10}$$

Choose $k = \int v_1 dx$ and let $w = v_1 - k$ $(m(\Omega) = 1)$. Then

$$\bar{d}_{11}\int_\Omega |\nabla w|^2 dx - \int_\Omega qw^2 dx = k\int_\Omega qw\,dx =$$

$$= k\int_\Omega qv_1 dx - k^2\int_\Omega q\,dx \leq$$

$$\text{(by (9), (10))} \leq 0.$$

Since w has zero average and $q \leq 1$

$$\bar{d}_{11}\int_\Omega |\nabla w|^2 dx \leq \int_\Omega w^2 dx \leq \mu_2\int_\Omega |\nabla w|^2 dx$$

and so we conclude that $v_1(x)$ is a constant. We distinguish two cases:

(a) $v_1 \equiv 0$

(b) $v_1 \not\equiv 0$.

If (b) holds then we obtain from (8) that $q \equiv 0$. It follows from

$$\bar{D}\Delta v = 0 \qquad \left.\frac{\partial v}{\partial n}\right|_{\partial\Omega} = 0$$

that $v_i = \text{constant}$ $(i = 2,\ldots,N)$.

Step III

We have from Part 1 that the ω-limit set consists of functions of two types: either $v(x) = (0, v_2(x), \ldots, v_N(x))$ with some x_0 such that $v_2^2(x_0) + \ldots + v_N^2(x_0) \neq 1$ or $v(x) = (c_1, \ldots, c_N)$ with $c_1^2 + \ldots + c_N^2 = 1$. We will show as a result of connectedness that both types can not coexist in the same ω-limit set.

Let us assume that the two sets are not separated in L^∞. Then there exists a sequence of solutions of (7) such that as $n\to\infty$ $(0,v_2^n(x),\ldots,v_N^n(x)) \xrightarrow{L^\infty} (0,c_2^*,\ldots,c_N^*)(c_2^*)^2+\ldots+(c_N^*)^2 = 1$.

We will arrive at a contradiction by showing that $S^{N-2} = \{(c_2)^2+\ldots+(c_N^2) = 1\}$ is isolated in the set of solutions

$$\begin{aligned} &\bar{d}_{22}\Delta v_2 +\ldots+ \bar{d}_{2N}\Delta v_N + (1-\langle v,v\rangle)v_2 = 0 \\ &\bar{d}_{N2}\Delta v_2 +\ldots+ \bar{d}_{NN}\Delta v_N + (1-\langle v,v\rangle)v_N = 0. \end{aligned} \tag{11}$$

$$\left.\frac{\partial v_i}{\partial n}\right|_{\partial\Omega} = 0,\ i = 2,\ldots,N. \quad v = \mathrm{col}(v_2,\ldots,v_N),\ \langle\ ,\ \rangle$$

inner product in $\mathbb{R}^{N-1}$.

For this purpose note first that by standard Schauder regularity theory the solutions of (11) are in $C^{2+\alpha}(\Omega)$, $\alpha > 0$. Furthermore a simple regularity argument reveals that it is enough to show that S^{N-2} is isolated in $C^{2+\alpha}$. To achieve this we will employ the implicit function theorem. Define

$$\begin{aligned} F(v_2,\ldots,v_N) = (&-\bar{d}_{22}\Delta v_2 -\ldots- \bar{d}_{2N}\Delta v_N - (1-\langle v,v\rangle)v_2,\ldots, \\ &-\bar{d}_{N2}\Delta v_2 -\ldots- \bar{d}_{NN}\Delta v_N - (1-\langle v,v\rangle)v_N) \end{aligned}$$

$$F:\ C^{2+\alpha} \longrightarrow C^\alpha.$$

We compute the Frechet derivative of F at an arbitrary point in S^{N-2} which without loss of generality may be taken $(1,0,\ldots 0)$.

First we introduce a decomposition of $C^{2+\alpha}$:

$$v = \begin{bmatrix} v_2(x) \\ \vdots \\ v_N(x) \end{bmatrix} = \begin{bmatrix} \int v_2 dx \\ 0 \\ \vdots \\ 0 \end{bmatrix} + \begin{bmatrix} 0 \\ \int v_3 dx \\ \vdots \\ \int v_N dx \end{bmatrix} + \begin{bmatrix} v_2(x) - \int v_2 dx \\ \vdots \\ v_N(x) - \int v_N dx \end{bmatrix} =$$

$$= w + y + z.$$

Define $G(w,y,z) = F(w + y + z)$. Then

$$\frac{\partial G}{\partial(w,z)} = \frac{\partial F}{\partial v}\,\frac{\partial v}{\partial(w,z)} .$$

The kernel of $\dfrac{\partial G}{\partial(w,z)}$ consists of $\mathrm{col}(c + h_2(x), h_3(x), \ldots, h_N(x))$ such that

$$\left[-D_1\Delta + \begin{bmatrix} 2 & 0 & \cdots & 0 \\ 0 & 0 & \cdots & 0 \\ \vdots & & & \\ 0 & 0 & \cdots & 0 \end{bmatrix} \right] \begin{bmatrix} c + h_2 \\ h_3 \\ \vdots \\ h_N \end{bmatrix} = 0 \tag{12}$$

where $\int_\Omega h_i(x)dx = 0$, c in $\mathbb{R}$ and $D_1 = (d_{ij})$, $i,j = 2,\ldots,N$.

Taking the inner product of (12) with $h = \mathrm{col}(c + h_2,\ldots,h_N)$ and integrating by parts

$$\int_\Omega \mathrm{tr}\{(\nabla h)^T D_1 (\nabla h)\}dx + 2\int_\Omega (c + h_2)^2 dx = 0$$

from which we obtain making use of the positivity of $\overline{D}_1$ that $c = 0$, $h \equiv 0$. Thus the implicit function theorem implies that near $(1,0,\ldots,0)$ the solutions of (11) be on an $N - 2$ manifold. This manifold necessarily then has to coincide locally with S^{N-2}. This established the desired contradiction.

Step IV

As it is evident from Step 3 the only case that needs further investigation is that in which the ω-limit set is a subset of S^{N-1}. In such an event we will show that actually the stronger result

$$\lim_{t\to\infty} \bar{v}(x,t) = c = \mathrm{col}(c_1,\ldots,c_N),\ c \text{ in } S^{N-1} \tag{13}$$

holds. The limit is understood to be uniform in x. First we linearize (7) in the setting employed in Step 3 at a point c in S^{N-1} and compute the spectrum. The linearized operator is

$$Lh = \overline{D}\Delta h - 2Ch$$

$$\left.\frac{\partial h}{\partial n}\right|_{\partial\Omega} = 0$$

where C is the matrix $(c_i c_j)$, $i,j = 1,\ldots,N$. It is well known that the spectrum consists of eigenvalues of finite multiplicity. Consider the eigenvalue problem

$$Lh = \lambda h, \quad \left.\frac{\partial h}{\partial n}\right|_{\partial\Omega} = 0.$$

Taking the inner product of both sides with h and integrating

$$-\int_\Omega \mathrm{tr}\{(\nabla h)^T \bar{D}(\nabla h)\}dx = 2\int_\Omega \langle Ch,h\rangle dx + \lambda\int_\Omega \langle h,h\rangle dx.$$

By the positivity of D and C we conclude that $\lambda \leq 0$. Moreover it is easily verified that 0 is an eigenvalue of multiplicity $N-1$. Since c sits on an $N-1$ dimensional manifold of equilibria, the set S^{N-1}, (which by an argument identical to that in Step III can be shown to be isolated in the set of solutions of (7)) the results of Hale and Massatt (cf Remark 3.2 in [6]) apply and (13) follows.

---

[1]See [7] for other nongeneric hypotheses that lead to the same conclusion.

REFERENCES

[1] Abrahams, E., Tsuneto, T., Time variation of the Giuzberg-Landau order parameter, Phys. Rev., 152, pp. 416-432, 1966.

[2] Brown, K. J., Dunne, P. C., Gardner, R. A., A Semilinear Parabolic System arising in the theory of Superconductivity, J. Differential Equations, 40, pp. 232-252, 1981.

[3] G. Nicolis and I. Prigogine, Selforganization in nonequilibrium systems: From dissipative structures to order through fluctuations, Wiley, New York (1977).

[4] Henry, D., Geometric Theory of Parabolic Equations, Springer-Verlag, Lecture Notes in Mathematics, 840, 1981.

[5] Alikakos, N. D., Remarks on Invariance in Reaction-Diffusion equations, Non, An., 5, pp. 593-614, 1981.

[6] Hale, J. K., Massatt, P., Asymptotic Behavior of Gradient-like Systems, preprint.

[7] Alikakos, N. D., Quantitative maximum principles and strongly coupled gradient like reaction-diffusion systems, Proc. Roy. Soc. Edin. A., to appear.

[8] Dafermos, C. M., Asymptotic behavior of solutions of evolution equations, in "Nonlinear Evolution Equations" (M. G. Crandall, Ed.) Academic Press, New York, 1978.

ACKNOWLEDGEMENT

I am indebted to K. Brown for a number of helpful discussions and for pointing out [2]. This work was done during the author's stay at Heriot-Watt University, the hospitality of which I am happy to acknowledge.

RELATION BETWEEN THE TRAPPED RAYS AND THE DISTRIBUTION OF THE EIGENFREQUENCIES OF THE WAVE EQUATION IN THE EXTERIOR OF AN OBSTACLE

Claude W Bardos

University of Paris, 13 Ave. J.B. Clement,
93/30 Villetaneuse and Ecole Normale Superieure,
45 Rue d'Ulm 75005, Paris, France.

In this paper which is a survey of a joint work with J.C. Guillot and J. Ralston, we extend for the exterior problem some recent results which connect the distribution of the eigenvalues and the geometry of the closed geodesics. This is related to the local decay of the solution of the wave equation. This property can be used for some nonlinear perturbations.

1. INTRODUCTION

This talk is a review of a joint paper with J.C. Guillot and J. Ralston [1]. In recent years many authors have developed the relations between the asymptotic behaviour of the eigenvalues for the Laplace operator in a bounded domain Ω (with Dirichlet or Neumann boundary condition) and the geometry of this open set Ω; in particular it turns out that the set of closed geodesics plays an important role.

For the exterior problem it appears that the eigenfrequencies (or poles of the S matrix) share in common many properties of the eigenvalues for the interior problem. We will try to deduce some of their properties in relation to the closed geodesics for the exterior problem.

2. NOTATIONS AND CLASSICAL RESULTS

Ω will denote the exterior of a compact set of $\mathbb{R}^n$ (n odd, $n \geqslant 3$), we assume that $\partial\Omega$ is smooth and we denote by r the radius of a ball containing $\mathbb{R}^n \backslash \Omega$. In Ω we consider the wave

J. M. Ball (ed.), Systems of Nonlinear Partial Differential Equations, 443–451.

equation with Dirichlet boundary condition

$$\frac{\partial^2 u}{\partial t^2} - \Delta u = 0 \quad , \quad u\big|_{\partial\Omega} = 0 \quad . \tag{1}$$

We denote by f the vector $(u, v = \frac{\partial u}{\partial t})$ and by A the operator:

$$A \begin{bmatrix} -u \\ v \end{bmatrix} = - \begin{bmatrix} 0 & I \\ \Delta & 0 \end{bmatrix} \begin{bmatrix} u \\ v \end{bmatrix} \tag{2}$$

and following Lax and Phillips we introduce the space $\mathcal{H} = H_0^1(\Omega) \times L^2(\Omega)$ which is constructed with the energy norm. The equation (1) can be written

$$\frac{d}{dt} f + Af = 0 \tag{3}$$

and A is the generator of a unitary group e^{tA} in $\mathcal{H}$ with domain $D(A) = (H^2(\Omega) \cap H_0^1(\Omega)) \times H_0^1(\Omega)$. (The space $H_0^1(\Omega)$ denotes the closure of $\mathcal{D}(\Omega)$ in the energy norm.)

For any compact K we denote by $E(t,K,u)$ the energy of the solution u contained in the compact K.

$$E(t,K,u) = \int_K (|\nabla u|^2 + |\frac{\partial u}{\partial t}|^2)\,dx \quad . \tag{4}$$

The first classical results concern the decay of the local energy. It has been proven (Lax and Phillips [4]) that the local energy always goes to zero (possibly arbitrarily slowly). On the other hand, combining several classical results (Lax and Phillips [4], Ralston [9], Melrose [8]) we have the following:

1. Theorem

For the wave equation in the exterior domain, with Dirichlet boundary conditions, the following assertions are equivalent

(i) there is no trapped ray

(ii) for any initial data with support in K *the local energy* $E(t,K,u)$ *decays exponentially in time.*

One of the ingredients of the proof of this Theorem is the introduction of a semigroup $Z_r(t)$ (r large enough such that

$K \subset \Omega \cap \{x \mid |x| < r\}$) which has the following properties

(i) $Z_r(t)$ is a contraction semigroup

(ii) for any f with $\operatorname{supp} f \subset B_r = \{x \mid |x| < r\}$ any $t \geq 0$ and any $x \in B_r$ we have

$$(Z_r(t)f)(x) = (e^{tA}f)(x) \quad . \tag{5}$$

Therefore $Z_r(t)$ describes the solution of the wave equation around the obstacle. Furthermore let B_r be the generator of $Z_r(t)$; then one can show that the spectrum of B_r, $\sigma(B_r)$, is strictly contained in the half plane $\operatorname{Re} z < 0$ and consists of a set of eigenvalues of finite multiplicity (independent of r) with no accumulation point. We will denote by $\mu_k = i\lambda_k$ these eigenvalues. The numbers λ_k are usually called the poles of the S matrix.

Finally one can show that if there is no trapped ray in Ω, the eigenvalues μ_k satisfy a relation of the following type:

$$|\operatorname{Re} \mu_k| \geq a + b \log |\operatorname{Im} \mu_k| \tag{6}$$

with a and b two positive constants.

3. A TRACE FORMULA FOR THE EXTERIOR PROBLEM

For the study of the distribution of the eigenvalue of the Laplacian one introduces the two following expressions;

$$\text{trace } e^{t\Delta} = \sum_k e^{-\lambda_k^2 t} \tag{7}$$

$$\text{trace } \cos t(\sqrt{-\Delta}) = \sum_k \cos t\lambda_k \tag{8}$$

where the λ_k^2 denote the eigenvalues of the Laplacian. It is known that information concerning the oscillation of the density of the eigenvalues does not appear in the formula (7) while it does appear in (8). Namely we have the following result (Colin de Verdiere [2], Chazarain [3] and others). The expression given by the right hand side of (8) is a distribution of finite order and the singular support of this distribution is contained in the union of the lengths of the closed geodesics, propagating in Ω and reflecting on the boundary $\partial\Omega$ according to the law of classical geometric optics.

For the exterior problem it is natural to replace the right hand side of (8) by the expression

$$\Sigma\, e^{\mu_k t} = \Sigma\, e^{i\lambda_k t}$$

which turns out to be the trace of the semigroup $Z_r(t)$.

With the finite speed of propagation property and some classical results of Lax and Phillips [4] (cf. Lemma 3.3 of Chapter 5) one can show that for $t > 2r$ the trace of $Z_r(t)$ is given by

$$\mathrm{Tr}\, Z_r(t) = \mathrm{Tr}\, (e^{tA} - e^{tA_o}) \tag{9}$$

where e^{tA_o} denote the propagator for the wave equation in $\mathbb{R}^n$ without obstacles. The right hand side of (9) can be analysed with the same method as for the interior problem and therefore we obtain

Theorem 2 (Bardos, Guillot and Ralston [1])
The sum $\Sigma\, e^{\mu_k t}$ *is for* $t > 2r$ *a distribution of order* $h + 1$ *and its singular support is contained in the union of the lengths of trapped rays.*

From the definition of the order of a distribution and the closed graph theorem one can deduce from Theorem 2 the

Corollary 1
For the poles of the S *matrix the following estimate is true:*

$$\sum_k e^{-2r|\mathrm{Im}\,\lambda_k|} \,/\, |\lambda_k|^{n+2} \leqslant C < +\infty \tag{10}$$

This corollary gives a crude description of the distribution of the poles in the upper half plane, namely that the number of poles contained in the half disk $|z| \leqslant \lambda$, $\mathrm{Im}\, z > 0$ is less than $Ce^{2r\lambda}\lambda^{n+2}$.

Remark 1

A sharper result has been obtained by Melrose [7] for the

equation $\frac{\partial^2 u}{\partial t^2} - \Delta u + Vu = 0$ and it may eventually be extended to the exterior problem. It claims that the trace of $Z(t)$ is for any $t > 0$ a well defined distribution of increasing order for t going to zero. This result implies that the number of poles contained in the half disk $|z| \leqslant \lambda$, $\mathrm{Im}\, z > 0$ is bounded by an expression of the following type: $C_\varepsilon\, e^{\varepsilon\lambda}$, where C_ε becomes arbitrarily large for ε going to zero.

The Theorem 2 is a generalisation, for the exterior of the first result (Chazarain [2] and Colin de Verdiere [3]) for the interior problem. On the other hand it has been noticed that when a bicharacteristic is isolated, and non degenerate one can compute its contribution to the singular support of the distribution trace $(\cos t\sqrt{-\Delta})$.

The same remark applies to the exterior problem and a simple example is given by an obstacle equal to the union of two non intersecting compact convex bodies; in this case there is only one trapped ray, and one can obtain the following:

Theorem 3 (Bardos, Guillot and Ralston [1])

Let Ω *be the exterior of two disjoint strictly convex and compact bodies contained in a ball of radius* r; *then for* $t > 2r$ *one has the formula*

$$\sum_k e^{i\lambda_k t} = 2d \sum_{m=1}^{\infty} \left|\det(P^m - I)\right|^{-\frac{1}{2}} \delta(t-2md) + h(t) \tag{11}$$

In the left hand side the λ_k denote the poles of the S matrix while in the right hand side $h(t)$ is a function belonging to $L^1_{loc}(2r,\infty)$, d is the distance between the two obstacles and P is the Poincaré map defined by the unique closed geodesic of length $2d$. The Poincaré map defined which appears in the formula (11) is non degenerate and it has $2(n-1)$ eigenvalues:

$$0 < \beta_1 \leqslant \beta_2 \ldots \beta_{n-1} < 1 < \frac{1}{\beta_{n-1}} \leqslant \frac{1}{\beta_{n-2}} \leqslant \ldots \leqslant \frac{1}{\beta_1})$$

and with the use of the classical Poisson formula one can obtain the relation:

$$\sum_{m=1}^{\infty} \left|\det(p^n - I)\right|^{-\frac{1}{2}} \delta(t-2md) = \sum_{\substack{m_o \in Z \\ m \in N^{n-1}}} e^{[2i\pi m_o + (\frac{1}{2}I+m)\cdot\nu]\frac{t}{2d}} \tag{12}$$

where ν denotes the vector:

$$\nu = (\log \beta_1, \log \beta_2, \ldots, \log \beta_{n-1}) \quad .$$

The formulas (11) and (12) relate the distribution of the poles of the S matrix and a set of numbers living in the upper half plane

$$\mu_k = (\tfrac{1}{2}I+m)\nu + 2i\pi m_o$$

which are called the pseudopoles. These pseudopoles have been used by Weinstein [11] to obtain some information concerning the distribution of the eigenvalues for the interior problem. Their use for the exterior problem is in some sense more natural because we use numbers which belong to the upper half plane to obtains information on the poles which also belong to the upper half plane.

With the formulas (11) and (12) and the estimate (10) one obtaines the following:

Theorem 4

In the situation of Theorem 3, *for any* $\varepsilon > 0$ *there exist an infinite number of poles in the region*

$$\operatorname{Im} z \leqslant \varepsilon \log|z| \quad . \tag{13}$$

Remark 2

In some sense the relation (13) deals with a situation (with only one trapped ray) which is the converse of the situation described in Theorem 1 where there is no trapped ray and where all the poles satisfy the relation:

$$|\operatorname{Im} \lambda_k| \geqslant a + b \log |\operatorname{Re} \lambda_k| \tag{14}$$

(cf. (6) with $\mu_k = i\lambda_k$).

<u>Remark 3</u>

On the other hand when there is at least one trapped ray there is no uniform exponential decay for the local energy. However in the present case when $\mathbb{R}^n \setminus \Omega$ is the union of two convex bodies Ikawa [5] has shown that there exists a number $\alpha > 0$ such that all the poles live in the half plane $\operatorname{Re} z \geqslant \alpha$. This result complements ours and all together we obtain a precise description of the distribution of the poles.

<u>Remark 4</u>

Due to the first result of Ralston [9] the semigroup $Z_r(t)$ does not decay exponentially in the presence of at least one trapped ray; therefore some part of the unit disk belongs to the spectrum of $Z_r(t)$ in the present example. Now a classical theorem of Hille and Phillips [12] says that one has for any semigroup

$$\operatorname{spec}(e^{tB}) \supset e^{t \operatorname{spec}(B)} \tag{15}$$

and that the two sides of (15) coincide with the possible exception of a) $\{0\}$, b) the continuous part of the spectrum of e^{tB}. Due to the result of Ikawa one has $\operatorname{spec}(B) \subset \{z | \operatorname{Re} z \leqslant -\alpha\}$ and this situation provides a physical example, for which the inclusion (15) is strict.

<u>Remark 5</u>

All the asymptotic results described in the present paper can be used for nonlinear perturbations of the wave equation. Indeed one can show (Klainerman [6], Shatah [10]) the following global existence result for the equation

$$\frac{\partial^2 u}{\partial t^2} - \Delta u = f(u) \quad \text{in } \mathbb{R}^n \quad . \tag{16}$$

Assume that f is a smooth function which satisfies for u small enough an estimate of the following type:

$$|f(u)| \leqslant C\,|u|^p \quad . \tag{17}$$

with p large enough with respect to the local decay property of the solution of the wave equation in $\mathbb{R}^n$; then for small enough initial data, the problem (16) has a smooth solution for all times. The present study of the semigroup $Z_r(t)$ may be used to extend this type of result for the solution of the wave equation

$$\frac{\partial^2 u}{\partial t} - \Delta u = f(u) \quad \text{in} \quad \Omega, \qquad u(x,t)\Big|_{\partial\Omega} = 0 \tag{18}$$

in the exterior of an obstacle.

REFERENCES

[1] Bardos, C., Guillot, J.C., and Ralston, J.: 1982, *La relation de Poisson pour l'equation des ondes daus un ouvert non borné application a la théorie de la diffusion*, Comm. in Partial Differential Equations, 7 (8), pp.905-958.

[2] Chazarain, J.: 1974, *Formule de Poisson pour le varietes riemanniennes*, Inventioner Math., 24, pp.65-82.

[3] Colin de Verdiere, V.: 1973, *Spectre du Laplacian et longueurs des geodesiques periodiques II*, Compositio Math., 27, pp.150-184.

[4] Lax, P., and Phillips, R.: 1967, *'Scattering theory'*, Academic Press.

[5] Ikawa, M.: *On the poles of the scattering matrix for two strictly convex obstacles*, to appear in J. of Math. of Kyoto Univ.

[6] Klainerman, S.: 1982. *Long time behaviour of solutions to nonlinear evolution equations*, Arch. Rat. Mech. and Anal., 78, pp.73-98.

[7] Melrose, R.: *Scattering theory and the trace of the wave group*, to appear.

[8] Melrose, R.: 1979, *Singularities and energy decay in acoustical scattering*, Duke Math. Journal, 16, pp.43-59.

[9] Ralston, J.: 1969, *Solution of the wave equation with localised energy*, Comm. Pure and Applied Maths., 22, pp.807-823.

[10] Shatah, J.: *Global existence for small solutions to non-linear evolution equations*, to appear.

[11] Weinstern, A.: 1972, *Fourier integral operators and the spectra of a Riemannian manifold*, Geometric Symplectique et Physique Mathematique Publication, no. 237 du CNRS.

[12] Hille, E., and Phillips, R.S.: *Functional Analysis*.

STABILIZATION PROPERTIES FOR NONLINEAR DEGENERATE PARABOLIC EQUATIONS WITH CUT-OFF DIFFUSIVITY

Piero de Mottoni

Istituto di Matematica Applicata, Università, L'Aquila

ABSTRACT: We prove stabilization results for a class of degenerate parabolic problems, in which the diffusivity vanishes beyond a given threshold. One characteristic feature is the onset of discontinuous stationary solutions, to which solutions of the evolutionary problem are shown to converge for large times. The basic tools are monotonicity, and a thorough analysis of the stationary solutions.

1. INTRODUCTION

The purpose of the present note is to prove asymptotic properties for a class of quasilinear degenerate parabolic problems, describing nonlinear diffusion processes, subject to a diffusion coefficient which is zero beyond a prescribed value of the density. We shall prove that nonnegative solutions of the evolutionary problem approach, as time tends to infinity, to a stationary solution which, under circumstances, shows up discontinuities. For this purpose we shall make use of a stabilization theorem valid for a broad class of quasilinear scalar initial-boundary value problems (Sect. 2 below), based on an extensive use of monotonicity methods: it has the advantage of being independent of compactness properties of the "trajectories" — a property which may be difficult to prove under general hypotheses (see the final remarks in [2]); its proof, together with some applications, can be found in a joint paper by A. Schiaffino, A. Tesei and the present author [8]. Another basic ingredient is a detailed analysis of the stationary solutions, i.e., the solutions of the associated stationary problem. It is precisely the structure of the set of stationary solutions, deriving from such analysis, which permits to draw from the stabilization theorem relatively "strong" asymptotic properties for the problem under consideration.

J. M. Ball (ed.), Systems of Nonlinear Partial Differential Equations, 453–460.

2. A STABILIZATION RESULT

The following general assumptions will be made throughout: Let G be an open bounded domain in $\mathbb{R}^n$, with $C^{2,\nu}$-boundary ∂G; h: $\bar{\mathbb{R}}^+ \to \bar{\mathbb{R}}^+$ a non-decreasing, continuous function, with $h(0)=0$; f: $\bar{\mathbb{R}}^+ \to \mathbb{R}$ a locally Lipschitz continuous function, with $f(0)=0$; λ a positive number. Moreover, let $u_o \in L^\infty(G)^+ := \{ u \in L^\infty(G);\ u \geq 0 \text{ a.e.}\}$. For $t > 0$, define Q_t as $[0, t] \times G$, and S_t as $(0, t] \times \partial G$. Consider the following (formal) parabolic problem:

$$\begin{aligned} u_t &= \Delta h(u) + \lambda f(u) \qquad \text{in } Q_\infty \\ u &= 0 \quad \text{in } S_\infty , \quad u = u_o \text{ in } \{0\}\times G \end{aligned} \tag{2.1}$$

A solution of (2.1) on (0,T) (T > 0) is a function u, $u \in L^\infty(Q_T) \cap C(0,T; L^1(G))$, such that

$$\int_G u(t)\phi(t) - \iint_{Q_t}(u\phi_t + h(u)\Delta\phi) = \int_G u_o\phi(0) + \lambda\iint_{Q_t} f(u)\phi \tag{2.2}$$

for all $t \in [0, T]$, and all $\phi \in C^2(Q_T)$, $\phi \geq 0$, $\phi = 0$ on S_T. A solution on $(0, \infty)$ is a solution on (0, T) for all $T > 0$. Upper, respectively lower solutions to (2.1) are defined similarly, replacing in (2.2) the equality sign with " $\geq$ ", respectively " $\leq$ ".

Likewise, stationary solutions, i.e., solutions of the stationary problem

$$\Delta h(u) + \lambda f(u) = 0 \text{ in } G, \quad u = 0 \text{ on } \partial G \tag{2.3}$$

are $L^\infty(G)$ - functions, satisfying

$$-\int_G h(u)\Delta\phi = \int_G f(u)\phi , \tag{2.4}$$

for all $\phi \in C^2(\bar{G})$, $\phi \geq 0$ in G, $\phi = 0$ on ∂G. Upper and lower solutions to (2.3) are defined similarly, starting from (2.4). Observe that stationary solutions always satisfy $h(u) \in C^{1,\mu}(\bar{G})$: if, in addition, h is strictly monotone and h^{-1} is Hölder continuous, then any stationary solution is "classical", in the sense that $u = h^{-1}(v)$, where $v \in C^{2,\mu}(\bar{G})$ is a classical solution of

$$\Delta v + \lambda g(v) = 0 \text{ in } G, \quad v = 0 \text{ on } \partial G, \quad \text{where} \tag{2.5}$$
$$g := f \circ h^{-1}.$$

We shall use the notation u_λ (resp., $v_\lambda = h(u_\lambda)$) for solutions of (2.3) (resp., (2.5)) whenever the dependence on the parameter λ is to be stressed. Before formulating the stabilization result, observe that, under the above hypotheses, the evolution problem (2.1) has a unique local solution for any given $u_o \in L^\infty(G)$, which we shall denote by $u(t;u_o)$: see e.g.[5]. We then have:

THEOREM 2.1. Assume the general hypotheses listed above and, moreover, either of the following additional properties of h:

H1) h is locally Lipschitz continuous;
H2) h is monotone increasing, and h^{-1} is locally Lipschitz continuous;

Then: (i) Suppose $u_o \in L^\infty(G)^+$ is a lower (respectively, upper) solution of the stationary problem (2.3). Then the corresponding solution $u(t;u_o)$ of (2.1) is a nondecreasing (respectively, nonincreasing) function of t, a.e. in G. (ii) Suppose there exist a lower solution u_1 and an upper solution u_2 of (2.3), satisfying $0 \leq u_1 \leq u_2$. Then the corresponding solutions $u(t;u_i)$ $(i=1, 2)$ are defined as $L^\infty(G)^+$ functions for any $t \geq 0$, and satisfy the estimate $u_1 \leq u(t;u_1) \leq u(t;u_2) \leq u_2$, for any $t \geq 0$; (iii) Under the assumptions of (ii), $u(t;u_1)$ and $u(t;u_2)$ converge, as $t \to \infty$, towards limits u_*, respectively u^*, which are solutions of the stationary problem, and satisfy $u_* \leq u^*$; the converge takes place in $L^\infty(G)$ if $n=1$, while $h(u(t;u_1))$ converges to $h(u_*)$ (and similarly $h(u(t;u_2))$ converges to $h(u^*)$) in $L^p(G)$ for any $p \geq 1$ if $n \geq 2$. Moreover, u_*, u^* are minimal, respectively maximal solutions of (2.3) within the interval set $\{u \in L^\infty(G)^+;\ u_1 \leq u \leq u_2$ a.e. in $G\}$.

A proof of (i), (ii) under the hypothesis H1) is contained in [2]; more general results of the same type are due to Damlamian (see [6]); a proof of Theorem 2.1 is contained in [8] .

3. EQUATIONS WITH CUT-OFF DIFFUSIVITY (*)

In this section we shall address equations (2.1), (3.3) under the following additional assumptions:

(HS) $\begin{cases} h \in C^1(\bar{\mathbb{R}}^+);\ \text{there is } a \in (0,1) \text{ such that } h(u) \text{ is strictly increasing for } 0 \leq u \leq a, \text{ and } h(u) = h(a) := \alpha \text{ for } u \geq a: \\ f \in C^1(\bar{\mathbb{R}}^+),\ f(0) = f(1) = 0;\ f \geq 0 \text{ on } (0,1),\ f < 0 \text{ for } u > 1. \\ \text{the function } v \to g(v)/v := f(h^{-1}(v))/v \text{ is decreasing on } (0, \alpha). \end{cases}$

In other words, the kinetic term f(u) is supposed to account for a logistic-type growth with carrying capacity 1 (in the absence of diffusion), while the diffusivity h' vanishes beyond the threshold value a of the density. The last assumption in (HS) was introduced for the sake of simplicity: a case where it is not satisfied will be briefly sketched below.

Simple examples of functions satisfying (HS) are:

$$f(u) = [u(1-u);\ h:\ h'(u) = [a-u]_+,\ h(0)=0,\ \tfrac{1}{2} < a < 1 \quad (3.1)$$

$$f \text{ as above, } h:\ h(0)=0,\ h'(u) = [u(a-u)]_+,\ \tfrac{2}{3} < a < 1 \quad (3.2)$$

note that in the last case the diffusivity is non-monotonic, and degenerates both at $u=0$ and for $u > a$.

(*) Early discussions with W. ALT on this subject are warmly acknowledged.

In the following we are going to characterize the non-negative solutions u_λ of (2.3) under the assumptions (HS), whence, by a straightforward application of the stabilization theorem, we shall prove attractivity properties of the solutions u_λ. We start by some auxiliary results:

LEMMA 3.1. Let u_λ be a non-negative solution of (2.3) such that the set $G'_\lambda := \{x \in G;\ u_\lambda(x) \geq a\}$ has positive measure. Then:

i) $G'_\lambda = \overline{G}'_\lambda \subset G$; $u_\lambda = 1$ a.e. on $\overline{G}'_\lambda$, $h(u_\lambda) = \alpha$ on $\overline{G}'$;

ii) for any ϕ as in (2.4) and for any $\eta > 0$ there is $\varepsilon < 0$ such that

$$|-\int_{\partial G'^{\varepsilon}_\lambda}(\partial h(u_\lambda)/\partial n) + \int_{G''^{\varepsilon}_\lambda}\Delta h(u_\lambda)\phi + \lambda\int_{G''^{\varepsilon}_\lambda} f(u_\lambda)\phi| < \eta \tag{3.3}$$

for any family $\{G'^{\varepsilon}_\lambda\}$ of subsets satisfying $\overline{G}'_\lambda \subset G'^{\varepsilon}_\lambda$, $\overline{G}'^{\varepsilon}_\lambda \subset G$, having smooth boundary $\partial G'^{\varepsilon}_\lambda$, and satisfying $\text{meas}(G'^{\varepsilon}_\lambda \sim G'_\lambda) < \varepsilon$; $\partial/\partial n$ denotes the normal (outward) derivative on $\partial G'^{\varepsilon}_\lambda$, and $G''^{\varepsilon}_\lambda := G \sim G'^{\varepsilon}_\lambda$.

Remark 3.2. If $\partial G'_\lambda$ is C^1 — as it certainly is, e.g., in the one-dimensional case, see below — then (3.3) amounts to

$$\begin{aligned} &\Delta h(u_\lambda) + \lambda f(u_\lambda) = 0 \text{ on } G''_\lambda := G \sim G'_\lambda ; \\ &u_\lambda = 0 \text{ on } \partial G,\ u_\lambda = a, \text{ and } \partial h(u_\lambda)/\partial n = 0 \text{ on } \partial G'_\lambda . \end{aligned} \tag{3.4}$$

Note that (3.4), and more generally (3.3) can be viewed as a free boundary problem, the unknowns being u_λ and G'_λ, or $\partial G'_\lambda$, on which two boundary conditions are imposed.

Furthermore, as it will be apparent from the proof, an equivalent formulation of (3.3) is

$$\int_{G''_\lambda} \nabla h(u_\lambda)\nabla\phi + \lambda\int_{G''_\lambda} f(u_\lambda)\phi = 0, \text{ for all } \phi \in C^1(\overline{G}),\ \phi = 0 \text{ on } \partial G \tag{3.5}$$

P r o o f of the Lemma 3.1: Let us drop, for definiteness, the suffix λ. First, since $h(u)$ is continuous (in fact, $C^{1,\mu}$, see Sect.1) and $h(u) = \alpha$ if and only if $u \geq a$, we see that G' is closed, and, as $h(u) = 0$ on ∂G, $G' \subset G$ immediately obtains. Picking test functions with support in $\overset{\circ}{G}'$, using (2.4) and the fact that $h(u)$ is constant on G', we infer $f(u) = 0$ a.e. on G', whence item i) easily follows. To prove ii), start from (2.4), and recall that $f(u) = 0$ a.e. on G': having chosen the subdomains G'^{ε} as in the statement of the Lemma yields, after an integration by parts:

$$-\int_{G'^{\varepsilon}\sim G''} \nabla h(u)\nabla\phi - \int_{G\sim G'} \nabla h(u)\nabla\phi + \lambda\int_{G'^{\varepsilon}\sim G'} f(u)\phi + \lambda\int_{G\sim G'} f(u)\phi = 0 \tag{3.6}$$

Since $\text{meas}(G'^{\varepsilon} \sim G') < \varepsilon$, and $h(u)$, $f(u)$ are bounded in G, formula (3.4) is easily established. Proceeding further, integration by parts of the second summand in (3.6) yields, reasoning as above, the required formula (3.3).

To proceed further, notice that the (non-negative, non-trivial) solutions of (2.3) whose maximum does not exceed a, exist and are uniquely determined for any λ "not too large": in fact, define λ_p as the principal eigenvalue of the Laplacian:

$$\Delta\phi_p + \lambda_p\phi_p = 0,\ \phi_p > 0 \text{ in } G,\ \phi_p = 0 \text{ on } \partial G,\ \max\phi_p = 1 \qquad (3.7)$$

and
$$\lambda_o := \begin{cases} \lambda_p/g'(0) & \text{if } g'(0) < \infty, \\ 0 & \text{if } g'(0) = \infty, \end{cases}$$

moreover, let g denote $f(h^{-1}(.))$ (defined on $[0,\alpha]$). We then have:

LEMMA 3.3. If $\lambda < \lambda_o$, no non-trivial non-negative solution of (2.3) exists, whose essential supremum does not exceed a. There is λ_1, $\lambda_o < \lambda_1 < \infty$ such that, for any $\lambda \in (\lambda_o, \lambda_1)$ a unique non-trivial non-negative solution u_λ of (2.3) exists, satisfying $u_\lambda <$ a. Such u_λ is a classical solution, and it approaches zero uniformly as $\lambda \to \lambda_o^+$. Moreover, $\lambda \to u_\lambda$ is (pointwise) non-decreasing, and $\max u_\lambda$ approaches a as $\lambda \to \lambda_1^-$.
The proof is straightforward: in fact, non-negative solutions of (2.3) not exceeding a are expressed as $u_\lambda = h^{-1}(v_\lambda)$, where v_λ is a solution of (2.5): in terms of such equation, the desired results easily follow from the assumptions on $g := f(h^{-1}(.))$, on the strength of well-known arguments (see e.g., [1],[7],[8]).

Note that λ_1 could be defined as $\sup\{\lambda > \lambda_o,\ v_\lambda < \alpha\}$; the fact that $\lambda_1 < \infty$ can be proved as follows: if not, we would have $0 < v_\lambda$ $v_\lambda \leq \alpha$ for all $\lambda > \lambda_o$, so that, on the one hand $g(v_\lambda)$ would approach zero as a distribution, for $\lambda \to \infty$, but on the other hand v_λ would converge pointwise towards a non-negative, non-trivial function, which contradicts the assumptions on f.

An important consequence of the arguments invoked in the above proof is the following

LEMMA 3.4. Let $\lambda_1 > \lambda_o$ be as above. Then for any non-negative non trivial solution u of (2.3) with $\lambda > \lambda_1$ the set $G'(u_\lambda) := \{x \in G;\ u_\lambda(x) \geq a\}$ has positive measure.

Sketch of the proof: If not, we could find a diverging sequence λ_n, and corrispondingly solutions $u_{\lambda n}$ such that ess sup $u_{\lambda n}$ $<$ a for all n. The u_n's would therefore be (regular) solutions of the type described in the previous Lemma, and, proceeding as in the proof of such Lemma, we would reach a contradiction.

Thus, solutions of (2.3) for $\lambda > \lambda_1$, if any, should enjoy the properties stated in Lemma 3.1: to complete our discussion, we have to proof that such solutions actually exist, and in fact they are uniquely determined. This will be achieved in the following

LEMMA 3.5. For any $\lambda > \lambda_1$ there is a unique non-negative non-trivial solution of (2.3), denoted by u_λ; moreover, $\lambda \to u_\lambda$ is (pointwise) non decreasing.

Proof: For proving existence we can make use of Theorem 2.1: to this end, suffice it to exhibit an upper and a lower solution. Now, $u=1$ is obviously an upper solution, as to lower solutions, it is easily seen that any solution u_ξ of (2.3) with λ replaced by ξ, and $\xi \leq \lambda$, is a lower solution to (2.3). Thus the existence of non-negative non-trivial solutions of (2.3) for $\lambda > \lambda_1$ follows from Lemma 3.1. As to uniqueness, observe first that it is sufficient to establish it within the class of comparable solutions. Suppose now that, for $\lambda > 0$, u, $\tilde{u}$ are two solutions of (2.3) satisfying $0 \leq u \leq \tilde{u}$, $0 \not\equiv u$, $u \not\equiv \tilde{u}$, and let $G' = \{x \in G,\ u(x) \geq a\}$, $\tilde{G}' = \{x \in G,\ \tilde{u}(x) \geq a\}$; of course, $\tilde{G}' \supseteq G'$, G' being possibly empty. If $\tilde{G}'$ is empty, then we fall in the case covered by Lemma 3.3. If not, write (3.5) for u, respectively $\tilde{u}$, choosing as test functions $h(\tilde{u})$, respectively $h(u)$ — which is allowed, since the latter quantities are both of class $C^{1,\mu}(\overline{G})$: we thus obtain:

$$- \int_{G \sim G'} \nabla h(u) \nabla h(\tilde{u}) + \lambda \int_{G \sim \tilde{G}'} f(u) h(\tilde{u}) = 0 \tag{3.8}$$

$$- \int_{G \sim \tilde{G}'} \nabla h(\tilde{u}) \nabla h(u) + \lambda \int_{G \sim \tilde{G}'} f(\tilde{u}) h(u) = 0 \tag{3.9}$$

Now the first integral in (3.8) actually extends to $G \sim \tilde{G}'$, so that, subtracting (3.9) from (3.8) we obtain

$$\int_{G \sim \tilde{G}'} f(\tilde{u}) h(u) - \int_{G \sim G'} f(u) h(\tilde{u}) = 0, \text{ that is}$$

$$\int_{G \sim \tilde{G}'} (f(\tilde{u}) h(u) - f(u) h(\tilde{u})) - \lambda\alpha \int_{\tilde{G}' \sim G'} f(u) = 0 .$$

From the above formula, since α, $f(u) \geq 0$, and $u \to f(u)/h(u)$ is strictly decreasing on $(0, a)$, we have a contradiction: thus, uniqueness is proved. That $\lambda \to u_\lambda$ is non-decreasing follows from the already observed fact that u_ξ is a lower solution to (2.3) whenever $\xi \leq \lambda$.

The above Lemmas can be summarized as follows:

THEOREM 3.6. Under the assumptions (HS), there are $\lambda_0 \geq 0$, $\lambda_1 \geq \lambda_0$ such that (i) for $\lambda < \lambda_0$ no non-negative solution of (2.3) exists, but zero; for $\lambda > \lambda_0$ there is a unique non-negative non-trivial solution u_λ of (2.3), which, for $\lambda < \lambda_1$, is continuous, does not exceed a, and is such that $h(u_\lambda) \in C^{2,\mu}(\overline{G})$; instead, for $\lambda > \lambda_1$, u_λ is discontinuous, equalling one on a non-zero measure set, G'_λ, and satisfying (3.3) (or (3.5)). Moreover, $\lambda \to u_\lambda$ is non-decreasing, and as $\lambda \to \infty$, $h(u_\lambda)$ converges to α uniformly on compact subsets of G.

Note that the last claim (which implies, by the way, that meas $G'_\lambda \to$ meas G as $\lambda \to \infty$), follows applying the same arguments used in the proof of Lemma 3.3. The above established properties can be visualized in the Figure 1, where, anticipating the results of Theorem 3.7 below, the "branch" of attractive solutions (relative to (2.1)) is represented by a solid line.

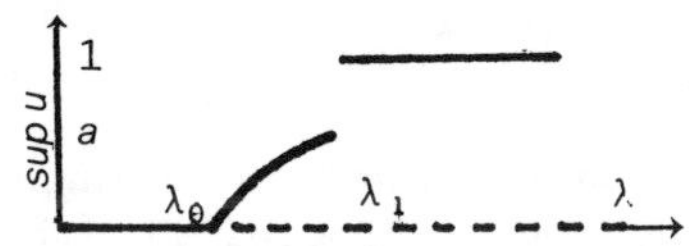

Fig. 1

For the sake of completeness, let us briefly sketch an example where the last assumption in (HS) is violated: to be specific, we take (3.1) with $a \in (0, \frac{1}{2})$, so that, as easily checked, $f(h^{-1}(v))/v$ is increasing, and the uniqueness proof given above (Lemma 3.5), while valid if ess sup $u < a$, fails if u exceeds a. To simplify the analysis, let us consider the case of one space dimension: $G = (0, L)$. Then an easy application of the maximum principle permits to show that G'_λ is an interval, and the solutions exceeding a are characterized by (3.4); a straightforward computation, based on "phase-energy" methods then shows that the bifurcation picture has the aspect depicted in Figure 2.

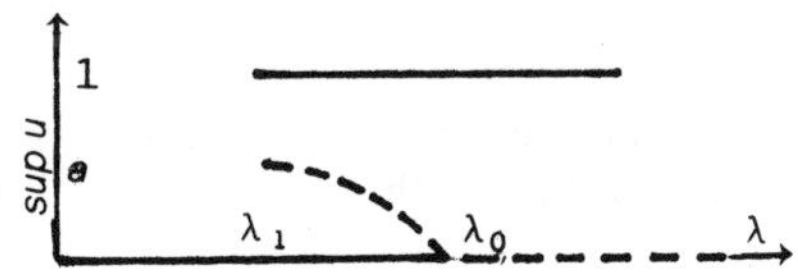

Fig. 2

Let us now turn to the asymptotics of solutions of (2.1) under the assumptions (HS). Since, as already observed, $u = 1$ is in any case an upper solution to (2.3), by virtue of the uniqueness of u_λ, it is sufficient to exhibit a lower solution u_1 of (2.3) in order to ensure (via Theorem 2.1) the convergence of $u(t;u_o)$ towards u_λ as $t \to \infty$, whenever $u_o \geq u_1$. Since lower solutions, if $\lambda > \lambda_o$ can be found (see e.g. [8]) of the form $h^{-1}(\sigma\phi_p)$, with $\sigma > 0$, $\sigma < \alpha$ and ϕ_p as in (3.7), the stabilization of $u(t;u_o)$ towards u_λ is granted, for $\lambda > \lambda_o$, provided

$$h(u_o) \geq \sigma\phi_p \text{ , for some } \sigma \text{ small enough (and positive)} \tag{3.10}$$

Now there are at least two possibilities to relax the condition on u_o expressed by (3.10). The first one is met, as suggested in [8], when

$$\lim_{v \to 0+} f(h^{-1}(v))/v = +\infty, \tag{3.11}$$

since in this case no "critical size" for G exists, and lower solutions can be found of the form $h^{-1}(\sigma\phi_p(D))$, $\sigma < \alpha$, where $\phi_p(D)$ is as in (3.7), but with G replaced by a subdomain D, $\bar{D} \subset G$ ($\phi_p(D)$ being zero on $G \sim D$). Thus stabilization towards u_λ of a solution of (2.1) with initial datum u_o occurs provided u_o fulfils

u_o has a positive infimum on some open non-empty subdomain of G. (3.12)

Another possibility is suggested by the Aronson-Bertsch-Peletier positivity theorem [3],[4]: in fact, if

$$h \in C^1(\bar{\mathbb{R}}^+) \cap C^3((0, s_o)), \text{ and}$$
$$\gamma < \frac{h(s)h''(s)}{h'(s)} < 1, \text{ for some } \gamma \in (0, 1),\ s_o \geq 0 \tag{3.11'}$$

then the assumptions of such theorem are met (note that our hypotheses imply $\liminf_{v \to 0+} f(v)/h(v) > -\infty$), and stabilization occurs under the same assumption (3.12) on u_o. Summarizing, we have the following stabilization result:

THEOREM 3.7. Assume (HS), and either (3.11) or (3.11'). Then, if $\lambda > \lambda_o$, any solution $u(t;u_o)$ of (2.1) with u_o satisfying (3.12) converges, in the sense specified by Theorem 2.1, as $t \to \infty$, to the unique, non-negative, non-trivial solution of (2.3).

Remark 3.7. If $\lambda \leq \lambda_o$, it can be easily proved that any non-negative solution of (2.1) converges to zero.

REFERENCES

1. Amann, H.: On the existence of positive solutions for nonlinear boundary value problems, 1971, Indiana Univ. Math. Journ., pp. 125-146.
2. Aronson, D.G., Crandall, M.G. and Peletier, L.A.: Stabilization of solutions of a degenerate nonlinear diffusion problem, 1981, M.R.C., T.S.R. n° 2220, Univ. of Wisconsin, Madison.
3. Aronson, D.G., Peletier, L.A.: Large time behaviour of solutions of the porous medium equation in bounded domains, 1981, Journ. Diff. Eqns. 39 (3), pp. 378-412.
4. Bertsch, M., Peletier, L.A.: A positivity property of solutions of nonlinear diffusion equations, Mathematical Institute of the University of Leiden, Preprint n° 1, Febr. 1982.
5. Brézis, H., Crandall, M.G.: Uniqueness of solutions of the initial-value problem for $u_t - \Delta\phi(u) = 0$, 1979, J. Math. pures appliquées 58, pp. 153-163.
6. Damlamian, A.: Some results on the multi-phase Stefan problem, 1977, Commun. in Part. Diff. Eqns. 2(10), pp. 1017-1044.
7. Lions, P.L.: On the existence of positive solutions in semilinear elliptic equations, 1981, M.R.C., T.S.R. n° 2209, Univ. of Wisconsin, Madison, WI.
8. de Mottoni, P., Schiaffino, A., and Tesei, A.: Attractivity properties of non-negative solutions for a class of nonlinear degenerate parabolic problems, 1982, Preprint (Universities of L'Aquila and Rome II).

DYNAMICS IN PARABOLIC EQUATIONS - AN EXAMPLE

Jack K. Hale

Division of Applied Mathematics
Brown University, Providence, R. I. 02912 U.S.A.

Abstract: For a parabolic equation, results are given about the orbits which connect equilibrium points. The approach is based on the theory of dynamical systems and the maximum principle.

1. The basic problem.

A system of nonlinear parabolic equations defined on a bounded domain generate a nonlinear semigroup of transformations $\{T_\mu(t), t \geq 0\}$ on some Banach space X. The parameter μ is supposed to represent the elliptic operators in the parabolic equation, the region Ω where the equation is defined, the nonlinear functions in the equation and the boundary conditions. The basic problem is to study how the qualitative properties of the orbits defined by $T_\mu(t)$ depend on the parameter μ.

Because $T_\mu(t)$ arises from parabolic equations on a bounded domain, one generally has

$$T_\mu(t),\ t > 0, \text{ is a completely continuous operator} \tag{1.1}$$

$$T_\mu(t) \text{ is one-to-one for all } t. \tag{1.2}$$

In particular, (1.1) implies that every bounded orbit belongs to a compact set. Thus, the ω-limit set of a bounded orbit is a nonempty, compact, connected set of X which is invariant under the semigroup $T_\mu(t)$. We say a set M in X is <u>invariant</u> for $T_\mu(t)$ if, for any $\varphi \in M$, there is a function $x(t,\varphi)$ defined for $t \in \mathbb{R}$, $x(0,\varphi) = 0$, $x(t,\varphi) \in M$, $t \in \mathbb{R}$, and, for any $\tau \in \mathbb{R}$, $T_\mu(t)x(\tau,\varphi) = x(t+\tau,\varphi)$, $t \geq 0$. We call $x(t,\varphi)$ a <u>backward extension</u> of φ and write $x(t,\varphi) = T_\mu(t)\varphi$ for $t \in \mathbb{R}$. Under

J. M. Ball (ed.), Systems of Nonlinear Partial Differential Equations, 461–472.

condition (1.2), backward extensions are unique. If a backward extension of φ exists and belongs to a compact set, then the α-limit set exists and is nonempty, compact, connected and invariant (see [Ha2]).

An important role in the theory is played by the set,

$$A_\mu = \{\varphi \in X: T_\mu(t)\varphi \text{ is defined and bounded for } t \in \mathbb{R}\}, \tag{1.3}$$

which contains the limit points of all bounded orbits as well as much more. In fact, one has the following elementary result.

Lemma 1.1. If A_μ is compact, then A_μ is the maximal compact invariant set of X. Furthermore, (1.2) implies that $T_\mu(t)$ is a group on A_μ.

One way to ensure that A_μ is compact is to assume the existence of invariant regions for the parabolic system (see [Sl]). Another weaker condition is to assume that the system is point dissipative; that is, there is a bounded set $B \subset X$ such that, for any $\varphi \in X$, there is a $t_0 = t_0(\varphi, B)$ such that $T_\mu(t)\varphi \in B$ for $t \geq t_0$. This condition implies the following result.

Theorem 1.2. If $T_\mu(t)$, $t > 0$, is completely continuous and point dissipative, then there is a maximal compact invariant set A_μ for $\{T_\mu(t), t \geq 0\}$, A_μ is uniformly asymptotically stable and attracts bounded sets of X; that is, for any bounded set U in X, $\text{dist}(T_\mu(t)U, A_\mu) \to 0$ as $t \to \infty$. If $T_\mu(t)$ is one-to-one, then $T_\mu(t)$ is a continuous group on A_μ.

The basic steps in the proof are as follows. One first proves that there is a compact set K in X such that for any compact set H in X, there is a neighborhood V of H such that $\text{dist}(T_\mu(t)V, K) \to 0$ as $t \to \infty$. This property implies there is a maximal compact invariant set A_μ which is uniformly asymptotically stable and attracts neighborhoods of compact sets. The compactness of $T_\mu(t)$ for $t > 0$ can be used to complete the proof.

The above theorem is actually true for much more general semigroups $\{T_\mu(t), t \geq 0\}$. One needs only that $T_\mu(t)$ is compact dissipative and an α-contraction for $t > 0$. This permits applications to a much broader class of problems including some hyperbolic systems (see [HA 3]). We do not consider such general cases here and refer the reader to [Ha 1] for a complete proof of the above result as well as historical references. The discussion is also contained in [Ha 2]. However, an error in the statement of one result in [Ha 2] requires a reordering of the material and this reordering can be found in [Ha 1].

We remark that it is not necessary to consider all of X to obtain an interesting invariant set A_μ. One could have a subset Y of X which is positively invariant with respect to $T_\mu(t)$. Point dissipative is then defined relative to the intersection of bounded sets with Y.

The basic problem in the qualitative theory of parabolic equations *is to study how* A_μ *and the flow on the invariant set* A_μ *change with the parameter* μ.

Before giving specific examples which illustrate the above remarks, let us state another interesting implication of point dissipativeness, a proof of which can be found in [HL 1].

Theorem 1.3. If $T_\mu(t)$ *is completely continuous for* $t > 0$ *and point dissipative, then there is an equilibrium point of* $T_\mu(t)$*; that is, there exists a* φ *in* X *such that* $T_\mu(t)\varphi = \varphi$ *for all* t.

When $T_\mu(t)$ is the semigroup generated by a system of parabolic equations, Theorem 1.3 implies the existence of a solution of the corresponding elliptic boundary value problem. For some types of problems, this approach could be easier to obtain the existence of solutions of the elliptic system (see [A1], [A2], [A3], [K1]).

2. *An example.* In this section, we consider the simplest nontrivial parabolic equation. In spite of its simplicity, there are several unanswered questions whose solution would lead to a better understanding of the role of diffusion in the dynamics.

Consider the scalar equation

$$\begin{aligned} &u_t = u_{xx} + \lambda f(u), \quad 0 < x < \pi, \\ &u = 0 \quad \text{at} \quad x = 0, \pi \end{aligned} \tag{2.1}$$

with $\lambda > 0$ being a real parameter and $f(u)$ being a given nonlinear function of u. If

$$V(\varphi) = \int_0^\pi [\varphi_x^2 - \lambda F(\varphi)]dx, \qquad F(u) = \int_0^u f, \tag{2.2}$$

and $u(t,x)$ is a solution of (2.1), then

$$\frac{d}{dt}V(u(t,x)) = -\int_0^\pi u_t^2 dx \leq 0 \tag{2.3}$$

Theorem 2.1. If

$$F(u) \to -\infty \quad \text{as} \quad u \to \pm\infty \tag{2.4}$$

then Eq. (2.1) generates a C_0-semigroup $T_\lambda(t)$, $t \geq 0$, on $X = H_0^1(0,\pi)$, each orbit is bounded and has ω-limit set as an equilibrium point. There is a maximal compact invariant set A_λ for $T_\lambda(t)$ which has the stability properties mentioned in Theorem 1.2. Finally, if $\varphi \in A_\lambda$, then the α-limit set of φ is an equilibrium point.

The equilibrium points of (2.1) are the solutions of the equation

$$\begin{aligned} u_{xx} + \lambda f(u) &= 0 \qquad 0 < x < \pi \\ u &= 0 \quad \text{at} \quad x = 0,\pi \end{aligned} \tag{2.5}$$

That (2.1) generates a C_0-semigroup is given in Henry [He 1]. The fact that the ω- and α-limit sets must be a single equilibrium point has been proved by a number of people (see Zelenyak [Z 1], Matano [M 2], Hale and Massat [HM 1]). Relation (2.4) implies the set of equilibrium points is bounded. Since every orbit approaches an equilibrium point, one obtains point dissipative.

An equilibrium point u_0 is hyperbolic if no eigenvalue of the operator $\partial^2/\partial x^2 + \lambda f'(u_0)$ on X is zero and it is called stable (hyperbolic) if all eigenvalues are negative. The unstable manifold $W^u(u_0)$ is the set of $\varphi \in X$ such that $T_\mu(t)\varphi$ is defined for $t \leq 0$ and $\to u_0$ as $t \to -\infty$. The stable manifold $W^s(u_0)$ is the set of $\varphi \in X$ such that $T_\mu(t)\varphi \to u_0$ as $t \to \infty$. The set $W^u(u_0)$ is an embedded submanifold of X of finite dimension m (m being the number of positive eigenvalues of the above operator). The set $W^s(u_0)$ is an embedded submanifold of codimension m (see [He 1]). These manifolds are tangent at u_0 to the stable and unstable manifolds of the linear operator $\partial^2/\partial x^2 + f'(u_0)$ on X.

The following remark is a simple but important consequence of Theorem 2.1.

Corollary 2.2. If (2.4) is satisfied and there are only a finite number of hyperbolic equilibrium points $\varphi_1, \varphi_2, \ldots, \varphi_k$ of (1.1) with each being hyperbolic, then

$$A_\lambda = \bigcup_{j=1}^{k} W^u(\varphi_j).$$

Corollary 2.2 states that A_λ is the union of a finite number of finite dimensional manifolds. The complete flow on A_λ seems to be difficult to describe in the general case. However, some nontrivial information is easily obtained if we make additional hypotheses.

Theorem 2.3. *If the conditions of Corollary 2.2 are satisfied and $\varphi_1(x) < \varphi_2(x) < \cdots < \varphi_k(x)$, $0 < x < \pi$, and $\dim W^u(\varphi_j) \leq 1$, then k is odd, $k = 2p+1$, φ_{2j+1} is stable hyperbolic and φ_{2j} is unstable for all j with the ω-limit set (α-limit set) of φ_{2j} being $\varphi_{2j+1}(\varphi_{2j-1})$.*

Proof: If φ_i is unstable, then $\dim W^u(\varphi_i) = 1$. Since $W^u(\varphi_i)$ is tangent at φ_i to the line spanned by the eigenfunction corresponding to the positive eigenvalue of $\partial^2/\partial x^2 + f'(\varphi_i)$ on X and this eigenfunction has no zeros on $(0,\pi)$, it follows that there is a solution $u(t,x)$ of (2.1) defined for $t \in \mathbb{R}$ with $\lim_{t\to-\infty} u(t,x) = \varphi_i(x)$ and $u(t,x) > \varphi_i(x)$ for t sufficiently negative. The maximum principle and our hypothesis on φ_{i+1} implies that $\varphi_i(x) < u(t,x) < \varphi_{i+1}(x)$ for $0 < x < \pi$. Thus, $\lim_{t\to\infty} u(t,x) = \varphi_{i+1}(x)$. The same argument shows there is a $\psi \in W^u(\varphi_i)$ such that the ω-limit set of ψ is φ_{i-1}. Since A_λ is stable, φ_1 and φ_k are stable hyperbolic and the theorem is proved.

Corollary 2.4. *Under the hypotheses of Theorem 2.3, the global dynamics of (2.1) is determined by the local bifurcation of equilibrium points.*

Theorem 2.3 has immediate application to positive solutions of Eq. (2.1). In fact, if $X^+ = \{\varphi \in X: \varphi(x) \geq 0,\ 0 \leq x \leq \pi\}$, and $T_\lambda(t): X^+ \to X^+$, then we get the maximal compact invariant set A_λ^+ in X^+. In this case, the $\dim W^u(\varphi) \leq 1$ for any equilibrium point. Thus, if the equilibrium points are hyperbolic, then the conclusion of Theorem 2.3 is true. For another discussion of this latter example using the Conley index, see Smoller [S 1].

Remark 2.5. Using the same proof as above, one observes that the conclusions of Theorem 2.3 are true if u_{xx} is replaced by Δu and $(0,\pi)$ is replaced by a bounded open set $\Omega \subset \mathbb{R}^N$.

We also remark that the strong conclusion in Theorem 2.3 and Corollary 2.4 are not consequences of only $\dim W^u(\varphi_j) \leq 1$. The parabolicity is used in an essential way. The reader is referred to Hale and Rybakowski [HR 1] for an example of a scalar gradient-like delay equation where the conclusions of Theorem 1.4 are not true even though the dimension of the unstable manifold of each equilibrium point is ≤ 1.

Let us now return to the general discussion. When the equilibrium points of (2.1 are hyperbolic, Corollary 2.2 gives A_λ as the union of the unstable manifolds of the equilibrium points. The complete dynamics on A_λ will only be known when we know the specific way in which the equilibrium points are connected to each other by orbits. In an effort to learn more about this, we introduce some other invariant sets of A_λ.

Let

$$A_\lambda^j = \{\varphi \in A_\lambda : \{T_\lambda(t)\varphi, t \in \mathbb{R} \text{ as well as its } \alpha\text{- and } \omega\text{-limit sets have exactly } j \text{ zeros in } (0,\pi)\}. \tag{2.6}$$

Each set A_λ^j is compact and invariant.

<u>Lemma 2.6. Suppose</u> $f(0) = 0$. <u>If</u> $\varphi \in A_\lambda \setminus \bigcup A_\lambda^p$ <u>with</u> α-<u>limit set in</u> A_λ^j <u>and</u> ω-<u>limit set in</u> A_λ^k, <u>then</u> $j > k$.

<u>Proof.</u> Suppose such a φ exists. Since, as $t \to -\infty$, $T_\lambda(t)\varphi \to \psi \in A_\lambda^j$, an equilibrium point with j zeros in $(0,\pi)$ and since the zeros of ψ are simple, it follows that $T_\lambda(t)\varphi$ has j zeros in $(0,\pi)$ for $t \leq -\tau$ with τ sufficiently large. Also, $T_\lambda(t)\varphi$ has $j + 1$ extreme values in $(0,\pi)$. Also, $T_\lambda(t)\varphi \to \eta \in A_\lambda^k$, with k simple zeros and $k + 1$ extreme values in $(0,\pi)$. Thus, if $\bar{\varphi} = T_\lambda(-\tau)\varphi$, then $T_\lambda(t)\bar{\varphi} \to \eta$ and the results of Matano [M 1] imply that the number of extreme values of $\bar{\varphi}$ must be $\geq$ the number of extreme values of η. This implies $j \geq k$. But since $j \neq k$, we have the result.

The condition $f(0) = 0$ is, in general, necessary to obtain the conclusion in Lemma 2.6. If $f(0) \neq 0$, then the best general result would be $j \geq k - 2$ (see [M 1]).

Let $M_i \in A_\lambda$, $i = 1,2,\dots,p$, be compact invariant sets in A_λ. Following Conley [C 1], we say $\{M_i\}$ is a <u>Morse decomposition</u> of A_λ if $\varphi \in A_\lambda \setminus \bigcup_i M_i$ implies there are integers $j > k$ such that the ω-limit set of φ is in M_k and the α-limit set of φ is in M_j.

Under the assumptions of Lemma 2.6, $f(0) = 0$ implies $u = 0$ is an equilibrium solution of (1.1). Thus, the set $\{0\}$ must be included with the A_λ^j in (2.6) in order to have any hope for a Morse decomposition. Since Lemma 2.6 says nothing about $\{0\}$, more detailed information is needed about the behavior of the solutions near $u = 0$. However, we can state the following general consequence of Lemma 2.6.

<u>Theorem 2.7. Suppose</u> $f(0) = 0$ <u>and the sets</u> A_λ^k, $k=0,1,\dots,p$, <u>are defined in (1.6) and let</u> $A_\lambda^{p+1} = \{0\}$. <u>If no orbit in</u> A_λ <u>has</u> ω-<u>limit set</u> A_λ^{p+1}, <u>then the sets</u> $\{A_\lambda^k, k=0,1,\dots,p+1\}$ <u>form a Morse decomposition for</u> A_λ.

Other variants of Theorem 2.7 can easily be given if one knows more details about the stable and unstable manifold of $u = 0$ in A_λ.

To introduce some other ideas, let us consider the special case of equation (2.1) studied extensively by Chafee and Infante [CI 1] and Henry [He 1], where

$$\begin{aligned} &f(0) = 0, \quad f'(0) = 1 \\ &\limsup f(u)/u \leq 0, \quad uf''(u) > 0 \text{ if } u \neq 0. \end{aligned} \tag{2.7}$$

Theorem 2.8. *If f satisfies (2.7) and $\lambda \in (n^2, (n+1)^2)$, n an integer, then there are exactly $2n+1$ equilibrium points $\alpha_\infty = 0$, α_j^+, α_j^-, $j=0,1\ldots,n-1$, where α_j^+, α_j^- have j zeros in $(0,\pi)$, $\dim W^u(\alpha_j^\pm) = j$, $0 \leq j \leq n-1$, $\dim W^u(\alpha_\infty) = n$ and*

$$A_\lambda = (\bigcup_j W^u(\alpha_j^\pm)) \cup W^u(\alpha_\infty).$$

With the A_λ^k defined as in (2.6) and $A_\lambda^\infty = \{\alpha_\infty = 0\}$, the set $\{A_\lambda^\infty, A_\lambda^k, k=0,1,2,\ldots,n-1\}$ is a Morse decomposition of A_λ.

Proof. The first part may be found in [He 1]. We prove that $\alpha_\infty = 0$ is completely unstable in A_λ. In fact, any function in $W^s(\alpha_\infty)$ except zero must have at least n zeros and $n+1$ extrema. The results of Matano [M 1] imply the assertion. Theorem 2.7 completes the proof.

Remark 2.9. The function $\alpha_j^+ [\alpha_j^-]$ is uniquely specified by requiring that $d\alpha_j^+(0)/dx > 0$ $[d\alpha_j^-(0)/dx < 0]$. Also, the properties of the orbits in the phase plane imply $\alpha_{2j+1}^+(\pi - x) = \alpha_{2j+1}^-(x)$.

Although Theorem 2.8 gives some information about orbits which connect equilibrium points, it is very imprecise. One would like to have a better understanding of exactly which connections exist. Henry [He 1] has given a complete answer to this problem for $n = 0,1,2,3$, for $f(u)\ldots f(-u)$. We will reprove this result in a slightly different way and, at the same time, give some more information which will hopefully point out the difficulties involved for larger n.

For the statement of the next result, some additional notation is needed. The set $W^s(\alpha_\infty)$ is an embedded submanifold of X (see [He 1,p.155 ff]) of codimension n and $W^s(\alpha_j^\pm)$ is an embedded submanifold of codimension j. The set $W^u(\alpha_\infty)$ is tangent at α_∞ to the linear manifold of X spanned by eigenfunctions of the operator $\partial^2/\partial x^2 + f'(\alpha_\infty)$ on X corresponding to the positive eigenvalues $0 < \lambda_{n-1,\infty} < \lambda_{n-2,\infty} < \ldots < \lambda_{0,\infty}$. The eigenfunction for $\lambda_{j,\infty}$ has j zeros in $(0,\pi)$ for each j. One can construct an imbedded $(n-j)$-dimensional submanifold $W_j^u(\alpha_\infty) \subset W^u(\alpha_\infty)$ for $j=0,1,\ldots,n-1$ which consists of all orbits which approach α_∞ as $t \to -\infty$ with an exponential rate $\leq \lambda_{j,\infty}$. Let

$$W_j^{u*}(\alpha_\infty) = W^u(\alpha_\infty) \setminus W_j^u(\alpha_\infty). \tag{2.8}$$

In the same way, for any $j \leq k$, $k = 0,1,\ldots,n-1$, one can define

$$W_j^{u*}(\alpha_k^{\pm}) = W^u(\alpha_k^{\pm}) \setminus W_j^u(\alpha_k^{\pm}). \tag{2.9}$$

The set $W_j^{u*}(\alpha_k^{\pm})$ is a $(k-j)$ dimensional submanifold of $W^u(\alpha_k^{\pm})$.

We also let

$$C_\lambda^{n,k} = \{\varphi \in A_\lambda : \lim_{t\to-\infty} T_\lambda(t)\varphi \in A_\lambda^n,\ \lim_{t\to\infty} T(t)\varphi \in A_\lambda^k\} \tag{2.10}$$

and let $\gamma(\varphi,\psi)$ designate an orbit whose α-limit set is φ and ω-limit set is ψ.

Lemma 2.10. *If* f *satisfies* (2.7), *is odd and* $\lambda \in (n^2,(n+1)^2)$, *then*

$$\begin{aligned} C_\lambda^{k,k} &= \emptyset \\ C_\lambda^{\infty,0} &= W_1^{u*}(\alpha_\infty) \supseteq \{\gamma(\alpha_\infty,\alpha_0^{\pm})\} \\ C_\lambda^{k,0} &= W_1^{u*}(\alpha_k^{+}) \cup W_1^{u*}(\alpha_k^{-}) \supseteq \{\gamma(\alpha_k^{\pm},\alpha_0^{\pm})\}\ . \end{aligned}$$

Proof. From the oddness of f and the existence of the Liapunov functional V in (2.2), it is easy to see that $C_\lambda^{k,k} = \emptyset$.

We now show that there is a special type of orbit from α_∞ to α_0^+. As remarked before, the set $W^u(\alpha_\infty)$ is tangent at α_∞ to the linear manifold of X spanned by the eigenfunctions of the operator $\partial^2/\partial x^2 + f'(\alpha_\infty)$ on X corresponding to the positive eigenvalues $0 < \lambda_{n-1,\infty} < \ldots < \lambda_{0,\infty}$. These eigenfunctions have, respectively, $n-1,\ldots,1,0$ zeros in $(0,\pi)$. Thus, there is a $\varphi \in X$ such that $T_\lambda(t)\varphi \to \alpha_\infty$ as $t \to -\infty$ and $T_\lambda(t)\varphi(x) > 0$ on $(0,\pi)$ for $t \leq -\tau$ for τ sufficiently large. The maximum principle implies $T_\lambda(t)\varphi(x) > 0$ for all t and, thus, the ω-limit set is an equilibrium point with no zeros in $(0,\pi)$; that is, α_0^+. The same type of argument shows that $\gamma(\alpha_\infty,\alpha_0^-)$ exists. Continuity, the Lyapunov functional, X_0 having no interior and

α_0^+, α_0^- are stable implies the assertion that $C_\lambda^{\infty,0} \supseteq W_1^{u*}(\alpha_\infty)$.

To show $\gamma(\alpha_j^+, \alpha_0^+)$, $j \geq 1$, exists, one proceeds in the same way as above to find a $\varphi \in X$ such that $T_\lambda(t)\varphi \to \alpha_j^+$ as $t \to -\infty$ and $T_\lambda(t)\varphi(x) > \alpha_j^+(x)$ on $(0,\pi)$. Then the ω-limit set of φ must be $> \alpha_j^+(x)$ on $(0,\pi)$. From the manner in which the $\alpha_j^\pm(x)$ are constructed from phase plane analysis, it follows that no equilibrium point is larger than $\alpha_j^+(x)$ on $(0,\pi)$ except α_0^+. In the same way, one obtains $\gamma(\alpha_j^+, \alpha_0^-), \gamma(\alpha_j^-, \alpha_0^+), \gamma(\alpha_j^-, \alpha_0^-)$. A use of an argument similar to the above shows that $C_\lambda^{k,0} \supseteq W_1^{u*}(\alpha_k^+) \cup W_1^{u*}(\alpha_k^-)$. It remains to show equality.

Since α_0^+, α_0^- are stable, it follows that the set

$$X_0 = X \setminus (W^s(\alpha_0^+) \cup W^s(\alpha_0^-)) \tag{2.11}$$

is invariant under $T_\lambda(t)$, has no interior and X_0 contains all the equilibrium points $\alpha_j^\pm$, $j \geq 1$, and α_∞ as well as their stable manifolds. Furthermore, if δ is any continuous curve joining α_0^+, to α_0^-, then $\delta \cap X_0 \neq \emptyset$.

Now suppose $C_\lambda^{\infty,0}$ contains an orbit $\tilde{\gamma}$ from α_∞ to α_0^+ which is not the ones constructed above. Then the oddness of f implies $-\tilde{\gamma}$ goes from α_∞ to α_0^-. The orbits $\tilde{\gamma}, -\tilde{\gamma} \in W_1^u(\alpha_\infty)$ where $W_1^u(\alpha_\infty)$ is defined in (2.8). The openness of $W^s(\alpha_0^+), W^s(\alpha_0^-)$ implies that, in a neighborhood U of α_∞, the set $X_0 \cap U$ is a proper subset of a submanifold of codimension 1. This contradicts $\delta \cap X_0 \neq \emptyset$ for any continuous curve δ joining α_0^+ to α_0^-. Doing the same argument for $\alpha_k^\pm$ completes the proof of the theorem.

<u>Lemma 2.11. Suppose f satisfies (2.7), $\lambda \in (n^2, (n+1)^2)$ and ψ is an equilibrium point with $j + k$ zeros and φ is an equilibrium point with j zeros in $(0,\pi)$. Then any orbit $\gamma(\psi,\varphi)$ going from ψ to φ must belong to $W_j^u(\psi)$.</u>

<u>Proof.</u> Let φ be an equilibrium point with $j + k$ zeros and let ψ be an equilibrium point with j zeros. Then the manner in which the equilibrium points are constructed from the phase plane implies that $\psi - \varphi$ has j zeros in $(0,\pi)$ and $j + 1$ extreme values. If $u = \varphi + v$ in (2.1), then $W^u(0)$ for the new equation has dimension $j + k$ with the basis for the tangent manifold at zero being given by functions with $0,1,\dots,j+k-1$ zeros in $(0,\pi)$. Since $\psi - \varphi$ is an equilibrium point for the new equation with $j + 1$ extreme values in $(0,\pi)$, one can apply

the results of Matano [M 1] to obtain the assertion in the Lemma.

To point out implications of this lemma, define

$$\begin{aligned} S_\lambda^0 &= A_\lambda, \\ S_\lambda^k &= S_\lambda^{k-1} \setminus [(W^s(\alpha_{k-1}^+) \cup W^s(\alpha_{k-1}^-)) \cap A_\lambda], \quad k=1,2,\ldots,n-1. \end{aligned} \tag{2.12}$$

The points α_0^+, α_0^- are stable in S_λ^0. Also, Lemma 2.11 (or Lemma 2.10 with $k=1$) imply that α_1^+, α_1^- are stable in S_λ^1. Furthermore, Lemma 2.10 implies these are the only stable points in S^1 and $\dim[W^u(\alpha_k^\pm) \cap S_\lambda^1] = k-1$, $k=2,3,\ldots,n-1$, $\dim[W^u(\alpha_\infty) \cap S_\lambda^1 = n-1$. Lemma 2.11 and the symmetry hypothesis implies
$C_\lambda^{2,1} = W_1^u(\alpha_2^+) \cup W_1^u(\alpha_2^-) \supseteq \{\gamma(\alpha_2^+,\alpha_1^+), \gamma(\alpha_2^+,\alpha_1^-), \gamma(\alpha_2^-,\alpha_1^+), \gamma(\alpha_2^-,\alpha_1^-)\}$.

In general, we can assert that α_j^+, α_j^- are stable in S_λ^j for every j. This is a consequence of Lemma 2.11. For $j=n-1$, this implies $\alpha_{n-1}^+, \alpha_{n-1}^-$ is stable in S_λ^{n-1}. The only other equilibrium point in S^{n-1} is α_∞ and, thus,
$C_\lambda^{\infty,n-1} = W_{n-2}^{u^*}(\alpha_\infty) = \{\gamma(\alpha_\infty,\alpha_{n-1}^+), \gamma(\alpha_\infty,\alpha_{n-1}^-)\}$.

For $j=2,3,\ldots,n-2$, we are unable to analyze the fine structure of the flow at this time. To see the difficulties, suppose $j = 2$. Then α_2^+, α_2^- are stable in S_λ^2. From what has been proved up to this point, there is the possibility that α_3^+, α_3^- are also stable in S_λ^2; that is, the two dimension unstable manifold of these points in S_λ^1 have ω-limit set $\{\alpha_1^+, \alpha_1^-\}$. If such things can occur, then the flow can be very complicated and can change its qualitative properties without going through a local bifurcation.

Summarizing the nicest part of what has been proved, we obtain the following result of Henry [He 1] .

<u>Theorem 2.12.</u> <u>If</u> f <u>satisfies</u> (2.7), <u>is odd and</u> $\lambda \in (n^2,(n+1)^2)$, $n=0,1,2,3$, <u>then</u>

$$\begin{aligned} C_\lambda^{m,k} &= W_k^{u^*}(\alpha_m^+) \cup W_k^{u^*}(\alpha_m^-) \supseteq \{\gamma(\alpha_m^\pm,\alpha_k^\pm)\}, \quad n-1 \geq m > k \geq 0 \\ C_\lambda^{\infty,k} &= W_k^{u^*}(\alpha_\infty) \geq \{\gamma(\alpha_\infty,\alpha_k^\pm)\}, \quad n-1 \geq k \geq 0 \\ A_\lambda^k &= \{\alpha_k^+,\alpha_k^-\} \\ A_\lambda &= \bigcup_{m>k} (A_\lambda^k \cup C_\lambda^{m,k}) \cup (A_\lambda^\infty \cup C_\lambda^{\infty,k}) \end{aligned}$$

Theorem 2.12 says that A is like a ball of n-dimensions if $\lambda \in (n^2,(n+1)^2)$, n=0,1,2,3. The pictures of A_λ and the flow on A_λ are shown below.

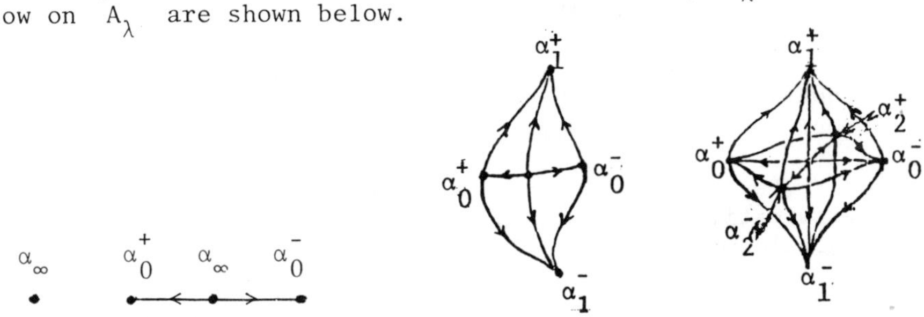

n = 0 n = 1 n = 2 n = 3

For $n \leq 3$, Theorem 2.12 asserts that the changes that occur in the dynamics on A_λ for $\lambda < 16$ come about only through local bifurcation. A proof that the same situation prevails for all λ or a counterexample would be equally interesting.

References

[A 1] H. Amann, Existence and multiplicity theorems for semi-linear elliptic boundary value problems. Math. Z. 150 (1976), pp.281-295.

[A 2] H. Amann, Periodic solutions of semilinear parabolic equations, Nonlinear Anal. pp.1-29, Academic Press, 1978.

[A 3] H. Amann, Existence and stability of solutions for semi-linear parabolic systems and applications to some diffusion reaction equations. Proc. Royal Soc. Edinburgh 81A(1978), pp.35-47.

[C 4] C. Conley, Isolated invariant sets and the Morse Index. NSF-CBMS Regional Conf. Series, No.38, Am. Math. Soc., Providence, R.I., 1978.

[CI 1] N. Chafee and E.F. Infante, A bifurcation problem for a nonlinear partial differential equation of parabolic type. Applicable Analysis, 4(1974), pp.17-37.

[Ha 1] J.K. Hale, Some recent results on dissipative processes. p.152-172 in Lecture Notes in Math. Vo. 799 (Ed.A.F. Izé), Springer-Verlag, 1980.

[Ha 2] J.K. Hale, Theory of Functional Differential Equations. Springer-Verlag, 1977.

[Ha 3] J.K. Hale, Topics in Dynamic Bifurcation Theory, NSF-CBMS Regional Conf. Series, No.47, Am. Math. Soc., Providence, R.I., 1981.

[He 1] D. Henry, Geometric Theory of Semilinear Parabolic Equations. Lecture Notes in Math. Vol. 840, Springer-Verlag, 1981.

[HL 1] J.K. Hale and O. Lopes, Fixed point theorems and dissipative processes, J. Differential Eqns. 13(1973), pp.391-402.

[HM 1] J.K. Hale and P. Massatt, Asymptotic behavior of gradient-like systems. Univ. Fla. Symp. Dyn. Systems, II, Academic Press, 1982.

[HR 1] J.K. Hale and K.P. Rybakowski, On a gradient-like integro-differential equation. Proc. Roy. Soc. Edinburgh, 92A(1982).

[K 1] Ju. S. Kolesov, Periodic solutions of quasilinear equations of the second order. Trans. Moscow Math. Soc. 21(1970), pp.114-146.

[M 1] H. Matano, Nonincrease of lap number of a solution for a one-dimensional semilinear parabolic equation. Pub. Fac. Sci. Univ. Tokyo. To appear.

[M 2] H. Matano, Convergence of solutions of one-dimensional semilinear parabolic equations. J. Math., Kyoto Univ. 18(1978), pp.221-227.

[S 1] C.C. Conley and J.A. Smoller, Algebraic and Topological Invariants for Reaction-Diffusion Equations. These Proceedings.

[Z 1] T.I. Zelenyak, Stabilization of solutions of boundary value problems for a second order parabolic equation with one space variable. Differential Equations 4 (1968), pp.17-22.

Added in proof: The details of the proof of that part of the proof of Lemma 2.10 which states that $C_\lambda^{\infty,0} \supseteq W_1^{u^*}(\alpha_\infty)$ are not as easily supplied as indicated. The same applies to $C_\lambda^{k,0}$. Thus, this must remain as a conjecture.

This research has been supported in part by the Air Force Office of Scientific Research under Contract #AF-AFOSR 81-0198, in part by the National Science Foundation under contract #MCS 79-05774-05 and in part by the U.S. Army Research Office under contract #ARO-DAAG-29-79-C-0161.

INDEX